A GUIDE TO ALGORITHM DESIGN

Paradigms, Methods, and Complexity Analysis

Chapman & Hall/CRC

Applied Algorithms and Data Structures Series

Series Editor

Samir Khuller

University of Maryland

Aims and Scopes

The design and analysis of algorithms and data structures form the foundation of computer science. As current algorithms and data structures are improved and new methods are introduced, it becomes increasingly important to present the latest research and applications to professionals in the field.

This series aims to capture new developments and applications in the design and analysis of algorithms and data structures through the publication of a broad range of textbooks, reference works, and handbooks. The inclusion of concrete examples and applications is highly encouraged. The scope of the series includes, but is not limited to, titles in the areas of parallel algorithms, approximation algorithms, randomized algorithms, graph algorithms, search algorithms, machine learning algorithms, medical algorithms, data structures, graph structures, tree data structures, and other relevant topics that might be proposed by potential contributors.

Published Titles

A Practical Guide to Data Structures and Algorithms Using Java
Sally A. Goldman and Kenneth J. Goldman

Algorithms and Theory of Computation Handbook, Second Edition – Two Volume Set
Edited by Mikhail J. Atallah and Marina Blanton

Mathematical and Algorithmic Foundations of the Internet
Fabrizio Luccio and Linda Pagli, with Graham Steel

The Garbage Collection Handbook: The Art of Automatic Memory Management
Richard Jones, Antony Hosking, and Eliot Moss

A Guide to Algorithm Design: Paradigms, Methods, and Complexity Analysis
Anne Benoit, Yves Robert, and Frédéric Vivien

Chapman & Hall/CRC
Applied Algorithms and Data Structures Series

A GUIDE TO ALGORITHM DESIGN

Paradigms, Methods, and Complexity Analysis

Anne Benoit, Yves Robert, and Frédéric Vivien

CRC Press is an imprint of the
Taylor & Francis Group an **informa** business
A CHAPMAN & HALL BOOK

CRC Press
Taylor & Francis Group
6000 Broken Sound Parkway NW, Suite 300
Boca Raton, FL 33487-2742

International Standard Book Number-13: 978-1-4398-2564-8 (Hardback)

Visit the Taylor & Francis Web site at
http://www.taylorandfrancis.com

and the CRC Press Web site at
http://www.crcpress.com

Contents

III Reasoning on problem complexity 239

List of exercises

Preface

Objective

YABA? *Yet Another Book on Algorithms?*

No thanks. There are so many good books on the design of algorithms that it is hard to choose and pick one. If asked to name our two favorite references, we would recommend *Introduction to Algorithms* by Cormen, Leiserson, Rivest, and Stein [27] and *Algorithms* by Dasgupta, Papadimitriou, and Vazirani [30]. For sure, this book does not intend to compete with such established monuments.

Instead, this book proposes a complementary perspective. It aims at guiding students and researchers who need to solve problems, either by finding optimal algorithms or by assessing new complexity results. In a nutshell, the main objective of this book is to outline the roadmap to follow, and to practice all the corresponding steps, in order to determine the complexity of a problem.

Intended audience and use

The target audience for this book is graduate students and postgraduate researchers in computer science and related fields.

This book does have prerequisites: We expect the reader to have some experience with the design of algorithms, maybe through following an undergraduate course in the field, or through reading a few chapters of the reference books quoted above. In particular, we assume that the reader is familiar with classic algorithms, such as comparison-based sorting (e.g., quick sort or merge sort), and has a good knowledge of elementary graph theory, including:

- traversals (depth-first, breadth-first, connected components);
- shortest paths (one-source, such as Dijkstra, all-pairs, such as Floyd–Warshall);
- maximum matchings in bipartite graphs.

In fact, one needs to know only that efficient algorithms exist to solve these graph theory problems, but, of course, it is better to understand how they work. Because excellent external sources already cover these topics, we refer to them.

Also, we assume that readers have already been exposed, at least up to some extent, to the basic paradigms of algorithm design: divide-and-conquer, greedy algorithms, dynamic programming, and amortized analysis. But here, rather than (or in addition to) referring to external sources, this book provides extensive material so that the readers can assess their skills by solving the many exercises in Part I.

Part II of this book can be used to teach an undergraduate or graduate class on NP-completeness, with a focus on polynomial reductions, and a survey of approaches that go beyond NP-completeness.

Part III of this book can be used to teach a graduate class on advanced algorithms, either in the form of a series of classes presenting the case studies, or in the form of projects assigned to students.

Book content and organization

The book is composed of three main parts:

- Part I: Polynomial-time algorithms: Exercises
- Part II: NP-completeness and beyond
- Part III: Reasoning on problem complexity

Part I aims at training the reader to design efficient algorithms. To do so, we provide a comprehensive set of problems to investigate. Problems are organized along the main design principles, which we each revisit through a brief introduction and a series of related exercises. This leads to five chapters:

1. Introduction to complexity
2. Divide-and-conquer
3. Greedy algorithms
4. Dynamic programming
5. Amortized analysis

All solutions to exercises are provided.

Part II deals with NP-completeness and beyond. Our coverage of NP-completeness focuses on polynomial reductions. We deliberately ignore Turing machines and the theoretic arsenal. The (small) price to pay is to admit Cook's theorem, the existence of the canonical NP-complete problem, formula satisfiability, a.k.a. SAT. In Part II, we also cover approaches that go beyond NP-completeness: identifying polynomial instances, approximation algorithms, linear programming, randomized algorithms, branch-and-bound, and backtracking. Part II consists of four chapters:

1. NP-completeness
2. Exercises on NP-completeness
3. Beyond NP-completeness
4. Exercises going beyond NP-completeness

All solutions to exercises are provided.

Part III constitutes the main originality of the book. It is devoted to case studies whose goal is to provide the reader with tools and techniques to assess problem complexity: which instances are polynomial, and which are NP-hard, and what do to for the latter. Part III consists of an introduction summarizing how to assess the complexity of a new problem, and it is illustrated with five case studies:

1. Chains-on-chains partitioning
2. Replica placement in tree networks
3. Packet routing
4. Matrix product, or tiling the unit square
5. Online scheduling

Thanks

The content of this book, or at least preliminary versions of it, has been used to teach courses at École Normale Supérieure de Lyon. We are grateful to the students for their feedback and suggestions. We also thank all our colleagues who helped gather the problems of Part I. The teaching assistants when Yves Robert was teaching the *Algorithms* course were (ordering by year) Odile Millet-Botta, Tanguy Risset, Alain Darte, Bruno Durand, Frédéric Vivien, Jean-Christophe Dubacq, Olivier Bodini, Daniel Hirschkoff, Matthieu Exbrayat, Natacha Portier, Emmanuel Hyon, Eric Thierry, Michel Morvan, and Yves Caniou. The teaching assistants when Anne Benoit took over were (ordering by year) Victor Poupet, Damien Regnault, Benjamin Depardon, Jean-François Pineau, Clément Rezvoy, Christophe Mouilleron, Fanny Dufossé, and Anne-Cécile Orgerie.

We also wish to thank the following people who have contributed to some of the content by their insightful suggestions, their own previously published work, or their help reviewing draft chapters: Guillaume Aupy, Marin Bougeret, Jean-Yves l'Excellent, Arnaud Legrand, Loris Marchal, Paul Renaud-Goud, Veronika Sonigo, and Bora Uçar.

Finally, a word of caution on bibliographical notes: Some exercises have appeared in many sources, and the references that we give may well not be the original ones. Also, the absence of any reference is not a claim for originality! However, all solutions are ours, and they have been tested and verified by the students at ENS Lyon, the teaching assistants, and ourselves (but we keep the sole responsibility for errors). We welcome comments and suggestions to our e-mail addresses.

Anne Benoit, Anne.Benoit@ens-lyon.fr
Yves Robert, Yves.Robert@ens-lyon.fr
Frédéric Vivien, Frederic.Vivien@inria.fr

Part I

Polynomial-time algorithms: Exercises

Chapter 1

Introduction to complexity

This chapter revisits basic notions on the cost of an algorithm and on the complexity of a problem. To illustrate these notions, in Section 1.1, we study the problem of computing x^n, given x and n (where n is a positive integer). Then, in Section 1.2, we recall the classical asymptotic notations O, o, Θ, and Ω. Finally, exercises are proposed in Section 1.3, with their solutions in Section 1.4.

1.1 On the complexity to compute x^n

We study the problem of computing x^n, given x and n (where n is a positive integer). Note that x is not necessarily a number; it can be a matrix, a polynomial with several unknowns, or any mathematical object for which the multiplication is defined.

We let $y_0 = x$, and we use the following "rule of the game": If I have already computed $y_1, y_2, \ldots, y_{i-1}$, then I can compute y_i as a product of any of two previous temporary results: $y_i = y_j \times y_k$, with $0 \leqslant j, k \leqslant i-1$. The goal is to reach x^n as soon as possible, i.e., to minimize the *cost* of the algorithm, expressed in the number of multiplications. The cost is the first index m such that $y_m = x^n$.

We define $Opt(n)$ as the minimum index m such that $y_m = x^n$, where the minimum is taken over all algorithms, i.e., all possible sequences of y_i. The cost of an algorithm, therefore, is always greater than or equal to $Opt(n)$. Formally,

$$Opt(n) = \min\left\{ m \;\middle|\; \begin{array}{l} \exists y_0 = x,\ y_1, y_2, \ldots, y_{m-1},\ y_m = x^n, \\ \forall i \in [1, m],\ \exists j, k \in [0, i-1],\ y_i = y_j \times y_k \end{array} \right\}.$$

In the following, we present four methods to compute x^n, and we compare their costs. Then we end the section with some complexity results that aim at providing bounds on $Opt(n)$.

1.1.1 Naive method

Let us consider the following naive algorithm: $y_i = y_0 \times y_{i-1}$. We have $y_{n-1} = x^n$, and thus a cost of $n - 1$.

1.1.2 Binary method

We can easily find a method more efficient than the naive algorithm:

$$x^n = \begin{cases} x^{n/2} \times x^{n/2} & \text{if } n \text{ is even,} \\ x^{\lfloor n/2 \rfloor} \times x^{\lfloor n/2 \rfloor} \times x & \text{if } n \text{ is odd.} \end{cases}$$

This algorithm can be formulated as follows: We write n in binary, and then we replace each "1" by SX and each "0" by S, and we remove the first SX. The word that we obtain gives a method to compute x^n. The i-th letter indicates how to compute y_i; letter S corresponds to a *squaring* operation ($y_i = y_{i-1} \times y_{i-1}$), while letter X corresponds to a *multiplying by x* operation ($y_i = y_{i-1} \times y_0$).

For instance, for $n = 23$ (n=10111), we obtain SX S SX SX SX, and after removing the first SX, we obtain the word SSXSXSX. Therefore, we compute, in order, $y_1 = y_0 \times y_0 = x^2$, $y_2 = y_1 \times y_1 = x^4$, $y_3 = y_2 \times y_0 = x^5$, $y_4 = y_3 \times y_3 = x^{10}$, $y_5 = y_4 \times y_0 = x^{11}$, $y_6 = y_5 \times y_5 = x^{22}$, and finally $y_7 = y_6 \times y_0 = x^{23}$.

The correction of the algorithm is easy to justify from the properties of binary decomposition. The cost is $\lfloor \log(n) \rfloor + \nu(n) - 1$, where $\nu(n)$ is the number of 1s in the binary writing of n. $\nu(n) - 1$ is thus the number of Xs, and $\lfloor \log(n) \rfloor$ is the number of Ss in the word. Logarithms are taken in base 2 here, and this will be the case throughout the book unless specified otherwise. In the example $n = 23$, there are four Ss and three Xs, and the cost is, therefore, 7. This value is also obtained with the formula.

Note that this binary method is not optimal; for instance, with $n = 15$, we get the word SXSXSX, leading to six multiplications, while one could notice that $15 = 3 \times 5$, that we need two multiplications to compute $z = x^3$ ($z = (x \times x) \times x$), and then three additional ones to compute $x^{15} = z^5$ (with the binary method: z^2, z^4, z^5).

1.1.3 Factorization method

This method is based on the factorization of n, that is applied recursively when $n \geqslant 2$:

$$x^n = \begin{cases} (x^p)^q & \text{if } p \text{ is the smallest prime factor of } n \ (n = p \times q), \\ x^{n-1} \times x & \text{if } n \text{ is a prime number.} \end{cases}$$

For instance, with this method, for $n = 15$, we obtain the computation described above, i.e., $x^{15} = (x^3)^5 = x^3 \times (x^3)^4$, leading to five multiplications: $y_1 = y_0 \times y_0 = x^2$, $y_2 = y_1 \times y_0 = x^3$, $y_3 = y_2 \times y_2 = (x^3)^2$, $y_4 = y_3 \times y_3 = (x^3)^4$, $y_5 = y_4 \times y_2 = (x^3)^5 = x^{15}$.

Note that if n is a power of 2, this method is identical to the binary method. Also, this factorization method is not optimal. For instance, with $n = 33$, we have seven multiplications ($x^{33} = (x^3)^{11} = x^3 \times (x^3)^{10} = x^3 \times ((x^3)^2)^5 = x^3 \times z \times z^4$, with $z = (x^3)^2$), while the binary method requires only six multiplications ($x^{33} = x \times x^{2^5}$). Note also that there is an infinity of numbers for which the factorization method is better than the binary method ($n = 15 \times 2^k$), and reciprocally ($n = 33 \times 2^k$).

However, we need to emphasize the fact that the cost of decomposing n into prime numbers is not accounted in this formulation, while this would be necessary to correctly quantify the cost of the factorization method. The problem is that we do not know, as of today, how to decompose n in polynomial time. This problem is indeed still open.

1.1.4 Knuth's tree method

The last method that we detail consists in using *Knuth's tree* [62], illustrated in Figure 1.1. The path from the root of the tree to n indicates a sequence of exponents from which we can compute efficiently x^n.

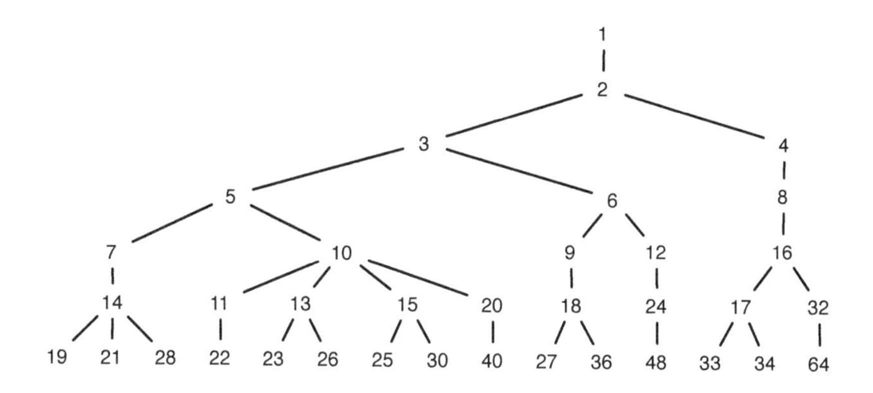

FIGURE 1.1: The first seven levels of Knuth's tree.

Building the tree. The root of the tree is 1. The tree is then built by induction. The $(k+1)$-th level of the tree is defined from the first k levels as follows. Consider each node j of the k-th level from the left to the right, and create nodes $j+1, j+a_1, j+a_2, \ldots, j+a_{k-1} = 2j$ at level $k+1$, as children of node j, in this order from left to right, where $1, a_1, \ldots, a_{k-1} = j$ is the path from the root to j. We do not add a node in the tree if there is already a node with the same value.

The algorithm. The algorithm simply consists of finding n in the tree (it appears only once by construction) and extracting nodes on the path from the root to n: $1, a_1, \ldots, n$. At each step of the algorithm, we compute $y_i = x^{a_i}$ as a product of two previous temporary results, which is possible by construction of the tree. The number of products to be done, i.e., the cost of the algorithm, is equal to the length of the path.

Statistics. Some interesting statistics are extracted from Knuth's book [62]. The smallest numbers for which the tree method is not optimal are $n = 77$, $n = 154$, and $n = 233$. The smallest number for which the tree method is better both to the binary and the factorization methods is $n = 23$. The smallest number for which the tree method is worse than the factorization method is $n = 19,879 = 103 \times 193$, and such cases are rare; for $n \leqslant 100,000$, the tree method is better than the factorization method 88,803 times, it is equivalent 11,191 times, and it is worse than the factorization method only 6 times.

At this point, we have several algorithms, but we do not know anything on the value of $Opt(n)$ yet. To assess the complexity of the problem, we have to provide bounds or asymptotic estimates for $Opt(n)$.

1.1.5 Complexity results

THEOREM 1.1. *For all integer $n \geqslant 1$, $Opt(n) \geqslant \lceil \log(n) \rceil$.*

Proof. Let us consider an algorithm that computes x^n in m steps. Recall that y_i is the intermediate result at step i of the algorithm and thus $y_m = x^n$. Let $\alpha(i)$ be the integer such that $y_i = x^{\alpha(i)}$, for $1 \leqslant i \leqslant m$. Then we prove by induction that $\alpha(i) \leqslant 2^i$.

Initially, we have $y_0 = x$, and thus $\alpha(0) = 1 \leqslant 1 = 2^0$.

For $1 \leqslant i \leqslant m$, there exist j and k $(0 \leqslant j, k < i)$ such that $y_i = y_j \times y_k$, by definition of the algorithm. Therefore, we have $\alpha(i) = \alpha(j) + \alpha(k)$, and we can apply the induction hypothesis on j and k, leading to $\alpha(j) \leqslant 2^j \leqslant 2^{i-1}$, and $\alpha(k) \leqslant 2^k \leqslant 2^{i-1}$. Finally, we have $\alpha(i) \leqslant 2^{i-1} + 2^{i-1} = 2^i$, which concludes the proof. □

Intuitively, the proof expresses the fact that we cannot do better at each step than doubling the exponent. Thanks to this theorem and to the study of the binary method, whose number of steps is bounded by $2\lfloor \log(n) \rfloor$ (recall that $\log(n)$ denotes $\log_2(n)$), we have the following result for all $n \geqslant 2$:

$$1 \leqslant \frac{Opt(n)}{\lceil \log(n) \rceil} \leqslant 2.$$

THEOREM 1.2. $\lim_{n \to \infty} \frac{Opt(n)}{\log(n)} = 1.$

Proof. The idea is to improve the binary method by applying it in base b. We let $b = 2^k$, where the value of k will be fixed later, and we write n in base b: $n = \alpha_0 b^t + \alpha_1 b^{t-1} + \cdots + \alpha_t$, where $t = \lfloor \log_b(n) \rfloor$, and $0 \leqslant \alpha_i \leqslant b-1$ (for $0 \leqslant i \leqslant t$). Then, we compute all x^d, for $1 \leqslant d \leqslant b-1$, with the naive method, in $b-2$ multiplications. Note that we do not necessarily need all these values (only the ones corresponding to the α_is), but they are computed on the fly and we can compute them without significant additional cost.

Then we successfully compute:

$$\begin{aligned} y_0 &= x^{\alpha_0}, \\ y_1 &= (y_0)^b \times x^{\alpha_1} &&= x^{\alpha_0 b + \alpha_1}, \\ y_2 &= (y_1)^b \times x^{\alpha_2} &&= x^{(\alpha_0 b + \alpha_1) b + \alpha_2}, \\ &\vdots \\ y_t &= (y_{t-1})^b \times x^{\alpha_t} &&= x^n. \end{aligned}$$

At each step i (for $1 \leqslant i \leqslant t$), we need $k+1$ computations (k squaring to compute $(y_{i-1})^b$, and one multiplication by x^{α_i}), and, therefore, we have a total cost of

$$\begin{aligned} t \times (k+1) + (b-2) &= \lfloor \log_b(n) \rfloor (k+1) + 2^k - 2 \\ &\leqslant (\log_b(n))\,(k+1) + 2^k = (\log(n))\,\frac{k+1}{k} + 2^k \end{aligned}$$

(recall that $\log_b(a) = \log_x(a)/\log_x(b)$).

We want k to be a function of n tending to infinity when n tends to infinity, so that we have $(k+1)/k$ tending to 1, and such that $2^k = o(\log(n))$ (see Section 1.2 for a definition of the o-notation). For instance, with $k = \lfloor \frac{1}{2} \log(\log(n)) \rfloor$, we have $2^k \leqslant \sqrt{\log(n)}$. (As we are interested only in the asymptotic behavior, we assume that $n > 16$; then $k \geqslant 1$ and $b \geqslant 2$.)

Therefore, we have $Opt(n) \leqslant (\log(n))\frac{k+1}{k} + \sqrt{\log(n)}$ and $\frac{k+1}{k} + \frac{1}{\sqrt{\log(n)}}$ tends to 1 when n tends to infinity. □

Note that this method is somewhat complicated, only to gain a factor 2 by comparison to the binary method.

Finally, we point out that the complexity of the problem of computing x^n is still open, i.e., we do not know whether there exists a polynomial-time method that performs the exact minimum number of operations. Formally, the underlying problem is that of *addition chains.* Starting with $a_0 = 1$, and given $a_0, a_1, \ldots, a_i$ for $i \geqslant 0$, we build a_{i+1} as $a_{i+1} = a_j + a_k$ where $0 \leqslant j \leqslant k \leqslant i$. The length of the chain is the smallest integer $\ell(n)$, if it exists, such that $a_{\ell(n)} = n$. Clearly, the a_is represent the exponents of the values x^{a_i}s that we compute to derive x^n. Given n, what is the complexity to derive an addition chain of minimal length?

An optimal method is easily derived from Knuth's tree. If we keep all possibilities in Knuth's tree, i.e., we add a node in the tree even if it already exists somewhere else, then we have an exhaustive method that always performs the minimum number of operations. However, this method clearly takes an exponential amount of time, and thus is not satisfying. In fact, to the best of our knowledge, the complexity of the problem is still open. There is a common misbelief that the problem of determining whether there exists an addition chain whose length does not exceed some bound is NP-complete. In fact, the result is known to be NP-complete only for a sequence of integers $n_1, n_2, \ldots, n_m$, but not for a single value n [32].

1.2 Asymptotic notations: O, o, Θ, and Ω

Let $f(n)$ be a function, where n is an integer. The asymptotic notations describe the complexity of the function for large values of n.

We say that $f(n) = O(g(n))$ if there exist positive constants c and n_0 such that for all $n \geqslant n_0$, $0 \leqslant f(n) \leqslant c\,g(n)$. The O-notation allows us to give an upper bound on the function, up to within a constant factor.

The o-notation expresses the fact that the upper bound is not asymptotically tight: $f(n) = o(g(n))$ if for any positive constant c, there exists a positive constant n_0 such that for all $n \geqslant n_0$, $0 \leqslant f(n) < c\,g(n)$.

The Ω-notation provides an asymptotic lower bound on the function: $f(n) = \Omega(g(n))$ if there exist positive constants c and n_0 such that for all $n \geqslant n_0$, $0 \leqslant c\,g(n) \leqslant f(n)$.

The Θ-notation is more accurate, since it bounds the function both from below and above: $f(n) = \Theta(g(n))$ if there exist positive constants c_1, c_2, and n_0 such that for all $n \geqslant n_0$, $0 \leqslant c_1\ g(n) \leqslant f(n) \leqslant c_2\ g(n)$. In other words, $f(n) = \Theta(g(n))$ if $f(n) = O(g(n))$ and $f(n) = \Omega(g(n))$.

1.3 Exercises

Exercise 1.1: Longest balanced section (solution p. 14)

Let F be an array of size $n \geqslant 1$ whose elements are 0 or 1. A section $[i..j]$ of consecutive elements of F, with $1 \leqslant i < j \leqslant n$, is *balanced* if it contains as many 0 as 1 elements:

$$\text{card}\{k \mid F[k] = 0,\ i \leqslant k \leqslant j\} = \text{card}\{k \mid F[k] = 1,\ i \leqslant k \leqslant j\}.$$

The length of a balanced section $[i..j]$ is its number of elements $j - i + 1$. The goal of this exercise is to find the longest balanced section of F.

1. Provide a solution whose complexity is $O(n^2)$.
2. Provide a solution whose complexity is $O(n)$.

The reader may want to think for a while before reading the following hint for linear-time complexity. Introduce an array $Q[-n..n]$ of size $2n+1$ and let $Q[b]$ be the first index j such that the imbalance of section $[1..j]$ in F is equal to b. Here the imbalance imbal(i,j) of section $[i..j]$ is defined as

$$\text{imbal}(i,j) = \text{card}\{k \mid F[k] = 1,\ i \leqslant k \leqslant j\} - \text{card}\{k \mid F[k] = 0,\ i \leqslant k \leqslant j\}.$$

Exercise 1.2: Find the star (solution p. 15)

In a group of n persons (numbered from 1 to n), a *star* is someone who does not know anybody else but who is known by all other persons. Our goal is to identify a star, if one exists, in the group. The only action that can be taken is to ask a question to any person i: "Do you know person j?" We assume that everybody tells the truth.

1. How many stars can exist in the group?
2. Design an algorithm to find the star (if any) that requires $O(n)$ questions.
3. Provide a lower bound on the complexity (in terms of number of questions) of any algorithm solving the problem. Prove that the best lower bound for this problem is $3n - \lfloor \log(n) \rfloor - 3$.

Exercise 1.3: Breaking boxes (solution p. 16)

The problem consists of finding the lowest floor of a building from which a box would break when dropping it. The building has n floors, numbered from 1 to n, and we have k boxes. There is only one way to know whether dropping a box from a given floor will break it or not. Go to that floor and throw a box from the window of the building. If the box does not break, it can be collected at the bottom of the building and reused.

The goal is to design an algorithm that returns the index of the lowest floor from which dropping a box will break it. The algorithm returns $n+1$ if a box does not break when thrown from the n-th floor. The cost of the algorithm, to be kept minimal, is expressed as the number of boxes that are thrown (note that re-use is allowed).

1. For $k \geqslant \lceil \log(n) \rceil$, design an algorithm with $O(\log(n))$ boxes thrown.
2. For $k < \lceil \log(n) \rceil$, design an algorithm with $O\left(k + \frac{n}{2^{k-1}}\right)$ boxes thrown.
3. For $k = 2$, design an algorithm with $O(\sqrt{n})$ boxes thrown.

Exercise 1.4: Maximum of n integers (solution p. 17)

The goal is to compute the maximum of n integers, and we study the complexity of the algorithms in terms of number of comparisons and number of assignments.

1. Write a naive algorithm to solve the problem. What is its complexity in the worst and best cases?

2. Is this algorithm optimal for the number of comparisons in the worst case?

3. What is its complexity in the average number of comparisons or assignments? To compute the average number of assignments, you may use the following reasoning. Let $P_{n,k}$ be the number of permutations σ of $\{1, \ldots, n\}$ such that on $T[1] = \sigma(1)$, $\ldots$, $T[n] = \sigma(n)$, the algorithm performs k assignments. Give a recurrence relation for $P_{n,k}$. Let $G_n(z) = \sum P_{n,k} z^k$. Prove that $G_n(z) = z(z+1)\cdots(z+n-1)$, and give a conclusion.

Exercise 1.5: Maximum and minimum of n integers (solution p. 20)

The goal is to compute simultaneously the maximum and the minimum of n integers, and we study the complexity of the algorithms in terms of number of comparisons in the worst case.

1. Design a naive algorithm and give its complexity.

2. One idea to improve the algorithm is to group elements by pairs, in order to decrease the number of comparisons that must be done. Design an algorithm based on this idea, and analyze its complexity.

3. Prove the optimality of such an algorithm by providing a lower bound on the number of comparisons. The idea is to use the *adversary* method. Let A be an algorithm that finds the maximum and minimum. For a given input, when the algorithm is executed, a *novice* is an element that has never been compared, a *winner* has been compared at least once and has always been superior in comparisons, a *loser* has been compared at least once and has always been inferior in comparisons, and the remaining elements are called *average* elements. The number of such elements is represented by a quadruplet of integers (i, j, k, l), with, of course, $i + j + k + l = n$. Give the value of this quadruplet at the beginning and at the end of the algorithm. Provide a strategy for the adversary, so as to maximize the duration of the execution of the algorithm. Conclude with a lower bound on the number of comparisons.

Exercise 1.6: Maximum and second maximum of n integers (solution p. 23)

The goal is to compute simultaneously the maximum and the second maximum of n integers, and we study the complexity of the algorithms in terms of the number of comparisons in the worst case.

1. Design a naive algorithm and give its complexity.

2. One idea to improve the algorithm is to compute the maximum following a tournament (as, for instance, a tennis tournament). If there are $n = 2^k$ numbers taking part in the tournament, how do we find the maximum and the second maximum once the tournament is over? What is the complexity of this algorithm? In the general case, how can we adapt the algorithm for any value of n?

3. Prove the optimality of this algorithm by providing a lower bound on the number of comparisons. The idea is to use *decision trees*. The decision tree of an algorithm is a tree that represents all the possible executions of the algorithm, on every possible input of size n. The internal nodes correspond to tests. In our case, the test is a comparison; if the answer is "yes" we move to the left child, otherwise to the right child, hence having a binary tree. The leaves correspond to the results of the different executions (several leaves may correspond to the same result). Each branch of the tree corresponds to an execution of the algorithm, and the number of comparisons is the height of the branch. The number of comparisons in the worst case is then obtained as the height of the tree.

   (a) Prove that any decision tree that computes the maximum of n integers has at least 2^{n-1} leaves.

   (b) Prove that any binary tree of height h and with f leaves is such that $2^h \geqslant f$.

   (c) Let A be a decision tree solving the problem. Give a lower bound on its number of leaves. Conclude with a lower bound on the number of comparisons in the worst case.

Exercise 1.7: Merging two sorted sets (solution p. 25)

The goal is to merge two sorted sets: a set A of size m and a set B of size n. The $m+n$ numbers to merge are all different and such that $A_1 < A_2 < \cdots < A_m$ and $B_1 < B_2 < \cdots < B_n$.

1. Prove that we need at least $\left\lceil \log \binom{m+n}{n} \right\rceil$ comparisons for the merge (recall that logarithms are taken in base 2).

2. Deduce that for $n = m$, there is a constant k such that, when n is sufficiently large, we need at least $2n - \frac{1}{2}\log(n) - k$ comparisons for the merge.

3. Recall briefly the usual merging algorithm and give its complexity.

4. Prove that for $n = m$, we cannot do better than the usual algorithm. Therefore, the lower bound of Question 2 cannot be matched.

Exercise 1.8: The toolbox (solution p. 26)

In a toolbox, there are n nuts, all of different sizes, and n corresponding bolts. However, everything is mixed up, and you wish to associate each nut with the corresponding bolt. The size differences are so small that it is not possible to decide if a nut (or a bolt) is larger than another one just by looking at them. The only way to proceed consists of trying one nut with one bolt, and each operation can lead to three possible answers: (i) the nut is strictly larger than the bolt, (ii) the bolt is strictly larger than the nut, and (iii) they correspond to each other.

1. Design a simple algorithm with $O(n^2)$ operations that associates each nut with the corresponding bolt.

2. Prove that the problem of finding the smallest nut and the corresponding bolt can be solved with no more than $2n - 2$ operations.

3. Prove that any algorithm solving the initial problem (i.e., associate each nut with the corresponding bolt) requires at least $\Omega(n\log(n))$ operations in the worst case.

Exercise 1.9: Sorting a small number of objects (solution p. 29)

This exercise investigates the complexity of sorting a small number of objects when the only possible operation is the comparison of two objects. For n elements, we know that the number of comparisons is at least $\lceil\log(n!)\rceil$ (see, for instance, Section 10.2 page 243). We ask whether this bound can be reached. Asymptotically, this is true because, for instance, the merge sort has a complexity in $O(n\log(n))$ in the worst case. We check if the bound can be exactly reached in terms of number of comparisons. In the following table, for $2 \leqslant n \leqslant 12$ objects, we indicate the lower bound on the number of comparisons ($\lceil\log(n!)\rceil$), the number of comparisons done by a merge-sort algorithm (merge-sort(n)), and the optimal number of comparisons (opt(n)).

n	2	3	4	5	6	7	8	9	10	11	12
$\lceil\log(n!)\rceil$	1	3	5	7	10	13	16	19	22	26	29
merge-sort(n)	1	3	5	8	11	14	17	21	25	29	33
opt(n)	1	3	5	7	10	13	16	19	22	26	30

Therefore, merge-sort is not reaching the bound as soon as $n \geqslant 5$. The goal of this exercise is to design ad hoc sorting algorithms for each value of n ($2 \leqslant n \leqslant 12$) that perform the optimal number of comparisons $\text{opt}(n)$.

Several techniques can be used:

- *Binary-search insertion*: If we want to insert an element in a sorted set of k elements, the cost is of r comparisons in the worst case if $2^{r-1} \leqslant k \leqslant 2^r - 1$. Therefore, it is less costly to insert an element in a set of three elements than in a set of two elements (two comparisons in both cases), and in a set of seven elements rather than between four and six elements (three comparisons), because the cost in the worst case is the same. In other words, insertion is the most cost effective when $k = 2^r - 1$.

- *Incremental sort* of n elements: We first sort $n - 1$ elements, and then we insert the last one with a binary-search insertion.

- *Divide-and-conquer*: To sort four elements, we create two pairs of two elements $(a \to b)$ and $(c \to d)$, where $(a \to b)$ means that $a \leqslant b$, and then we compare the two largest elements to obtain, for instance, $(a \to b \to d)$. Finally, we insert c with a binary search. The following figure illustrates this technique:

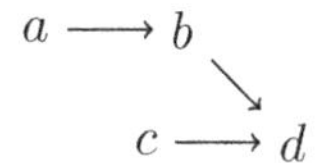

For instance, for $n = 3$, we compare two elements, hence obtaining $(a \to b)$ with one comparison, and then we compare the third element to a and b with two more comparisons, obtaining $3 = \text{opt}(3)$ comparisons. For $n = 4$, we can use the incremental sort to obtain $(a \to b \to c)$ with three comparisons, and then we insert the last element with a binary search, with two comparisons, hence a total of $3 + 2 = 5 = \text{opt}(4)$ comparisons.

1. Provide another technique for $n = 4$, based on divide-and-conquer.

2. Following the previous ideas, provide algorithms for any value $5 \leqslant n \leqslant 11$ that perform $\text{opt}(n)$ comparisons.

3. For $n = 12$, provide a method with 30 comparisons. Indeed, it is impossible to succeed with $\lceil \log(12!) \rceil = 29$ comparisons; researchers have tested all possible algorithms with the brute force method, and it took two hours of computation in 1990 (and it was a real challenge at that time!).

1.4 Solutions to exercises

Solution to Exercise 1.1: Longest balanced section

Quadratic complexity. There are many possible solutions. Obviously, a balanced section $[i..j]$ is such that $\text{imbal}(i,j) = 0$. One method is to fix the origin i of a section and to compute iteratively $\text{imbal}(i,j)$ for all values of j such that $i+1 \leqslant j \leqslant n$. This requires scanning the section $[i..n]$ only once, and one can record on the fly the largest value of j (if any) such that $\text{imbal}(i,j) = 0$. The complexity is $O(n)$ for each value of the origin i, hence a solution whose cost is $O(n^2)$.

Linear complexity. The hint gives the main idea. We initialize the $2n+1$ elements of Q to the value -1. We scan the array F by letting the index i vary from 1 to n, computing iteratively (hence, in constant time) the value $b = \text{imbal}(1,i)$. Now there are two cases:

- Either the value b is met for the first time, which means we had $Q[b] = -1$ before reaching i. Then we store the value of i by letting $Q[b] = i$.
- Or the value b had already been met, meaning that $Q[b] \neq -1$. Here $Q[b]$ was the first index such that $b = \text{imbal}(1, Q[b])$. But, we also have $b = \text{imbal}(1,i)$, which means that the section $[(Q[b]+1)..i]$ is balanced, and we record its length $i - Q[b]$ if it exceeds the largest value currently found.

Altogether, we are led to Algorithm 1.1, whose complexity is indeed $O(n)$:

```
1  max ← 0
2  for b = −n to n do
3  └ Q[b] ← −1
4  Q[0] ← 0 /* the empty section is balanced */
5  b ← 0 /* b is the current imbalance */
6  for i = 1 to n do
7  │ if F[i] = 1 then b ← b + 1 else b ← b − 1
8  │ if Q[b] = −1 then Q[b] ← i
9  │ else if i − Q[b] > max then
10 └ └ max ← i − Q[b]
11 return max
```

ALGORITHM 1.1: Longest balanced section.

Solution to Exercise 1.2: Find the star

1. There is, at most, one star in a group because a star does not know any other person; thus, no other person can be known by everybody in the group.

2. When we ask the question: Does i know j?, we obtain the following result:

   - If the answer is "yes," i is not a star but j may be one;
   - If the answer is "no," j is not a star but i may be one.

   Therefore, each test identifies one person as not being a star. We use this property to build a linear-time algorithm, Algorithm 1.2. This algorithm first scans the group of persons while memorizing at each step the only potential star candidate among the persons tested so far (*while* loop at Step 2). Then it checks that the star is known by all other group members (*while* loop at Step 6, optimized to take into account all knowledge gathered by the previous loop). Finally, it checks that the candidate star does not know any group member (*while* loop at Step 10).

   Each of three *while* loops executes at most n iterations, hence a complexity in $O(n)$.

```
1  i ← 1 and j ← 2
2  while j ⩽ n do
3  |  if "Does j know i?" then j ← j + 1
4  |  else i ← j and j ← j + 1
5  star_found ← TRUE and j ← 1
6  while j < i and star_found do
7  |  if "Does j know i?" then j ← j + 1
8  |  else star_found ← FALSE
9  k ← 1
10 while k ⩽ n and star_found do
11 |  if k ≠ i and "Does i know k?" then star_found ← FALSE
12 |  else k ← k + 1
13 if star_found then return "The star is i."
14 else return "There is no star."
```

ALGORITHM 1.2: Algorithm identifying a star in a group.

3. First of all, we remark that the worst case is reached when the group of persons contains a star. To identify a star, it is necessary that each

person, but the star, is involved in a question that identifies him or her as not being a star. This requires $n-1$ questions. It is also necessary that, if the group contains a star i, all $n-1$ questions: Does i know j? and all $n-1$ questions: Does j know i? are asked. However, these two sets of questions are not independent.

As noted earlier, each question can, at most, invalidate one candidate as being a star. It is suboptimal for an algorithm to ask a question involving a person who is known not to be a star when there are still at least two star candidates. Indeed, in the worst case, this question will not provide any new knowledge. This is obviously the case if we already know that the two persons are not stars. Otherwise, the candidate involved will not be invalidated by the answer to the question but will eventually turn out not to be a star. In other words, in an optimal algorithm, each of the first $n-1$ questions should involve two persons who can still be stars, when taking into account the answers to all questions asked so far.

Then, after $n-1$ questions, there remains exactly one person who can be a star. This person has already been involved in a certain number of questions, say k. To complete the algorithm, one needs to ask the remaining $2n-2-k$ questions involving the candidate star. Overall, $3n-3-k$ questions will be necessary in the worst case to identify the star. If we represent the set of the initial $n-1$ questions by a tree, k is exactly the length of the path in the tree from its root to the leaf corresponding to the candidate (the length of the path being the number of edges visited). Therefore, in the worst case, the number of questions needed is $3n-3-h$, where h is the length of the shortest branch of the tree. Then, a lower bound for this value is obtained with trees for which this value is minimal, that is, trees of minimum height. For these trees, $h = \lfloor \log(n) \rfloor$, and thus an overall minimal bound is

$$3n - 3 - \lfloor \log(n) \rfloor.$$

Solution to Exercise 1.3: Breaking boxes

1. The complexity in $O(\log(n))$ is a hint: One should use a binary search. Indeed, if we have $k \geqslant \lceil \log(n) \rceil$, we know the result for the floors whose indices range from i to j by dropping a box from the m-th floor where $m = \lfloor \frac{i+j}{2} \rfloor$ and then by iterating with floors i to $m-1$ if the box broke, and by iterating with floors m to j otherwise. The principle of the binary search guarantees that we will obtain the desired result (when $i = j$) and in at most $\lceil \log(n) \rceil$ steps, and, thus, after having broken at most $\lceil \log(n) \rceil$ boxes.

2. As we have only $k < \lceil \log(n) \rceil$ boxes, we cannot directly apply a binary search. We, however, will solve this problem in a simple way. We apply

the binary search using $k-1$ boxes in order to narrow as much as possible the search interval around the desired floor. We then use the last box to scan the remaining interval floor by floor, from the lowest to the highest. After throwing $k-1$ boxes, if the target floor has not been identified, there are at most $n/2^{k-1}$ floors in the search interval, hence a worst-case complexity of $O(k + n/2^{k-1})$.

3. When $k = 2$ we do not want to have to test each floor one after the other, thereby ending up with a linear complexity. We, therefore, will adapt the idea of narrowing the search interval. We partition the set of floors into "slices" of $\sqrt{n}$ floors (assume that n is a square without loss of generality, or use ceil functions). Then we throw the first box from the first floor of each slice until it breaks, starting with the lowest slice. Then, we return to the last tested floor, m, from which the box was dropped but did not break. We then test one by one the floors using the second box. We start with floor $m+1$ and, in the worst case, we go up to the floor from which the first box broke. There are, by construction, $\sqrt{n}$ floors in that slice. Therefore, we have two series of tests, with $O(\sqrt{n})$ tests in each of them. Hence, the overall complexity is $O(\sqrt{n})$.

Solution to Exercise 1.4: Maximum of n integers

1. Algorithm 1.3 is a naive algorithm to compute the maximum over n values.

```
1 max ← T[1]
2 for i = 2 to n do
3 └ if T[i] > max then max ← T[i]
4 return max
```

ALGORITHM 1.3: Maximum over n values.

Complexity in the number of comparisons: Whatever the instance, Algorithm 1.3 performs exactly $n-1$ comparisons.

Complexity in the number of assignments:

- worst case: n (when values are sorted in nondecreasing order);
- best case: 1 (if the first element is the maximum).

2. We provide two different proofs that $n-1$ comparisons are requested to determine the maximum among n values (thus establishing the optimality of Algorithm 1.3). A classical error in establishing this result is

to write that any nonmaximum value must have been compared to the maximum. This is true only by transitivity.

First proof. A value that never was the smaller one in a comparison is, potentially, the maximum. Therefore, for the maximum to be determined, all $n-1$ values that are not the maximum must have been the smaller one in at least one comparison, which requires at least $n-1$ comparisons.

Second proof. We identify the values with the vertices of a graph. Initially, the graph has no edge and thus contains n connected components, one per vertex. For each comparison, we add to the graph an edge between the two corresponding vertices. Adding an edge either decreases the number of connected components by one or keeps it unchanged. If the graph contains two distinct connected components at the end, then there has been no comparison between the two sets of values; each one may contain the maximum, and one cannot determine where the maximum value is. Therefore, to be able to determine the maximum, the graph must contain a single connected component, and at least $n-1$ comparisons must have been performed.

3. Average complexity in the number of comparisons. Whatever the instance, the algorithm always performs $n-1$ comparisons, and the average complexity is equal to the best-case and worst-case complexity.

   Average complexity in the number of assignments. The average complexity of an algorithm A on data of size n is defined as:

$$avg_A(n) = \sum_{\text{d data of size } n} p(d)\ cost_A(d),$$

   where $p(d)$ is the probability that d is an entry of algorithm A and $cost_A(d)$ is the complexity reached by algorithm A on entry d. Here the entry data are the permutations of $\{1,\ldots,n\}$. The $n!$ possible permutations are supposed to be equiprobable. The cost of algorithm A, in this question, is the number of assignments that it performs. This cost is equal to k for each of the permutations taken into account in $P_{n,k}$. Therefore, the average complexity of algorithm A can be expressed as:

$$avg_A(n) = \frac{1}{n!} \sum_{T \text{ permutation of } \{1,\ldots,n\}} cost_A(T) = \frac{1}{n!} \sum_{k=1}^{n} k P_{n,k}.$$

   We compute $P_{n,k}$ using a recursion. The maximum element is either the last one (being stored in $T[n]$) or not:

   - If the maximum is in $T[n]$, $T[n]$ is the cause of one assignment and, to reach a total of k assignments, the first $n-1$ values must be responsible of $k-1$ assignments. Therefore, in that case, the number of permutations leading to k assignments is equal to the number of permutations leading to $k-1$ assignments when processing an input of size $n-1$: $P_{n,k} = P_{n-1,k-1}$.

- If the maximum is not in $T[n]$, then all the k assignments are due to the first $n-1$ values. Furthermore, there are $n-1$ permutations of $\{1,\ldots,n\}$ that give rise to the same permutation of $\{1,\ldots,n-1\}$, because the rank of the value $T[n]$ (when the values are sorted) can be anything except that of the largest value. Therefore, in that case, the number of permutations of $\{1,\ldots,n\}$ leading to k assignments is equal to $P_{n,k} = (n-1)\ P_{n-1,k}$.

Gathering the previous two results, we obtain: $P_{n,k} = P_{n-1.k-1} + (n-1)\ P_{n-1,k}$. The limit cases are, for any integer n, $P_{n,0} = 0$, $P_{n,k} = 0$ if $k > n$, and $P_{n,n} = 1$. To explicit $P_{n,k}$, we use the generating function: $G_n(z) = \sum_{k=1}^{n} P_{n,k}\ z^k$.

$$\begin{aligned}
G_{n+1}(z) &= \sum_{k=1}^{n+1} P_{n+1,k}\ z^k = \sum_{k=1}^{n+1} (P_{n,k-1} + nP_{n,k})\ z^k \\
&= \left(\sum_{k=1}^{n+1} zP_{n,k-1}\ z^{k-1}\right) + \left(n\sum_{k=1}^{n+1} P_{n,k}\ z^k\right) \\
&= z\left(\sum_{k=0}^{n} P_{n,k}\ z^k\right) + n\left(\sum_{k=1}^{n+1} P_{n,k}\ z^k\right)
\end{aligned}$$

However, $P_{n,0} = 0$ and $P_{n,n+1} = 0$. Therefore,

$$G_{n+1}(z) = z\left(\sum_{k=1}^{n} P_{n,k}\ z^k\right) + n\left(\sum_{k=1}^{n} P_{n,k}\ z^k\right) = (z+n)G_n(z).$$

Since $G_1(z) = P_{1,1}z = z$, we obtain $G_n(z) = z(z+1)\cdots(z+n-1)$. We now need to link the value that we want to compute, namely the average complexity of Algorithm 1.3, to this function $G_1(z)$. We remark that

$$G_n'(z) = \sum_{k=1}^{n} kP_{n,k}\ z^{k-1}\ ,\ G_n'(1) = \sum_{k=1}^{n} kP_{n,k}\ ,\text{and } G_n(1) = \sum_{k=1}^{n} P_{n,k} = n!$$

Therefore,

$$\begin{aligned}
avg_A(n) &= \frac{G_n'(1)}{G_n(1)} = [\ln(G(z))]'(1) \\
&= \left[\sum_{i=0}^{n-1} \ln(z+i)\right]'(1) \\
&= 1 + \frac{1}{2} + \cdots + \frac{1}{n} = H_n
\end{aligned}$$

where H_n is the n-th partial sum of the diverging harmonic series (or the n-th harmonic number). Therefore, $avg_A(n) = O(\ln(n))$.

Solution to Exercise 1.5: Maximum and minimum of n integers

1. Algorithm 1.4 is a naive algorithm to compute both the maximum and the minimum from a set of n values. This algorithm performs $n-1$ comparisons to find the maximum, and as many to find the minimum. Hence, the complexity in the number of comparisons is equal to $2n-2$.

```
1 max ← T[1]
2 min ← T[1]
3 for i = 2 to n do
4   if T[i] > max then max ← T[i]
5 for i = 2 to n do
6   if T[i] < min then min ← T[i]
7 return (max,min)
```

ALGORITHM 1.4: Naive algorithm to compute the minimum and the maximum of a set of n values.

2. In Algorithm 1.4, we considered one new value at each step. Here we will consider two new values at each step. We first compare them, and then we compare the largest one to the current maximum and compare the smallest one to the current minimum. Algorithm 1.5 presents such an algorithm.

   If n is even, there are $\frac{n}{2}$ pairs and thus $\frac{n}{2}$ comparisons of pair elements. Then, there are $\frac{n}{2}-1$ additional comparisons to compute the maximum and as many for the minimum. Hence, the complexity is in $\frac{3n}{2}-2$.

   If n is odd, there are $\lfloor \frac{n}{2} \rfloor$ pairs and thus $\lfloor \frac{n}{2} \rfloor$ comparisons of pair elements. Then, there are $\lfloor \frac{n}{2} \rfloor - 1$ additional comparisons in the loop to compute the maximum and as many for the loop for the minimum. Finally, there are one or two additional comparisons to handle the value $T[n]$, hence a maximum number of comparisons of $3\lfloor \frac{n}{2} \rfloor$. However, when n is odd, $3\lfloor \frac{n}{2} \rfloor = \lceil \frac{3n}{2} \rceil - 2$.

   The complexity of Algorithm 1.5 in number of comparisons is thus $\lceil \frac{3n}{2} \rceil - 2$.

3. Let A be any algorithm. The quadruplet (i, j, k, l) represents the cur-

```
1  for i = 1 to ⌊n/2⌋ do
2  ⌊ if T[2i − 1] > T[2i] then exchange T[2i − 1] and T[2i]
3  max ← T[2]
4  for i = 2 to ⌊n/2⌋ do
5  ⌊ if T[2i] > max then max ← T[2i]
6  min ← T[1]
7  for i = 2 to ⌊n/2⌋ do
8  ⌊ if T[2i − 1] < min then min ← T[2i − 1]
9  if n is odd then
10 |  if T[n] > max then max ← T[n]
11 ⌊  else if T[n] < min then min ← T[n]
```

ALGORITHM 1.5: Algorithm that groups elements by pairs to compute the minimum and the maximum of a set of n values.

rent knowledge, that is, respectively, the number of novices, of winners, of losers, and of average elements (obviously $i + j + k + l = n$). The algorithm starts with no knowledge, that is, with the configuration $(i, j, k, l) = (n, 0, 0, 0)$. When the algorithm completes, there is only one winner, only one loser, and all the other elements must have won and lost at least one comparison each. The quadruplet corresponding to that situation is $(i, j, k, l) = (0, 1, 1, n - 2)$.

The aim of the adversary is to slow down as much as possible the progression of the algorithm. In other words, the aim of the adversary is to minimize the knowledge gained by the algorithm from each comparison. For instance, assume that the situation is (i, j, k, l) and that the algorithm compares a novice N and a winner W. Whatever the output of the comparison, once it is made, N is no longer a novice element. There are nevertheless two cases:

- $W < N$: Then N becomes a winner and W an average element (it just lost a comparison and had previously won at least one as it was labeled a winner). The new situation is then $(i - 1, j, k, l + 1)$.
- $W > N$: Then W remains a winner and N becomes a loser. The new situation is then $(i - 1, j, k + 1, l)$.

The adversary chooses the latter case as it leads to a situation that is further from the completion of the algorithm than the former case. Indeed, in the former case the knowledge gain is that one element (W) can be neither the minimum nor the maximum and thus can be safely ignored. In the latter case, the knowledge gain is just that there is one more element that cannot be the maximum.

Table 1.1 presents the choices made by the adversary depending on the elements compared by the algorithm. In that table, N designates a novice, W a winner, L a loser, and A an average element. The symbol "/" represents a case where the integer is left unchanged. The symbol "open" means that the adversary can pick any of the two possible outcomes except when results of previous comparisons dictate the outcome. For instance, if we had $L[2] < L[5]$ and $L[5] < L[4]$, then $L[2] < L[4]$ (this can happen when comparing two average elements).

TABLE 1.1: Strategy of the adversary in order to maximize the number of comparisons needed before an algorithm A can complete.

comparison	choice	i	j	k	l
$N : N$	open	$i-2$	$j+1$	$k+1$	/
$N : W$	$W > N$	$i-1$	/	$k+1$	/
$N : L$	$L < N$	$i-1$	$j+1$	/	/
$N : A$	$A > N$	$i-1$	/	$k+1$	/
	$N > A$	$i-1$	$j+1$	/	/
$W : W$	open	/	$j-1$	/	$l+1$
$W : L$	$W > L$	/	/	/	/
$W : A$	$W > A$	/	/	/	/
$L : L$	open	/	/	$k-1$	$l+1$
$L : A$	$L < A$	/	/	/	/
$A : A$	open	/	/	/	/

We now need to derive a lower bound on the number of comparisons needed, in the worst case. Each novice must be compared at least once. Then, the most efficient comparison type is the $N : N$ comparison, which decreases the number of novices by two. Therefore, we need at least $\lceil \frac{n}{2} \rceil$ comparisons to get rid of all novices. Furthermore, at the end we must have $n-2$ average elements. A comparison creates at most one average element. Hence, we need at least $n-2$ additional comparisons to reach the desired number of average elements. Finally, we remark that no comparison implying a novice leads to an increase in the number of average elements. Therefore, we can add the two lower bounds (i.e., the one to get rid of all novices and the one to create the required number of average elements) to obtain a global lower bound. A lower bound on the number of comparisons is thus:

$$n - 2 + \left\lceil \frac{n}{2} \right\rceil = \left\lceil \frac{3n}{2} \right\rceil - 2.$$

Since this lower bound is equal to the complexity of Algorithm 1.5, this algorithm is optimal.

Solution to Exercise 1.6: Maximum and second maximum of n integers

1. Algorithm 1.6 is a naive algorithm that first scans all values looking for the maximum one and then scan all values, except the maximum one, to look for the second maximum. Its first loop performs $n-1$ comparisons and the second one $n-2$ for an overall complexity in the number of comparisons of $2n-3$.

```
1  max_1 ← T[1]
2  posmax_1 ← 1
3  for i = 2 to n do
4      if T[i] > max_1 then
5          max_1 ← T[i]
6          posmax_1 ← i
7  if posmax_1 ≠ 1 then max_2 ← T[1] else max_2 ← T[2]
8  for i = 2 to n with i ≠ posmax_1 do
9      if T[i] > max_2 then  max_2 ← T[i]
10 return (max_1, max_2)
```

ALGORITHM 1.6: Algorithm computing the first and second maximum of a set of n values.

2. We first consider the case where $n = 2^k$. We compute the first maximum using a classical tournament scheme. At each round, we partition the values into pairs, perform one comparison per pair, and keep the largest value for the next round. Figure 1.2 presents the comparison tree (to be read bottom up). In this tree, the dotted lines mark the trajectory of the maximum value. $n-1$ comparisons are performed to determine the maximum. The second maximum is one of the values that lost their comparison against the maximum. Therefore, it is one of the values tagged by a question mark in Figure 1.2. There are k such values, and, therefore, $k-1 = \log(n)-1$ comparisons are needed to determine the second maximum. Hence, the overall complexity is of $n+\log(n)-2$ comparisons.

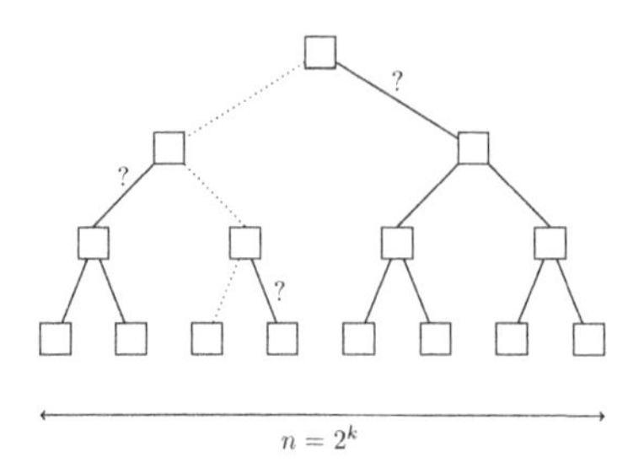

FIGURE 1.2: Tree of comparisons to determine the maximum of a set of $n = 2^k$ values.

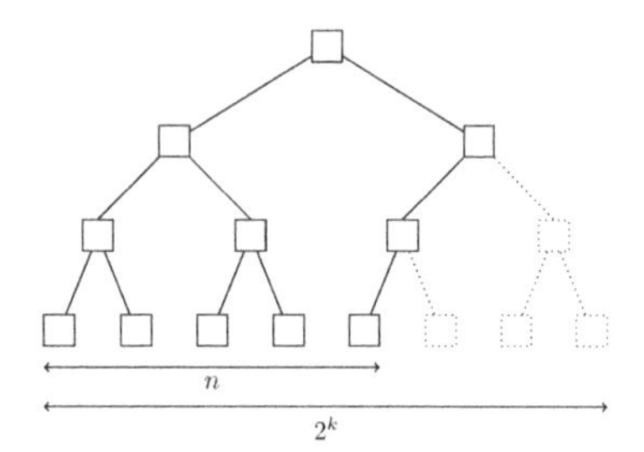

FIGURE 1.3: Tree of comparisons to determine the maximum of a set of $n \neq 2^k$ values.

For the general case, we want to fall back to the case where n is a power of 2. Therefore, we take for k the value $\lceil \log(n) \rceil$. The algorithm is then illustrated by Figure 1.3. As previously, we need $n-1$ comparisons to identify the maximum. Then, the longest branch in the tree has a length of $\lceil \log(n) \rceil$. Therefore, we need at most $\lceil \log(n) \rceil - 1$ comparisons to identify the second maximum, for an overall complexity of $n + \lceil \log(n) \rceil - 2$.

3. (a) We have seen in Exercise 1.4 that we need to perform at least $n-1$ comparisons to determine the maximum among n values. This means that any branch of the decision tree from the root to a leaf contains at least $n-1$ internal nodes that each have two children. Indeed, internal nodes that have a single child correspond to useless comparisons. We can suppress each such node from the tree (directly linking its parent to its child). We then obtain a tree whose first n levels constitute a complete binary tree. Therefore, the decision tree contains at least 2^{n-1} leaves.

   (b) We prove the result by induction on the height h of the tree. If $h = 0$, the tree has exactly one leaf and the result holds. We now suppose the result holds true for any tree whose height is at most h, and we consider a tree of height $h+1$ that has f leaves. There are two cases:

   - If the root has a single child, all the leaves belong to the subtree rooted at the single child of the root node. This subtree is a tree of height h. Using the induction hypothesis, we have $f \leqslant 2^h$. Therefore, $f \leqslant 2^{h+1}$.
   - If the root has two children, let f_1 and f_2 be the number of the leaves of the subtrees rooted at the left and right children of the root. Let h_1 and h_2 be the respective heights of these trees. Then, according to the induction hypothesis, $f_1 \leqslant 2^{h_1}$ and $f_2 \leqslant 2^{h_2}$. Therefore, $f = f_1 + f_2 \leqslant 2^{h_1} + 2^{h_2} \leqslant 2^h + 2^h = 2^{h+1}$.

(c) Figure 1.4 presents a decision tree for the maximum and the second maximum among four values.

We partition the leaves of the decision tree A with respect to the value of the (first) maximum. A_i is the subtree obtained from A by pruning exactly the branches that did not end in a leaf concluding that $T[i]$ is the maximum. We then remove from A_i the internal nodes that perform a test on $T[i]$. These nodes obviously had a single child each by construction of A_i. A_i is then a decision tree to compute the second maximum when the first maximum is known (this is $T[i]$). Therefore, $A[i]$ is a decision tree for the computation of the maximum among $n-1$ values. According to Question 3a, each tree $A[i]$ contains at least 2^{n-2} leaves. Since, by construction, the $A[i]$s define a partition of the leaves of A, A contains at least $n \cdot 2^{n-2}$ leaves. Then, according to Question 3b, the height h of the tree A must satisfy:

$$2^h \geqslant n \cdot 2^{n-2} \quad \Rightarrow \quad h \geqslant n - 2 + \lceil \log(n) \rceil.$$

Therefore, any algorithm that computes the first and second maximum must perform, in the worst case, at least $n - 2 + \lceil \log(n) \rceil$ comparisons.

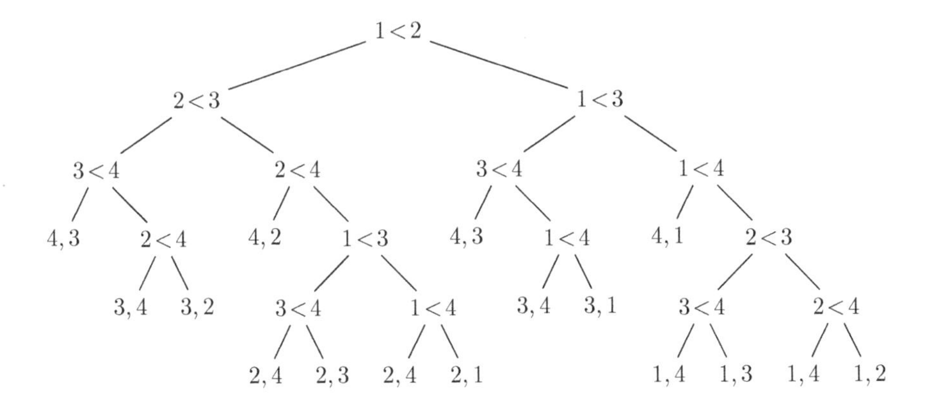

FIGURE 1.4: Decision tree for computing the first and second maximum among four values.

Solution to Exercise 1.7: Merging two sorted sets

1. The merged list contains $n + m$ elements. The relative order of the elements of A is fixed, because A is an ordered set. The same is true for B. Therefore, merging the two lists is equivalent to determining

the positions in the merged list that contain elements of B. Therefore, merging the two sorted lists can lead to $\binom{m+n}{n}$ different solutions.

The idea is then to consider the *decision tree* of any given merging algorithm (see Exercise 1.6 for the definition of decision trees). Each node in the tree corresponds to a comparison. Each leave corresponds to a possible output (several leaves can correspond to the same output). From what precedes, any decision tree must have at least $\binom{m+n}{n}$ leaves, and its height must be at least $\lceil \log \binom{m+n}{n} \rceil$.

2. Here we consider the case $m = n$. By definition, we have $\binom{2n}{n} = \frac{(2n)!}{(n!)^2}$. We then use Stirling's approximation of the factorial function: $n! \approx \sqrt{2\pi n}\left(\frac{n}{e}\right)^n$. We have $2 \leqslant \sqrt{2\pi} \leqslant 3$. Therefore, for n sufficiently large, we have $2\sqrt{n}\left(\frac{n}{e}\right)^n \leqslant \sqrt{2\pi n}\left(\frac{n}{e}\right)^n \leqslant 3\sqrt{n}\left(\frac{n}{e}\right)^n$, and

$$\frac{(2n)!}{(n!)^2} \geqslant \frac{2\sqrt{2n}\left(\frac{2n}{e}\right)^{2n}}{\left(3\sqrt{n}\left(\frac{n}{e}\right)^n\right)^2} = \frac{2\sqrt{2}}{9}\frac{1}{\sqrt{n}}\,2^{2n}.$$

Finally, for n sufficiently large,

$$\left\lceil \log \binom{2n}{n} \right\rceil \geqslant \log\left(\frac{2\sqrt{2}}{9}\right) - \frac{1}{2}\log(n) + 2n \geqslant 2n - \frac{1}{2}\log(n) - 2.$$

3. Algorithm 1.7 merges two sorted lists. This algorithm stops performing any comparison as soon as all the elements of one of the two input sorted sets have been stored in the resulting array. Furthermore, each time a comparison is done, one element from one of the two input sorted sets is stored in C. Therefore, in the worst case, after $(n-1)+(m-1)$ comparisons, there remains one element in input set A and one in B. Then, one final comparison is needed for the algorithm to be able to complete its task. The worst-case complexity of Algorithm 1.7 is thus of $n + m - 1$ comparisons.

4. Here, we once again consider the case $m = n$. Let us consider an instance such that we have $A[1] \leqslant B[1] \leqslant A[2] \leqslant B[2] \leqslant A[3] \leqslant B[3] \leqslant ... \leqslant A[n] \leqslant B[n]$. Then, for $1 \leqslant i \leqslant n-1$, $B[i]$ must be compared to both $A[i]$ and $A[i+1]$. Furthermore, $B[n]$ must be compared to $A[n]$. Overall we need to perform at least $2(n-1)+1 = 2n-1$ comparisons. Hence, the lower bound cannot be reached in that case (the distance to the bound being at least equal to $\log(n)$).

Solution to Exercise 1.8: The toolbox

1. To associate bolts and nuts, a simple solution is to pick one bolt and to test it with each nut. In at most n tests, the associated nut is identified.

```
1  i ← 1 /*  index to scan the set A */
2  j ← 1 /*  index to scan the set B */
3  k ← 1 /*  index to scan the resulting array C */
4  while i ⩽ n and j ⩽ m do
5  |  if A[i] < B[j] then
6  |  |  C[k] ← A[i]
7  |  |  i ← i + 1
8  |  else
9  |  |  C[k] ← B[j]
10 |  |  j ← j + 1
11 |  k ← k + 1
12 for l = i to n do
13 |  C[k] ← A[l]
14 |  k ← k + 1
15 for l = j to m do
16 |  C[k] ← B[l]
17 |  k ← k + 1
```

ALGORITHM 1.7: Merge the two sorted sets A and B.

Then, one repeats the process with the remaining $n-1$ bolts and $n-1$ nuts, and so on. All bolts and nuts are then associated in at most $\frac{n(n-1)}{2} = O(n^2)$ tests.

2. The idea is to number, in an arbitrary order, the bolts and the nuts from 1 to n. We associate one counter to each set. While we scan the sets, after each comparison we leave unchanged the counter corresponding to the smaller object and we increase the other counter. If the tested objects fit together, we memorize the association and we increment one of the counters, always the same one (for instance, the nut counter). Algorithm 1.8 realizes this scheme.

   We now show that Algorithm 1.8 enables one to solve the problem. At each iteration of the loop, exactly one among the nut counter and the bolt counter is incremented, except for the case $i = j = n$ at which none is incremented and the algorithm completes. First, we remark that once the smallest bolt is encountered, the bolt counter j stays constant (Step 11 is the only one that ever increments this counter). Therefore, when the algorithm completes, bolt j is the smallest bolt. Note also that, because of this property on the smallest bolt, the conditional of the while loop does not require a condition $j \leqslant n$.

```
1  ass_nut ← 0
2  ass_bolt ← 0
3  i ← 1
4  j ← 1
5  while (i ⩽ n) do
6      if i = j = n then
7          if ass_bolt = n then smallest_nut ← ass_nut
8          else smallest_nut ← n
9          break out of the loop
10     if nut.i = bolt.j then ass_nut ← i;  ass_bolt ← j;  i ← i + 1
11     if nut.i < bolt.j then j ← j + 1
12     if nut.i > bolt.j then i ← i + 1
13 if i = n + 1 then smallest_nut ← ass_nut
14 smallest_bolt ← j
15 return (smallest_nut, smallest_bolt)
```

ALGORITHM 1.8: Find the smallest nut and the smallest bolt.

The nut counter i does not stay constant once the smallest nut has been encountered. This is because of the special processing for the case $nut.i = bolt.j$. But, right before increasing the nut counter i after finding its corresponding bolt, we memorize the index of the corresponding nut and bolt (Step 10).

If the while loop ends with the case $i = j = n$, then, as we have already remarked, the smallest bolt is bolt n. If $ass\_bolt = n$, then the smallest nut has already been encountered and is ass_nut. Otherwise, $ass\_bolt \neq n$, the smallest nut has not been encountered as of yet and it is then the nut n. Therefore, in the case $i = j = n$, one can identify the smallest nut and the corresponding bolt without performing any further comparison.

If the algorithm does not complete on the condition $i = j = n$, it completes on the condition $i = n + 1$. Then, the smallest nut had been encountered because all nuts have been visited. The smallest nut is ass_nut (and, as always, the smallest bolt is j, which is here equal to ass_bolt).

If the while loop ends with the case $i = j = n$, $n - 1$ comparisons were necessary to increase i from 1 to n, and $n - 1$ to increase j from 1 to n, for a total of $2n - 2$ comparisons. In any other case, the algorithm completes with $i = n + 1$ and $j \leqslant n - 1$. To reach such a state requires at most $n + (n - 2) = 2n - 2$ comparisons. Hence, in the worst case, the algorithm performs $2n - 2$ comparisons. (In the best case, when the smallest bolt is of index 1, Algorithm 1.8 performs only n comparisons.)

3. As in the two previous exercises, we consider the decision tree of any algorithm. Such a decision tree is a ternary tree. Indeed, there are three possible outcomes for any comparison:

   (a) The nut is smaller than the bolt;
   (b) The bolt is smaller than the nut;
   (c) The bolt and the nut correspond.

   There are $n!$ possible ways to associate n bolts with n nuts and thus $n!$ possible outputs for the algorithm. Therefore, the decision tree of any algorithm contains at least $n!$ leaves. Let h be the height of the decision tree of a given algorithm. A ternary tree of height h contains at most 3^h leaves (see below). Hence:

$$3^h \geqslant n! \quad \Leftrightarrow \quad h \geqslant \log_3(n!) \sim n \log_3(n) = \Theta(n \log(n)).$$

   We prove by induction that a ternary tree of height h contains at most 3^h leaves. When $h = 0$, the tree is reduced to a single leaf and the result holds. We assume that the result holds for some value h and we consider a tree of height $h + 1$. Each of the three children of the root define a subtree of the root. These three subtrees of the root are ternary trees of height at most h. Therefore, using the induction hypothesis, each of these three trees contains at most 3^h leaves. Therefore, overall the tree contains at most $3 \times 3^h = 3^{h+1}$ leaves.

Solution to Exercise 1.9: Sorting a small number of objects

1. To sort four numbers, we just follow the divide-and-conquer principle described in the text of the exercise. We create two pairs of elements $(a \to b)$ and $(c \to d)$ with one comparison for each pair; then we compare the two largest elements with an additional comparison; finally, we insert c with a binary search in the sorted list containing a and b (we already know that $c \leqslant d$), which requires two more comparisons. Hence, we sort four numbers in $2 \times 1 + 1 + 2 = 5$ comparisons, which is optimal.

2. **Sorting 5 numbers.** Sorting four numbers and then inserting the fifth one with a binary search would cost: $5 + 3 = 8$ comparisons, which would be suboptimal. Therefore, we proceed otherwise.
   We start, as previously, by creating two pairs of two elements $(a \to b)$ and $(c \to d)$ with one comparison for each pair. Then we compare the two largest elements with an additional comparison. After three comparisons, we obtain the same configuration as previously:

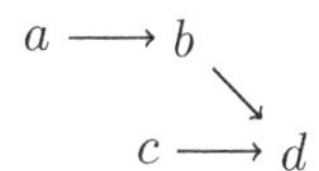

Then, we insert the fifth element, e, in the chain $a \to b \to d$ with two comparisons. Finally, we are left only with inserting c in the sorted chain made of the three elements a, b, and e, which costs two additional comparisons (we already know that $c \leqslant d$). Overall, we sort five numbers in $2 \times 1 + 1 + 2 + 2 = 7$ comparisons.

Sorting 6 numbers. We sort five of the numbers and then insert the sixth one using a binary search in $7 + 3 = 10$ comparisons.

Sorting 7 numbers. We sort six of the numbers and then insert the seventh one using a binary search in $10 + 3 = 13$ comparisons.

Sorting 8 numbers. We sort seven of the numbers and then insert the eighth one using a binary search in $13 + 3 = 16$ comparisons.

Sorting 9 numbers. We start by creating four sorted pairs (and an element is left alone) with four comparisons. Then we sort the greatest elements of the pairs in five comparisons. We then obtain the following configuration:

$$\begin{array}{ccccccccc} b & \longrightarrow & d & \longrightarrow & f & \longrightarrow & h & & i \\ \uparrow & & \uparrow & & \uparrow & & \uparrow & & \\ a & & c & & e & & g & & \end{array}$$

The trick is then to insert the elements c, e, g, and i in the chain $a \to b \to d \to f \to h$ in such a way as to minimize the number of comparisons required in the worst case. First, we insert e in $a \to b \to d$ (we already know that $e \leqslant f$) with two comparisons. Then, if $e \geqslant d$, we insert c in $a \to b$ (we already know that $c \leqslant d$). Otherwise, we insert c in $\{a, b, e\}$. In both cases, c is inserted in a sorted chain of at most three elements, which costs two comparisons. So far, we have sorted the set $\{a, b, c, d, e, f, h\}$. We insert i in this set of six elements using three comparisons. All that is left is then to insert g in the resulting set. In fact, because $g \leqslant h$, we need only to insert g in a set of six elements if $h \leqslant i$, or of seven elements if $i < h$. Such an insertion costs three additional comparisons. Overall, we sort nine numbers in $4 \times 1 + 5 + 2 + 2 + 3 + 3 = 19$ comparisons.

Sorting 10 numbers. To sort 10 numbers, we proceed as previously. We start by creating five sorted pairs with five comparisons. Then we sort the greatest elements of the pairs in seven comparisons. We then obtain the following configuration:

$$\begin{array}{ccccccccc} b & \longrightarrow & d & \longrightarrow & f & \longrightarrow & h & \longrightarrow & j \\ \uparrow & & \uparrow & & \uparrow & & \uparrow & & \uparrow \\ a & & c & & e & & g & & i \end{array}$$

Next we insert e in $a \to b \to d$ with two comparisons. Then, if $e \geqslant d$, we insert c in $a \to b$. Otherwise, we insert c in $\{a, b, e\}$. In both cases, c is inserted with, at most, two comparisons. So far, we have sorted the set $\{a, b, c, d, e, f, h, j\}$. We then insert i in the set of seven elements $\{a, b, c, d, e, f, h, \}$ using three comparisons (we already know that $i \leqslant j$). All that is left is to insert g in the resulting set. In fact, because $g \leqslant h$, we need only to insert g in a set of six elements if $h \leqslant i$, or of seven elements if $i < h$. Such an insertion costs three additional comparisons. Overall, we sort 10 numbers in $5 \times 1 + 7 + 2 + 2 + 3 + 3 = 22$ comparisons.

Sorting 11 numbers. We sort 10 of the numbers and then insert the 11th one using a binary search in $22 + 4 = 26$ comparisons.

3. To sort 12 numbers, we first sort 11 of them and then insert the 12th number using a binary search, in $26 + 4 = 30$ comparisons.

1.5 Bibliographical notes

Section 1.1 is inspired from the book by Knuth [62] (another monument). Several exercises come from the wonderful book by Rawlins [92], but the solutions are ours. This includes Exercise 1.3 (breaking boxes), Exercise 1.5 (maximum and minimum of n integers), and Exercise 1.8 (the toolbox). Exercise 1.1 (longest balanced section) comes from the book by Dijkstra [31]. Exercise 1.2 (find the star) is the classical problem to find whether there is a sink in a directed graph with n vertices, i.e., a vertex with in-degree $n - 1$ and out-degree 0; see, for instance, [17, Example 15]. Exercise 1.6 (maximum and second maximum of n integers) and Exercise 1.9 (sorting a small number of objects) come from the book by Froidevaux, Gaudel, and Soria [37].

Chapter 2

Divide-and-conquer

This chapter revisits the divide-and-conquer paradigms and explains how to solve recurrences, in particular, with the use of the "master theorem." We first illustrate the concept with Strassen's matrix multiplication algorithm (Section 2.1) before explaining the master theorem (Section 2.2) and finally providing techniques to solve recurrences (Section 2.3). These techniques are further illustrated in the exercises of Section 2.4, with solutions found in Section 2.5.

2.1 Strassen's algorithm

The classical matrix multiplication algorithm computes the product of two matrices of size $n \times n$ with $Add(n) = n^2(n-1)$ additions and $Mult(n) = n^3$ multiplications. Indeed, there are n^2 coefficients to compute, each of them corresponding to a scalar product of size n, thus with n multiplications, $n-1$ additions, and one affectation. Can we do better than this?

Note that the question was raised at a time when it was mainly interesting to decrease the number of multiplications, even though this would imply computing more additions. The pipelined architecture of today's processors allows us to perform, in steady-state mode, one addition or one multiplication per cycle time.

Strassen introduced a new method in his seminal paper [101]. Let us compute the product of two 2×2 matrices:

$$\begin{pmatrix} r & s \\ t & u \end{pmatrix} = \begin{pmatrix} a & b \\ c & d \end{pmatrix} \times \begin{pmatrix} e & f \\ g & h \end{pmatrix}$$

We first compute seven intermediate products

$$\begin{aligned}
p_1 &= a(f-h)\\
p_2 &= (a+b)h\\
p_3 &= (c+d)e\\
p_4 &= d(g-e)\\
p_5 &= (a+d)(e+h)\\
p_6 &= (b-d)(g+h)\\
p_7 &= (a-c)(e+f)
\end{aligned}$$

and then we can write

$$\begin{aligned}
r &= p_5+p_4-p_2+p_6\\
s &= p_1+p_2\\
t &= p_3+p_4\\
u &= p_5+p_1-p_3-p_7.
\end{aligned}$$

If we count operations for each method, we obtain the following:

Classic	*Strassen*
$Mult(2) = 8$	$Mult(2) = 7$
$Add(2) = 4$	$Add(2) = 18$

Strassen's method gains one multiplication, but at the price of 14 extra additions, thus being worse on modern processors than the classical method for 2×2 matrices. However, it is remarkable that the new method does not require the commutativity of multiplication, and, therefore, it can be used, for instance, with matrices instead of numbers. We can readily use it with matrices of even size n, say $n = 2m$. We consider that $a, b, c, d, e, f, g, h, r, s, t$, and u are matrices of size $m \times m$. So, let $n = 2m$, and use the previous approach with submatrices of size $m \times m$. To compute each p_i $(1 \leqslant i \leqslant 7)$ with the classic matrix multiplication algorithm, we need m^3 multiplications, thus a total $Mult(n) = 7m^3 = 7n^3/8$. For the additions, we need to add the additions performed in the seven matrix multiplications to form the intermediate products p_i, namely $7m^2(m-1)$, with the number of additions required to form the auxiliary matrices, namely $18m^2$. Indeed, there are 10 matrix additions to compute the p_is, and then 8 other matrix additions to obtain r, s, t, and u. Therefore, we have a total of $Add(n) = 7m^3 + 11m^2 = 7n^3/8 + 11n^2/4$.

Asymptotically, the dominant term is in $\frac{7}{8}n^3$ for $Mult(n)$ as for $Add(n)$, and the new method is interesting for n large enough. The intuition is the following: Multiplying two matrices of size $n \times n$ requires $O(n^3)$ operations (both for pointwise multiplications and additions), while adding two matrices of size $n \times n$ requires only $O(n^2)$ operations. For n large enough, matrix additions have a negligible cost in comparison to matrix multiplications (and the main source of pointwise additions is within these matrix multiplications). That was not the case for real numbers, hence, the inefficiency of the method for 2×2 matrices.

Strassen's algorithm is the recursive use of the decomposition explained above. We consider the case in which n is a power of 2, i.e., $n = 2^s$. Otherwise, we can extend all matrices with zeroes so that they have a size that is the first power of 2 greater than n, and replace in the following $\log(n)$ by $\lceil \log(n) \rceil$:

$$(X) \longrightarrow \begin{pmatrix} X & 0 \\ 0 & 0 \end{pmatrix} .$$

Let us consider matrices of size $n \times n$, where $n = 2^s$. We proceed by induction. We use the method recursively to compute each of the matrix products p_i, for $1 \leqslant i \leqslant 7$. We stop when matrices are of size 1 or, better, when Strassen's method is more costly than the classical method, for matrix sizes below a "crossover point." In practice, this crossover point is highly system dependent. By ignoring cache effects, we can obtain crossover points as low as $n = 8$ [48], while [29] determines the crossover points by benchmarking on various systems, and it ranges from $n = 400$ to $n = 2150$.

In the following, we stop the recursion when $n = 1$, and:

- $M(n)$ is the number of multiplications done by Strassen's algorithm to multiply two matrices of size $n \times n$;
- $A(n)$ is the number of additions done by Strassen's algorithm to multiply two matrices of size $n \times n$.

For the multiplications, we have:

$$\begin{cases} M(1) = 1 \\ M(n) = 7 \times M(n/2) \end{cases} \Longrightarrow M(n) = 7^s = 7^{\log(n)} = n^{\log(7)}.$$

As before, additions come from two different sources: the additions that are done in the 7 matrix multiplications (recursive call) and the 18 matrix additions (construction of the p_i's and of r, s, t, and u). We finally have:

$$\begin{cases} A(1) = 0 \\ A(n) = 7 \times A(n/2) + 18 \times (n/2)^2 \end{cases} \Longrightarrow A(n) = 6 \times (n^{\log(7)} - n^2). \quad (2.1)$$

We explain in Section 2.3 how this recurrence can be solved. Note that the recursive approach has improved the order of magnitude of the total computation cost, not just only the constant (previously, we had only $\frac{7}{8}n^3$ instead of n^3). The new order of magnitude is $O(n^{\log(7)})$ and $\log(7) \approx 2.81$.

Finally, we conclude by saying that Strassen's algorithm is not widely used because it introduces some numerical instability. Also, there are some algorithms with a better complexity. At the time of this writing, the best algorithm is the Coppersmith–Winograd algorithm, in $O(n^{2.376})$ [26]. The problem of establishing the complexity of matrix product is still open. The only known lower bound is a disappointing $O(n^2)$; we need to touch each coefficient at least once.

Strassen's algorithm provides, however, an excellent illustration of the divide-and-conquer paradigm, which we formalize in the next section through the master theorem.

2.2 Master theorem

Before formulating the master theorem, we need to formalize the divide-and-conquer paradigm that was illustrated in the previous section through the Strassen's algorithm.

DEFINITION 2.1 (Divide-and-conquer). Consider a problem of size n. In order to solve the problem, divide it into a subproblems of size n/b that will allow us to find the solution. The cost of this divide-and-conquer algorithm is then

$$S(n) = a \times S\left(\frac{n}{b}\right) + R(n) \tag{2.2}$$

where $R(n)$ is the cost to reconstruct the solution of the problem of size n from the solutions of the subproblems; it is often equal to $R(n) = c \times n^\alpha$, for some constants c and α. Initially, we often have $S(1) = 1$ (or equal to another constant value).

For instance, with Strassen's algorithm, if we consider the number of additions to be executed in a matrix product, we have $a = 7$, $b = 2$, $\alpha = 2$, and $c = \frac{18}{4}$. Indeed, the product of two matrices of size $n \times n$ is performed by first computing 7 products of matrices of size $n/2 \times n/2$ and reconstructing the solution through 18 additions of matrices of size $n/2 \times n/2$; therefore, $R(n) = 18(n/2)^2 = \frac{18}{4}n^2$. In this case, the initial cost is $S(1) = 0$.

Let us assume that there exists $k \in \mathbb{N}$ such that $n = b^k$; thus, $k = \log_b(n)$ and $a^k = n^{\log_b(a)}$. If we develop the formula in equation (2.2), we obtain the following:

$$\begin{aligned} S(n) &= a \times S(\tfrac{n}{b}) \quad + R(n) \\ &= a^2 \times S(\tfrac{n}{b^2}) + a \times R(\tfrac{n}{b}) + R(n) \\ &= \cdots \\ &= a^k \times S(1) \quad + \textstyle\sum_{i=0}^{k-1} a^i \times R(\tfrac{n}{b^i}). \end{aligned}$$

We consider the most usual case in which $R(n) = c \times n^\alpha$, and, therefore, we have $\sigma = \sum_{i=0}^{k-1} a^i \times R(\frac{n}{b^i}) = c \times n^\alpha \sum_{i=0}^{k-1} (a/b^\alpha)^i$.

We then distinguish several cases:

1. $(a > b^\alpha)$: $\sigma = \Theta(n^\alpha \times (\frac{a}{b^\alpha})^k) = \Theta(a^k) \Longrightarrow S(n) = \Theta(n^{\log_b(a)})$;

2. $(a = b^\alpha)$: $\sigma = \Theta\left(k \times n^\alpha\right) \Longrightarrow S(n) = \Theta\left(n^\alpha \times \log(n)\right)$;

3. $(a < b^\alpha)$: $\sigma = \Theta\left(n^\alpha \times \frac{1}{1-\frac{a}{b^\alpha}}\right) \Longrightarrow S(n) = \Theta(n^\alpha)$.

We have proved the following theorem:

THEOREM 2.1 (Master theorem). *The cost of a divide-and-conquer algorithm such that* $S(n) = a \times S(\frac{n}{b}) + c \times n^\alpha$ *is the following:*

(i) if $a > b^\alpha$, *then* $S(n) = \Theta(n^{\log_b(a)})$;

(ii) if $a = b^\alpha$, *then* $S(n) = \Theta(n^\alpha \times \log(n))$;

(iii) if $a < b^\alpha$, *then* $S(n) = \Theta(n^\alpha)$.

A fully detailed proof of Theorem 2.1 is given in [27]. Let us come back to Strassen's algorithm. We divided the matrices into four blocs of size $n/2$, and we would like to investigate a solution in which we would rather divide matrices into nine blocs of size $n/3$:

$$\left(\begin{array}{c|c|c} & & \\ \hline & & \\ \hline & & \end{array}\right) \times \left(\begin{array}{c|c|c} & & \\ \hline & & \\ \hline & & \end{array}\right) .$$

We would then have $b = 3$ and $\alpha = 2$ (the reconstruction cost is still in n^2). Let us assume that we are in case (i) of the master theorem. Then, this new algorithm would become better than Strassen's if and only if:

$$\begin{array}{rrcl} & \log_3(a) & < & \log(7) \\ \iff & \log(a) & < & \log(7) \times \log(3) \\ \iff & a & < & 7^{\log(3)} \approx 21.8. \end{array}$$

This is an open problem; one knows a method with $a = 23$ subproblems [69] but not with $a = 21$!

2.3 Solving recurrences

In this section, we detail how to solve recurrences that occur in the cost analysis of divide-and-conquer algorithms but that are slightly more complex than in the application case of the master theorem. We start with homogeneous recurrences and then consider the most general case of recurrences with a second member.

2.3.1 Solving homogeneous recurrences

A homogeneous linear recurrence with constant coefficients has the form $p_0 \times s_n + p_1 \times s_{n-1} + \cdots + p_k \times s_{n-k} = 0$, where each p_i is a constant and $(s_i)_{i \geqslant 0}$ is an unknown sequence. It is said to be homogeneous because the second member is null, i.e., the linear combination is set equal to zero. Solving such recurrences requires finding all the roots of the polynomial $P = \sum_{i=0}^{k} p_i \times X^{k-i}$, together with their multiplicity order. However, we see that P is a polynomial of degree k, and no algebraic method can find the roots of arbitrary polynomials of degree 5 or higher. Therefore, we need additional information, such as trivial roots, for high-degree recurrences.

Let us assume that we can find the k roots of P, $r_1, \ldots, r_k$. If these roots are distinct, then the general form of the solution is $s_n = \sum_{i=1}^{k} c_i \times {r_i}^n$, where the c_is are some constants that depend upon the first values of the sequence. Otherwise, let q_i be the order of multiplicity of root r_i, for $1 \leqslant i \leqslant \ell$ (with $\ell < k$ distinct roots). Then we have $s_n = \sum_{i=1}^{\ell} P_i(n) \times {r_i}^n$, where $P_i(n)$ is a polynomial of degree $q_i - 1$. Here again, the coefficients of the $P_i(n)$s are computed using the initial values of the recurrence.

2.3.2 Solving nonhomogeneous recurrences

In the general case, the recurrence may have a nonzero right-hand side, for instance, $s_n - 2s_{n-1} = 2^{n+1}$. Such recurrences are called *nonhomogeneous*. To explain how to solve them, we start by introducing a few notations. A sequence is represented by writing down its n-th element formula in curly brackets, for instance $\{3^n\}$ represents the sequence $1, 3, 9, 27, \ldots$ (starting at $n = 0$).

Then we introduce E, an operator that transforms a sequence by shifting it and leaving out its first element. In our example, $E\{3^n\} = 3, 9, 27, 81, \ldots = \{3^{n+1}\}$. More generally, $E\{s_n\} = \{s_{n+1}\}$.

We then define the following operations on sequences:

$$
\begin{aligned}
c\{s_n\} &= \{cs_n\}, \\
(E_1 + E_2)\{s_n\} &= E_1\{s_n\} + E_2\{s_n\}, \\
(E_1E_2)\{s_n\} &= E_1(E_2\{s_n\}).
\end{aligned}
$$

For instance, $(E - 3)\{s_n\} = \{s_{n+1} - 3s_n\}$, and $(2 + E^2)\{s_n\} = \{2s_n + s_{n+2}\}$.

We are looking for annihilators of the sequences. That is, we are looking for operators $P(E)$ such that $P(E)\{s_n\} = \{0\}$. For our example, $(E - 3)\{3^n\} = \{3^{n+1} - 3 \times 3^n\} = \{0\}$. We provide a few more examples, where $Q_k(n)$ is a polynomial in n of degree k:

sequence	annihilator
$\{c\}$	$E - 1$
$\{Q_k(n)\}$	$(E - 1)^{k+1}$
$\{c^n\}$	$E - c$
$\{c^n \times Q_k(n)\}$	$(E - c)^{k+1}$

The first three lines are special cases of the fourth line; therefore, we need to prove only the last relation. We prove it by induction on k. We start with $k = 0$, writing $Q_0(n) = q$:

$$
(E - c)\{c^n \times Q_0(n)\} = qE\{c^n)\} - c\{qc^n\} = \{qc^{n+1}\} - \{qc^{n+1}\} = \{0\}.
$$

Now by induction for $k \geqslant 1$, writing $Q_k(n) = a_0 n^k + Q_{k-1}(n)$:

$$\begin{aligned}(E-c)^{k+1}\{c^n \times Q_k(n)\} &= (E-c)^{k+1}\{c^n \times (a_0 n^k + Q_{k-1}(n))\} \\ &= (E-c)^k[(E-c)\{c^n(a_0 n^k + Q_{k-1}(n))\}] \\ &= (E-c)^k\{c^{n+1}(a_0(n+1)^k + Q_{k-1}(n+1)) \\ &\qquad - \; c^{n+1}(a_0 n^k + Q_{k-1}(n))\} \\ &= (E-c)^k[c^{n+1} \times R_{k-1}(n)],\end{aligned}$$

where $R_{k-1}(n)$ is a polynomial in n of degree $k-1$, because both $(n+1)^k - n^k$ and $Q_{k-1}(n+1) - Q_{k-1}(n)$ are polynomials of degree $k-1$. With the induction hypothesis, we obtain the result: $(E-c)^{k+1}\{c^n \times Q_k(n)\} = \{0\}$.

2.3.3 Solving the recurrence for Strassen's algorithm

We focus on the recurrence for the number of additions (see equation (2.1)):

$$A(n) = 7 \times A\left(\frac{n}{2}\right) + \frac{18}{4} \times n^2.$$

We have $n = 2^s$, and we consider the sequence $\{A_s\}$ such that $A_s = A(2^s)$. Thus, we have $A_{s+1} = 7 \times A_s + \frac{18}{4} \times (2^{s+1})^2 = 7 \times A_s + 18 \times 4^s$, and the annihilator is $(E-4)(E-7)$:

$$(E-4)(E-7)\{A_s\} = (E-4)\{A_{s+1} - 7A_s\} = (E-4)\{18 \times 4^s\} = \{0\}.$$

We have found an annihilator for the sequence, namely, $(E-4)(E-7)$, and, even better, it is in decomposed form, so we immediately have its two distinct roots, 4 and 7. From the previous result, we know the general form of the solution, namely,

$$A_s = k_1 \times 7^s + k_2 \times 4^s \ .$$

From the initial conditions $A_0 = 0$ and $A_1 = 18$, we obtain the values $k_1 = 6$ and $k_2 = -6$, and, finally,

$$A(n) = 6 \times 7^s - 6 \times 4^s \ .$$

2.4 Exercises

Exercise 2.1: Product of two polynomials (solution p. 42)

The goal of this exercise is to multiply two polynomials efficiently. An n-polynomial is a polynomial with a degree strictly less than n, thus with n coefficients.

Let $P = \sum_{i=0}^{n-1} a_i X^i$ and $Q = \sum_{i=0}^{n-1} b_i X^i$ be two n-polynomials. Their product $R = P \times Q$ is a $(2n-1)$-polynomial. We denote by $M(n)$ (resp. $A(n)$) the number of multiplications (resp. number of additions) done by an algorithm to multiply two n-polynomials.

1. Compute $M(n)$ and $A(n)$ for the usual algorithm to multiply two n-polynomials.

2. We assume that n is even, $n = 2 \times m$. We can then write $P = P_1 + X^m \times P_2$ and $Q = Q_1 + X^m \times Q_2$. What is the degree of the polynomials P_1, P_2, Q_1, and Q_2?

3. Let $R_1 = P_1 \times Q_1$, $R_2 = P_2 \times Q_2$, and $R_3 = (P_1 + P_2) \times (Q_1 + Q_2)$. Can you express $R = P \times Q$ as a function of R_1, R_2, and R_3? What is the degree of these three new polynomials? Compute $M(n)$ and $A(n)$, assuming that we use the classical multiplication algorithm to compute R_1, R_2, and R_3.

4. We assume now that $n = 2^s$ and we apply recursively the previous algorithm. Compute $M(n)$ and $A(n)$ for this algorithm.

Exercise 2.2: Toeplitz matrices (solution p. 44)

A *Toeplitz matrix*, or diagonal-constant matrix, named after Otto Toeplitz, is an $n \times n$ matrix with $(a_{i,j})$ coefficients $(1 \leqslant i, j \leqslant n)$ and such that $a_{i,j} = a_{i-1,j-1}$ for $2 \leqslant i, j \leqslant n$.

1. Let A and B be two Toeplitz matrices. Is the sum $A + B$ a Toeplitz matrix? And the product $A \times B$?

2. Give an algorithm to add two Toeplitz matrices in $O(n)$.

3. We assume here that $n = 2^k$. How can we compute the product of an $n \times n$ Toeplitz matrix M by a vector T of length n? What is the complexity of the algorithm?

   Hint: Decompose M as a matrix of blocks of size 2^{k-1}, decompose T accordingly:

   $$M = \begin{pmatrix} A & B \\ C & A \end{pmatrix} \quad \text{and} \quad T = \begin{pmatrix} X \\ Y \end{pmatrix}$$

   and consider the three matrices $U = (C + A)X$, $V = A(Y - X)$, and $W = (B + A)Y$.

Exercise 2.3: Maximum sum (solution p. 45)

Let T be a table of n relative integers. We want to find the maximum sum of contiguous elements, namely, two indices i and j $(1 \leqslant i \leqslant j \leqslant n)$ that maximize $\sum_{k=i}^{j} T[k]$.

1. If the values in the table are $T[1] = 2$, $T[2] = 18$, $T[3] = -22$, $T[4] = 20$, $T[5] = 8$, $T[6] = -6$, $T[7] = 10$, $T[8] = -24$, $T[9] = 13$, and $T[10] = 3$, can you return the two indices and the corresponding optimal sum?

2. Design an algorithm that returns the maximum sum of contiguous elements with a divide-and-conquer algorithm.

3. Design a linear-time algorithm that solves the problem through a single scan of the array.

Exercise 2.4: Boolean matrices: The Four-Russians algorithm

(solution p. 49)

The goal in this exercise is to multiply two $n \times n$ Boolean matrices, A and B. All matrix elements are either 0 or 1, and the sum and product correspond respectively to the *or* and *and* operations on Booleans.

1. Can we easily apply Strassen's algorithm to compute the product $A \times B$?

2. Apart from the classical multiplication algorithm, another way to view the product consists of multiplying columns of A with rows of B. Give an expression of $A \times B$, using $A_c[\ell]$, the ℓ-th column of A, and $B_r[\ell]$, the ℓ-th row of B.

3. To optimize the matrix product, the idea is to partition the columns of A and the rows of B into n/k equal-sized groups of size k (we can assume, for simplicity, that k divides n; otherwise, the last group is smaller). Therefore, for $1 \leqslant i \leqslant n/k$, A_i is an $n \times k$ matrix with k columns of A ($A_c[(i-1) \times k + 1], \dots, A_c[(i-1) \times k + k]$), and, similarly, B_i is a $k \times n$ matrix with k rows of B ($B_r[(i-1) \times k + 1], \dots, B_r[(i-1) \times k + k]$). Give an expression of $A \times B$, using the matrices A_i and B_i.

4. Provide a method to compute $C_i = A_i \times B_i$ in time $O(n^2)$ for all $1 \leqslant i \leqslant n/k$. (*Hint:* Show that each row of C_i can take only 2^k different values, precompute all possible values and store them in a table. What is the size of the table, i.e., the additional space required to run the algorithm? What is the time required to build the table?)

5. Building upon the previous method, provide an algorithm to compute $A \times B$, and give its complexity, in terms of k and n.

6. Which value of k would be most suited for this algorithm? What is the complexity of this matrix product algorithm? Compare with Strassen's algorithm.

Note that this algorithm is known as the *Four-Russians algorithm*, and it is due to Arlazarov et al. [3, 75].

Exercise 2.5: Matrix multiplication and inversion (solution p. 50)

Let $M(n)$ be the complexity of multiplying two square matrices of size n and $I(n)$ be the complexity of inverting a (square) matrix of size n. The functions $M(n)$ and $I(n)$ are not known, but the goal of this exercise is to show the following: If we assume that $M(n) = \Theta(n^\alpha)$ and $I(n) = \Theta(n^\beta)$, then $\alpha = \beta$. In other words, the complexity of both operations is of the same order under our hypothesis.

1. Prove that $2 \leqslant \alpha, \beta \leqslant 3$.

2. Prove that $\alpha \leqslant \beta$; matrix multiplication is not more complex than matrix inversion (which is intuitive).

3. Prove that $\beta \leqslant \alpha$; reciprocally, matrix inversion is not more complex than matrix multiplication (which is less intuitive).
   (*Hint*: Show that we can reduce the problem to inverting symmetric and positive definite matrices A whose size is an exact power of 2, and use the Schur complement $S = D - CB^{-1}C^T$ to recursively compute the inverse of $A = \begin{pmatrix} B & C^T \\ C & D \end{pmatrix}$. Note that B and D are symmetric and positive definite, too.)

2.5 Solutions to exercises

Solution to Exercise 2.1: Product of two polynomials

1. With the usual algorithm to multiply n-polynomials:

$$M(n) = n^2 \text{ and } A(n) = n^2 - \underbrace{(2n-1)}_{assignments} = (n-1)^2.$$

   Indeed, we multiply each of the n coefficients of P with each of the n coefficients of Q. Then, the number of additions is equal to the number of multiplication results minus the number of results computed, and there are $2n-1$ coefficients in the computed polynomial.

2. P_1, P_2, Q_1, and Q_2 are m-polynomials and of degree $m-1$.

3. We have $R = R_1 + (R_3 - R_2 - R_1) \times X^m + R_2 \times X^{2m}$. R_1, R_2, and R_3 are polynomials of degree $2m-2 = n-2$, and thus $(n-1)$-polynomials.

   Following this computation scheme, $M(n) = 3M(\frac{n}{2}) = \frac{3n^2}{4}$, as the computation of R_1, R_2, and R_3 each requires $M(m) = M(\frac{n}{2})$ multiplications. There are four types of additions: (1) those involved in the computations of R_1, R_2, and $R3$; (2) those involved in computing the polynomials $P_1 + P_2$ and $Q_1 + Q_2$ needed by R_3; (3) those involved in computing $R_3 - R_2 - R_1$; and (4) those needed to compute R from the partial results. We, therefore, obtain:

$$A(n) = \underbrace{3A(m)}_{R_1,\ R_2,\ \text{and}\ R_3} + \underbrace{2\ m}_{R_3\ \text{arguments}} + \underbrace{2(2m-1)}_{R_3 - R_2 - R_1} + \underbrace{(2m-2)}_{\text{building}\ R}.$$

Indeed, R is defined as follows:

$$\begin{aligned} R &= R_1 + (R_3 - R_1 - R_2) \times X^m + R_2 \times X^{2m} \\ &= \underbrace{\sum_{i=0}^{2m-2} r_{1,i} \times X^i}_{X^0 \to X^{2m-2}} + \underbrace{\sum_{i=0}^{2m-2} z_i \times X^{i+m}}_{X^m \to X^{3m-2}} + \underbrace{\sum_{i=0}^{2m-2} r_{2,i} \times X^{i+2m}}_{X^{2m} \to X^{4m-2}} \end{aligned}$$

where, for any $1 \leqslant j \leqslant 3$, $R_j = \sum_{i=0}^{2m-2} r_{j,i} X^i$ and, for any $0 \leqslant i \leqslant 2m-2$, $z_i = r_{3,i} - r_{1,i} - r_{2,i}$. Therefore, each of the terms in the second sum, except z_{m-1}, is added either to a term of the first sum or to a term of the third sum. Hence, there are a total of $(2m-1) - 1 = 2m - 2$ additions.

Therefore,

$$A(n) = 3A\left(\frac{n}{2}\right) + 4n - 4 = 3\left(\frac{n}{2} - 1\right)^2 + 4n - 4 = \frac{3}{4}n^2 + n - 1.$$

One can then check that under this scheme, the number of multiplications is always decreased, and the number of additions decreases as soon as $n \geqslant 12$.

4. We recursively apply the above scheme when $n = 2^s$. We have:

$$\begin{cases} M(1) = 1 \\ M(n) = 3 \times M(\frac{n}{2}) \end{cases} \implies M(n) = 3^s = n^{log_2(3)} \approx n^{1.58}.$$

$$\begin{cases} A(1) = 0 \\ A(n) = 3 \times A\left(\frac{n}{2}\right) + 4n - 4 \end{cases} \implies A(n) = 6n^{log_2(3)} - 8n + 2.$$

The expression for $A(n)$ is obtained using the method explained in Section 2.3 (p. 37). Indeed, from $A(n) = 3A\left(\frac{n}{2}\right) + 4n - 4$, we define $A_s = 3A_{s-1} + 4 \times 2^s - 4$. Then, using the notations of Section 2.3, $(E-1)(E-2)(E-3)\{A_s\} = \{0\}$ and, thus, there exist k_1, k_2, and k_3 such that $A_s = k_1 \times 3^s + k_2 \times 2^s + k_3$. Using $A_0 = 0$, $A_1 = 4$, and $A_2 = 24$, we obtain $k_1 = 6$, $k_2 = -8$, and $k_3 = 2$.

Solution to Exercise 2.2: Toeplitz matrices

1. The sum of two Toeplitz matrices is a Toeplitz matrix. However, the product of two Toeplitz matrices is not a Toeplitz matrix:

$$\begin{pmatrix} 1 & 0 \\ 1 & 1 \end{pmatrix} \times \begin{pmatrix} 1 & 1 \\ 0 & 1 \end{pmatrix} = \begin{pmatrix} 1 & 1 \\ 1 & 2 \end{pmatrix}.$$

2. We need to perform only the addition of the first row of A with the first row of B, and the first column of A with the first column of B, hence, $2n - 1$ additions. A Toeplitz matrix indeed can be fully represented by its first row and its first column.

3. We compute the matrix product $M \times T$ according to the decomposition proposed in the hint:

$$M \times T = \begin{pmatrix} A & B \\ C & A \end{pmatrix} \times \begin{pmatrix} X \\ Y \end{pmatrix} = \begin{pmatrix} A \times X + B \times Y \\ C \times X + A \times Y \end{pmatrix}.$$

We remark that A, B, and C are Toeplitz matrices. With the notations $U = (C + A)X$, $V = A(Y - X)$, and $W = (B + A)Y$, we have:

$$M \times T = \begin{pmatrix} W - V \\ U + V \end{pmatrix}.$$

The motivation of this approach is that we went from four to three multiplications of $2^{k-1} \times 2^{k-1}$ matrices. To assess the potential gain of this method, we look at the complexity of this multiplication scheme.

Computing the complexity. We denote, respectively, by $M(n)$ and $A(n)$ the number of multiplications and of additions performed for the product of two matrices of size $n \times n$. The number of multiplications, $M(n)$, is defined as follows:

$$\begin{cases} M(1) & = 1 \\ M(2^k) & = 3 \times M(2^{k-1}). \end{cases}$$

The computation of $A(n)$ is more involved:

$$\begin{cases} A(1) & = 0 \\ A(2^k) & = \underbrace{3 \cdot A(2^{k-1})}_{\substack{\text{from multiplications} \\ \text{in } U,\, V, \text{ and } W}} + \underbrace{2 \cdot (2 \cdot 2^{k-1} - 1)}_{(C+A) \text{ and } (B+A)} + \underbrace{2^{k-1}}_{Y-X} + \underbrace{2 \cdot 2^{k-1}}_{(W-V) \text{ and } (U+V)} \\ & = 3 \cdot A(2^{k-1}) + 2 \cdot (2^k - 1) + 3 \cdot 2^{k-1}. \end{cases}$$

To solve the recursions, we let $M_s = M(2^s) = M(n)$ and $A_s = A(2^s) = A(n)$, and we follow the method and notation of Section 2.3 (p. 37). For the number of multiplications performed, $M_0 = 1$ and $M_s = 3M_{s-1}$.

Therefore, $(E-3)\{M_s\} = \{0\}$, and thus $M_s = k_1 \times 3^s$. From $M_0 = 1$, we obtain $k_1 = 1$ and, finally, $M(n) = 3^{\log(n)} = n^{\log(3)}$.

This result also can be established using the following method:

$$M(2^k) = 3 \times M(2^{k-1}) \quad \Leftrightarrow \quad \frac{M(2^k)}{2^k} = \frac{3}{2}\frac{M(2^{k-1})}{2^{k-1}}.$$

We then let $u_k = \frac{3}{2}u_{k-1}$ and $u_0 = 1$, which leads to $u_k = \left(\frac{3}{2}\right)^k$. Finally, $M(2^k) = \left(\frac{3}{2}\right)^k \times 2^k = 3^k$.

We now compute the number of additions required. $A_0 = 0$ and $A_s = 3A_{s-1} + 7 \times 2^{s-1} - 2$. Therefore,

$$\begin{array}{ll} A_s - 3A_{s-1} = 7 \times 2^{s-1} - 2 & \Rightarrow \\ (E-3)\{A_s\} = 7 \times 2^s - 2 & \Rightarrow \\ (E-2)(E-3)\{A_s\} = -2 & \Rightarrow \\ (E-1)(E-2)(E-3)\{A_s\} = \{0\}. & \end{array}$$

Therefore, there exists k_1, k_2, and k_3 such that $A_s = k_1 3^s + k_2 2^s + k_3$. As $A_0 = A(1) = 0$, $A_1 = A(2) = 5$, and $A_2 = A(4) = 27$, to find k_1, k_2, and k_3, we have to solve the system:

$$\begin{cases} k_1 + k_2 + k_3 = 0 \\ 3k_1 + 2k_2 + k_3 = 5 \\ 9k_1 + 4k_2 + k_3 = 27 \end{cases}$$

and we obtain $k_1 = 6$, $k_2 = -7$, and $k_3 = 1$. Finally,

$$A(n) = 6 \times n^{\log(3)} - 7 \times n + 1.$$

Solution to Exercise 2.3: Maximum sum

1. The algorithm should return the indices 4 and 7, which leads to a sum of 32.

2. We want to divide the array T into halves. Then, three cases are possible:

   (a) The interval of maximum sum is included in the first half;
   (b) The interval of maximum sum is included in the second half;
   (c) The interval of maximum sum contains elements from both halves.

   The first two cases are solved through simple recursive calls on the halves. For the third case, we compute the interval of maximum sum that starts at the first element of the second half (this interval is then included in the second half). Symmetrically, we compute the interval of maximum sum that ends with the last element of the first half. The union of these two intervals is the interval of maximum sum that contains

```
1  if b = e then return (b, b, A[b])
2  middle ← ⌊(e−b)/2⌋
3  (start_1, end_1, sum_1) ← LS(b, middle)
4  (start_2, end_2, sum_2) ← LS(middle + 1, e)
5  LeftHalfSum ← A[middle];        LeftLimit ← middle
6  s ← A[middle]
7  for i = middle − 1 downto b do
8  |   s ← s + A[i]
9  |   if s > LeftHalfSum then
10 |   |_ LeftHalfSum ← s;    LeftLimit ← i
   |_
11 RightHalfSum ← A[middle + 1];        RightLimit ← middle + 1
12 s ← A[middle + 1]
13 for i = middle + 2 to e do
14 |   s ← s + A[i]
15 |   if s > RightHalfSum then
16 |   |_ RightHalfSum ← s;    RightLimit ← i
   |_
17 sum_3 ← LeftHalfSum + RightHalfSum
18 if sum_1 = max{sum_1, sum_2, sum_3} then
19 |_ return (start_1, end_1, sum_1)
20 if sum_2 = max{sum_1, sum_2, sum_3} then
21 |_ return (start_2, end_2, sum_2)
22 return (LeftLimit, RightLimit, sum_3)
```

ALGORITHM 2.1: LS(b, e): Find the maximum sum of contiguous elements of the array T between the indices b and e (included).

elements from both halves. Algorithm 2.1 presents a simple realization of this divide-and-conquer principle.

A call to Algorithm 2.1 with an array of size n leads to two recursive calls on arrays of size $n/2$. Dividing the array has a constant cost. However, computing the overall results from the subresults requires scanning the whole array and, hence, has a cost of $\Theta(n)$. Therefore, the complexity is given by:

$$C(n) = C\left(\frac{n}{2}\right) + \Theta(n).$$

Using the master theorem (p. 36), we have $a = 2$, $b = 2$, and $\alpha = 1$, and thus

$$C(n) = \Theta(n \log n).$$

In order to lower the complexity of the divide-and-conquer solution, we do not want to scan the entire array to compute the interval of maximum sum that contains elements of both halves. In other words, we want to be able to compute such an interval in constant time. To solve this problem, we just have to remark that the maximum-sum interval starting with the first element of an array either is fully contained in the first half of this array or includes all of this first half and then is equal to the whole first half plus the maximum-sum interval starting with the first element of the second half of the array. If the recursive calls on the two halves identify:

- The maximum-sum interval starting with the first element;
- The maximum-sum interval ending with the last element;
- The sum of all the elements in the array

we will be able to compute these values in constant time for the original array. Algorithm 2.2 exactly realizes this scheme.

A call to Algorithm 2.2 with an array of size n leads to two recursive calls on arrays of size $n/2$. Dividing the array and computing all the needed results from the subresults has a constant cost. Therefore, the complexity is:

$$C(n) = C\left(\frac{n}{2}\right) + \Theta(1).$$

Using the master theorem (p. 36), we have $a = 2$, $b = 2$, and $\alpha = 0$, and thus

$$C(n) = \Theta(n).$$

3. This new solution to the maximum-sum problem relies on several properties:
   - If none of the elements in the array is positive, then the interval of maximum sum is reduced to a single element, the maximum array element.
   - No strict prefix or suffix of the interval of maximum sum has a negative sum. We prove this result by contradiction, assuming that the interval of maximum sum starts at index i, ends at k ($k > i$), and admits a prefix of negative sum that ends at j ($i \leqslant j < k$). Then, $\sum_{l=i}^{k} T_l = \left(\sum_{l=i}^{j} T_l\right) + \left(\sum_{l=j+1}^{k} T_l\right) < \sum_{l=j+1}^{k} T_l$. As a consequence, if at least one element in the array is positive, there exists an interval of maximum sum whose first and last elements are positive.
   - If (j, k) is an interval of maximum sum, then, whatever the index $i < j$, the interval $(i, j-1)$ has a nonpositive sum. To prove this, we remark that as (j, k) is an interval of maximum sum, its sum is greater than or equal to that of the interval (i, k). Therefore:

$$\sum_{l=i}^{k} T[l] \leqslant \sum_{l=j}^{k} T[l] \quad \Leftrightarrow \quad \sum_{l=1}^{j-1} T[l] \leqslant 0.$$

```
1  if b = e then return (b, T[b], b, T[b], T[b], b, b, T[b])
2  middle ← ⌊(f−d)/2⌋
3  (start_1, sum_start_1, end_1, sum_end_1, sum_1, b_1, e_1, s_1)
   ← LSR(b, middle)
4  (start_2, sum_start_2, end_2, sum_end_2, sum_2, b_2, e_2, s_2)
   ← LSR(middle+1, e)
5  if sum_start_1 ⩾ sum_1 + sum_start_2 then
6      sum_start ← sum_start_1
7      start ← start_1
8  else
9      sum_start ← sum_1 + sum_start_2
10     start ← start_2
11 if sum_end_2 ⩾ sum_2 + sum_end_1 then
12     sum_end ← sum_end_2
13     end ← end_2
14 else
15     sum_end ← sum_2 + sum_end_1
16     end ← end_1
17 sum ← sum_1 + sum_2
18 b_3 ← end_1
19 e_3 ← start_2
20 s_3 ← sum_end_1 + sum_start_2
21 if s_3 = max{s_1, s_2, s_3} then
22     return (start, sum_start, end, sum_end, sum, b_3, e_3, s_3)
23 if s_2 = max{s_1, s_2, s_3} then
24     return (start, sum_start, end, sum_end, sum, b_2, e_2, s_2)
25 return (start, sum_start, end, sum_end, sum, b_1, e_1, s_1)
```

ALGORITHM 2.2: LSR(b, e): Find the maximum sum of contiguous elements of the array T between the indices b and e (included).

Algorithm 2.3 computes an interval of maximum sum that is based on the above properties. First, it looks for the first positive element in the array. If none is found, the solution is reduced to the largest element in the array. Otherwise, starting at the first positive element, it scans the array computing the sum of the current interval and updating the maximum sum if needed. Once an interval of nonpositive sum is encountered, using the last of the above properties, the algorithm skips the elements so far. Indeed, if a suffix of the current interval was a prefix of an interval of maximum sum, then the current interval would have a nontrivial, nonpositive prefix and the algorithm would have skipped it.

```
1  start ← 1;        sum_max ← T[1]
2  max_elem ← 1
3  while T[start] < 0 and start < n do
4  |  start ← start + 1
5  |_ if T[start] > T[max_elem] then max_elem ← start
6  if start = n then return (max_elem, max_elem, T[max_elem])
7  sum_max ← T[start]
8  end ← start
9  local_max ← sum_max;        local_start ← start
10 for i = start + 1 to n do
11 |  if local_max + T[i] < 0 then
12 |  |  local_max ← 0
13 |  |_ local_start ← i + 1
14 |  else
15 |  |  if local_max > sum_max then
16 |  |  |  sum_max ← local_max
17 |  |  |  start ← local_start
18 |_ |_ |_ end ← i
19 return (start, end, sum_max)
```

ALGORITHM 2.3: Find the maximum sum of contiguous elements of an array through a single array scan.

Solution to Exercise 2.4: Boolean matrices: The Four-Russians algorithm

1. No, because subtraction is not defined on booleans. A trick allows us to use Strassen's algorithm, by considering every bit as an integer modulo $n + 1$, where n is the size of the matrices, and then use the rules of addition, multiplication, and subtraction on these integers (see [75]).

2. $A \times B = \sum_{\ell=1}^{n} A_c[\ell] \times B_r[\ell]$.

3. $A \times B = \sum_{i=1}^{n/k} A_i \times B_i$.

4. Each row of C_i is a Boolean sum of the rows of B_i according to the corresponding row of A_i; if $k = 3$ and the j-th row of A_i is 1 0 1, then we add the first and the third rows of B_i to obtain the j-th row of C_i. The rows of A_i can take only 2^k different values, hence, the first result.

The table contains 2^k rows, each of length n, hence, a total size in $O(n \times 2^k)$, and it must be constructed for each B_i. The row can be addressed by coding the corresponding possible entry of A_i as an integer (i.e., in our example, the j-th row of C_i would be accessible as the 6th entry of the table because $1\ 0\ 1 = 5$).

We precompute the tables by induction, with at most one addition of a row for each entry. Because it takes n Boolean additions to add a row, this precomputing is done with $O(n \times 2^k)$ operations.

Finally, to compute C_i, for each of the n rows, it takes a time $O(1)$ to find the row in the table, and copying this row to the appropriate row in C_i takes a time $O(n)$, hence, a total computation time in $O(n^2)$.

5. To compute $A \times B$, we need to compute the n/k products C_i, following the method of the previous question. Each such product takes $O(n^2)$ time, and constructing the table takes $O(n \times 2^k)$ time. Therefore, the total running time of the algorithm is $O(n^3/k + n^2 \times 2^k/k)$.

6. A reasonable value would be $k = \log(n)$, so that the complexity of the algorithm is $O\left(\frac{n^3}{\log(n)}\right)$. This complexity is worse than that of Strassen's algorithm, which is $O(n^{\log(7)})$.

Solution to Exercise 2.5: Matrix multiplication and inversion

The first question is to prove that if such α and β exist, then they are between 2 and 3. The lower bound is because we need to access each matrix element at least once. The upper bound comes from the usual algorithm for matrix multiplication (no need for Strassen's algorithm here), and from the Gauss–Jordan algorithm for matrix inversion. Next, we show both inequalities as follows:

(i) Multiplication is not more complex than inversion. We aim at computing the product of two matrices A and B of size n. Let Z be the matrix of size $3n$ defined as

$$Z = \begin{pmatrix} I & A & 0 \\ 0 & I & B \\ 0 & 0 & I \end{pmatrix}.$$

We have

$$Z^{-1} = \begin{pmatrix} I & -A & A.B \\ 0 & I & -B \\ 0 & 0 & I \end{pmatrix}.$$

Hence, inverting Z requires computing the product AB, and $M(n) \leq I(3n)$; thus, $\alpha \leqslant \beta$.

(ii) Inversion is not more complex than multiplication. We aim at computing the inverse of a matrix A of size n. We proceed in two steps.

If A is symmetric and positive definite, and of dimension $n = 2^k$, then we write $A = \begin{pmatrix} B & C^T \\ C & D \end{pmatrix}$, where B, C, and D are of size 2^{k-1}. We obtain that

$$A^{-1} = \begin{pmatrix} B^{-1} + B^{-1}C^T S^{-1} C B^{-1} & -B^{-1}C^T S^{-1} \\ -S^{-1}CB^{-1} & S^{-1} \end{pmatrix}$$

where $S = D - CB^{-1}C^T$.

To compute A^{-1}, we have to compute the matrices

$$B^{-1}, CB^{-1}, (CB^{-1})C^T, S^{-1}, S^{-1}(CB^{-1}), \text{ and } (C.B^{-1})^T(S^{-1}(CB^{-1})).$$

The hint tells us that B and S are symmetric and positive definite as well (see [40] for a reference), hence, we can use the method recursively and

$$I(n) = 2I(n/2) + 4M(n) + O(n^2).$$

We deduce that $I(n) = O(M(n))$ from the master theorem and the fact that $2 \leqslant \alpha, \beta \leqslant 3$.

General case. If A is of dimension n, we augment it with zeros and build $\tilde{A} = \begin{pmatrix} A & 0 \\ 0 & I \end{pmatrix}$, where I is the identity matrix and the dimension of $\tilde{A}$ is $2^{\lceil \log n \rceil} \leqslant 2n$. Then we let $B = \tilde{A}^T \tilde{A}$, which is symmetric and positive definite and of dimension an exact power of 2. We compute the inverse of B by the previous method and write

$$I = B^{-1}.B = B^{-1}(\tilde{A}^T \tilde{A}) = (B^{-1}.\tilde{A}^T)\tilde{A},$$

which shows that $\tilde{A}^{-1} = B^{-1}\tilde{A}^T$. We derive that $I(n) = 2M(n) + O(M(n)) + O(n^2) = O(M(n))$. Finally, $\beta \leqslant \alpha$.

2.6 Bibliographical notes

Section 2.3 on recurrences comes from the book by Kronsjö [66]. Exercise 2.1 (product of two polynomials) is the well-known algorithm of Karatsuba used in many computer algebra systems. Exercises 2.2 (Toeplitz matrices) and 2.5 (matrix multiplication and inversion) come from the book by Cormen, Leiserson, Rivest, and Stein [27]. Exercise 2.3 (maximum sum) comes from the book by Bentley [14]. Exercise 2.4 (the Four-Russians algorithm) comes from the book by Manber [75].

Chapter 3

Greedy algorithms

This chapter explains the reasoning in finding optimal greedy algorithms. The main feature of a greedy algorithm is that it builds the solution step by step, and, at each step, it makes a decision that is locally optimal. Throughout Sections 3.1 to 3.3, we illustrate this principle with several examples and also outline situations where greedy algorithms are not optimal; taking a good local decision may prove a bad choice in the end! In Section 3.4, we also cover matroids, a (mostly theoretical) framework to prove the optimality of greedy algorithms. All of these techniques are then illustrated with a set of exercises in Section 3.5, with solutions found in Section 3.6.

3.1 Motivating example: The sports hall

Problem. Let us consider a sports hall in which several events should be scheduled. The goal is to have as many events as possible, given that two events cannot occur simultaneously (only one hall). Each event i is characterized by its starting time s_i and its ending time e_i. Two events are compatible if their time intervals do not overlap. We would like to solve the problem, i.e., find the maximum number of events that can fit in the sports hall, with a greedy algorithm.

A first greedy algorithm. The first idea consists of sorting events by increasing durations $e_i - d_i$. At each step, we schedule an event into the sports hall if it fits, i.e., if it is compatible with events that have already been scheduled. The idea is that we will be able to accommodate more shorter events than longer ones. However, we make local decisions at each step of the algorithm (this is a greedy algorithm!), and it turns out that we can make decisions that do not lead to the optimal solution. For instance, in the example of Figure 3.1, the greedy algorithm schedules only the shortest event i, while the two compatible events j and k would lead to a better solution.

A second greedy algorithm. In order to avoid the problem encountered in the previous example, we design a new algorithm that sorts events by

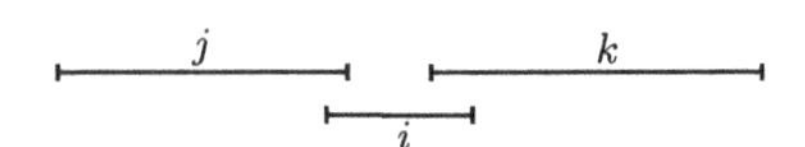

FIGURE 3.1: The first greedy algorithm is not optimal.

starting times s_i and then proceeds similarly to the first greedy algorithm. In the example of Figure 3.1, this greedy algorithm returns the optimal solution. However, the local decisions that are made may not be the optimal ones, as shown in the example of Figure 3.2. Indeed, the algorithm schedules event i at the first step, and then no other event can be scheduled, while it would be possible to have eight compatible events. Note that the first greedy algorithm would return the optimal solution for this example.

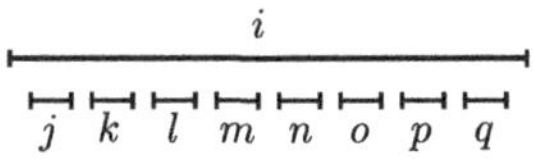

FIGURE 3.2: The second greedy algorithm is not optimal.

A third greedy algorithm. Building upon the first two algorithms, we observe that it is always a good idea to select first events that do not intersect with many other events. In the first example, events j and k intersect with only one other event, while event i intersects with two events and is chosen later; therefore, the new algorithm finds the optimal solution. Similarly in the second example, event i intersects eight other events, and it is the only event not to be scheduled. However, this greedy algorithm is still not optimal. We can build an example in which we force the algorithm to make a bad local decision. In the example of Figure 3.3, event i is the first to be chosen because it has the smallest number of intersecting events. However, if we schedule i, we can have only three compatible events, while we could have a solution with four compatible events, j, k, l, and m.

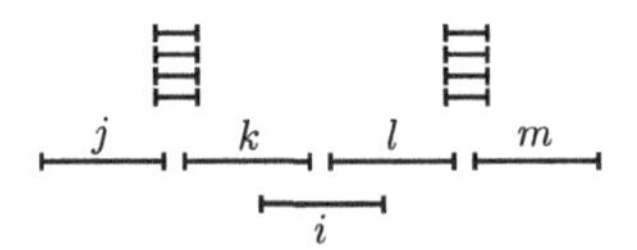

FIGURE 3.3: The third greedy algorithm is not optimal.

An optimal greedy algorithm. Even though many greedy choices do not lead to an optimal solution, as observed with the preceding algorithms, there is a greedy algorithm that solves the sports hall problem in polynomial time. The idea is to sort the events by increasing ending times e_i and then to greedily schedule the events. This way, at each step we fit the maximum number of events up to a given time, and we never make a bad decision. We now prove the optimality of this algorithm.

Let f_1 be the event with the smallest ending time. We prove first that there exists an optimal solution that schedules this event. Let us consider an optimal solution $O = \{f_{i_1}, f_{i_2}, \ldots, f_{i_k}\}$, where k is the maximum number of events that can be scheduled in the sports hall and where events are sorted by nondecreasing ending times. There are two possible cases: Either (i) $f_{i_1} = f_1$, the optimal solution schedules f_1, and nothing needs to be done, or (ii) $f_{i_1} \neq f_1$. In this second case, we replace f_{i_1} with f_1 in solution O. We have $e_1 \leqslant e_{i_1}$ by definition of event f_1, and $e_{i_1} \leqslant s_{i_2}$ because O is a solution to the problem (f_{i_1} and f_{i_2} are compatible). Therefore, $e_1 \leqslant s_{i_2}$ and, thus, f_{i_2} is compatible with f_1. The new solution is still optimal (the number of events remain unchanged), and event f_1 is scheduled.

The proof works by induction, following the previous reasoning. Once f_1 is scheduled, we consider only events that do not intersect with f_1, and we iterate the reasoning on the remaining events to conclude the proof.

Finally, we emphasize that there can be many optimal solutions, and not all of them will include the first event f_1 selected by the greedy algorithm, namely, the event with the smallest end time. However, schedules that select f_1 are *dominant*, meaning that there exists an optimal solution that includes f_1.

3.2 Designing greedy algorithms

The example of the sports hall gives a good introduction to the design principles of greedy algorithms. Actually, the binary method to compute x^n in Section 1.1.2 also is a greedy algorithm, in which we decide at each step which computation to perform. We can formalize the reasoning to find greedy algorithms as follows:

1. Decide on a greedy choice that allows us to optimize the problem locally;

2. Search for a counterexample that shows that the algorithm is not optimal (and go back to step 1 if a counterexample is found), or prove its optimality through steps 3 and 4;

3. Show that there is always an optimal solution that performs the greedy choice of step 1;

4. Show that if we combine the greedy choice with an optimal solution of the subproblem that we still need to solve, then we obtain an optimal solution.

We say that a greedy algorithm is a *top-down* algorithm because at each step we make a local choice, and we then have a single subproblem to solve, given this choice. On the contrary, we will see in Section 4 that dynamic programming algorithms are *bottom-up*; we will need results of multiple subproblems to make a choice.

3.3 Graph coloring

In this section, we further illustrate the principle of greedy algorithms through the example of graph coloring. The problem consists of coloring all vertices of a graph using the minimum number of colors while enforcing that two vertices, which are connected with an edge, are not of the same color. Formally, let $G = (V, E)$ be a graph and $c : V \rightarrow \{1..K\}$ be a K-coloring such that $(x, y) \in E \Rightarrow c(x) \neq c(y)$. The objective is to minimize K, the number of colors.

3.3.1 On coloring bipartite graphs

We start with a small theorem that allows us to define a bipartite graph, defined as a graph that can be colored with only two colors.

THEOREM 3.1. *A graph can be colored with two colors if and only if all its cycles are of even length.*

Proof. Let us first consider a graph G that can be colored with two colors. Let $c(v) \in \{1, 2\}$ be the color of vertex v. We prove by contradiction that all cycles are of even length. Indeed, if G has a cycle of length $2k+1$, $v_1, v_2, \ldots, v_{2k+1}$, then we have $c(v_1) = 1$, say, which implies that $c(v_2) = 2$, $c(v_3) = 1$, until $c(v_{2k+1}) = 1$. However, since it is a cycle, there is an edge between v_1 and v_{2k+1}, so they cannot be of the same color, which leads to the contradiction.

Now, if all the cycles of the graph G are of even length, we search for a 2-coloring of this graph. We assume that G is connected (the problem is independent from one connected component to another). The idea consists of performing a breadth-first traversal of G.

Le $x_0 \in G$, $X_0 = \{x_0\}$ and $X_{n+1} = \bigcup_{y \in X_n} N(y)$, where $N(y)$ is the set of nodes connected to y, but not yet included in a set X_k, for $k \leqslant n$. Each vertex appears in one single set, and we color with color 1 the elements from sets X_{2k}, and with color 2 the elements from sets X_{2k+1}.

This 2-coloring is valid if and only if two vertices connected by an edge are of different colors. If there is an edge between $y \in X_i$ and $z \in X_j$, where i and j are both either even or odd, then we have a cycle $x_0, \ldots, y, z, \ldots, x_0$ of length $i + j + 1$, and this value is even, leading to a contradiction. The coloring, therefore, is valid, which concludes the proof. □

In a bipartite graph, if we partition vertices into two sets according to the colors, all edges go from one set to the other. We retrieve here the usual definition of bipartite graphs, namely, graphs whose vertices are partitioned into two sets and with no edge inside these sets. We now consider colorings of general graphs, and we propose a few greedy algorithms to solve the problem.

3.3.2 Greedy algorithms to color general graphs

The first greedy algorithm takes the vertices in a random order, and, for each vertex v, it colors it with the smallest color number that has not been yet given to a neighbor of v, i.e., a node connected to v.

Let $K_{greedy1}$ be the total number of colors needed by this greedy algorithm. Then we have $K_{greedy1} \leqslant \Delta(G) + 1$, where $\Delta(G)$ is the maximal degree of a vertex (number of edges of the vertex). Indeed, at any step of the algorithm, when we color vertex v, it has at most $\Delta(G)$ neighboring vertices and, therefore, the greedy algorithm never needs to use more than $\Delta(G) + 1$ colors.

Note that this algorithm is optimal for a fully connected graph (a clique), since we need $\Delta(G) + 1$ colors to connect such a graph (one color per vertex). However, this algorithm is not optimal in general; on the following bipartite graph, if the order of coloring is 1 and then 4, we need three colors, while the optimal coloring uses only two colors.

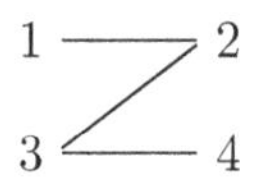

In order to improve the previous algorithm, one idea is to order vertices in a smart way and then to proceed as before, i.e., color each vertex in turn with the smallest possible color.

Let $n = |V|$ be the number of vertices and d_i be the degree of vertex v_i. We have $K_{greedy2} \leqslant \max_{1 \leqslant i \leqslant n} \min(d_i + 1, i)$. Indeed, when we color vertex v_i, it has at most $\min(d_i, i - 1)$ neighbors that have already been colored, and thus its own color is at most $1 + \min(d_i, i - 1) = \min(d_i + 1, i)$. To obtain the result, we take the maximum of these values on all vertices.

This result suggests that it would be smart to color vertices with a high degree first, so that we have $\min(d_i + 1, i) = i$. Therefore, the second greedy algorithm sorts the vertices by nonincreasing degrees.

Once again, the algorithm is not optimal. On the following bipartite graph, we choose to color vertex 1, then vertex 4, which imposes the use of three colors instead of the two required ones.

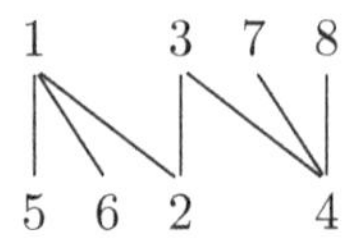

Based on these ideas, several greedy algorithms can be designed. In particular, a rather intuitive idea consists of giving priority to coloring vertices that have already many colored neighbors. We define the color-degree of a vertex as the number of its neighbors that are already colored. Initially, the color-degree of each vertex is set to 0, and then it is updated at each step of the greedy algorithm.

The following greedy algorithm is called the *Dsatur* algorithm in [20]. The ordering is done by (color-degree, degree); we choose a vertex v with maximum color-degree, and such that its degree is the largest among the vertices with maximum color-degree. This vertex v is then colored with the smallest possible color, and the color-degrees of the neighbors of v are updated before proceeding to the next step of the algorithm. We illustrate this algorithm on the following example:

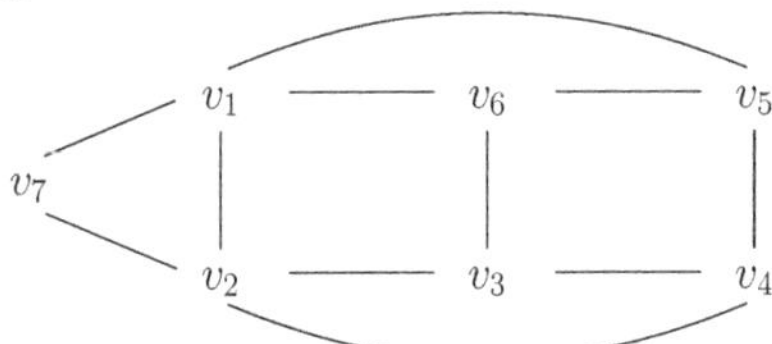

We first choose a vertex with maximum degree, for instance v_1, and it is colored with color 1. The color-degree of v_2, v_5, v_6, and v_7 becomes 1, and we choose v_2, which has the maximum degree (between these four vertices); it is assigned color 2. Now, v_7 is the only vertex with color-degree 2; it is given the color 3. All remaining noncolored vertices have the same color-degree 1 and the same degree 3; we arbitrarily choose v_3 and color it with 1. Then, v_4, with color-degree 2, receives color 3. Finally, v_5 is colored in 2 and v_6 in 3; the graph is 3-colored, and it is an optimal coloring.

The name *Dsatur* comes from the fact that maximum color-degree vertices are saturated first. We prove below that *Dsatur* always returns an optimal coloring on bipartite graphs; however, it may use more colors than needed on arbitrary graphs.

THEOREM 3.2. *The* Dsatur *algorithm is optimal on bipartite graphs, i.e., it always succeeds to color them with two colors.*

Proof. Consider a connected bipartite graph $G = (V, E)$, where $V = B \cup R$ and each edge in E is connecting a vertex in B (color 1 is blue) and a vertex in R (color 2 is red). Note first that the first two greedy algorithms may fail. Let G be such that $B = \{b_1, b_2, b_3\}$, $R = \{r_1, r_2, r_3\}$, and $E = \{(b_1, r_2), (b_2, r_3), (b_3, r_1), (b_i, r_i) | 1 \leqslant i \leqslant 3\}$, as illustrated below.

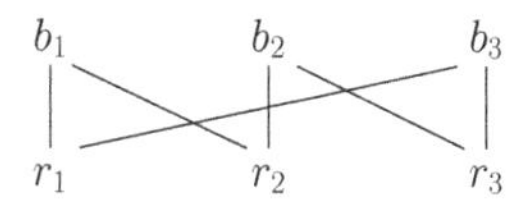

All vertices have a degree 2. If we start by coloring a vertex of B, for instance b_1, and then a nonconnected vertex of R, r_3, with the same color 1, it is not possible to complete the coloring with only two colors. The use of the color-degree prevents us from such a mistake, since once b_1 has been colored, we need to color either r_1 or r_2 with the color 2 and finish the coloring optimally.

In the general case, with *Dsatur*, we first color a vertex, for instance from B, with color 1 (blue). Then we have to color a vertex of color-degree 1, that is, a neighboring vertex. This neighboring vertex belongs necessarily to R. It is colored with color 2 (red). We prove by induction that at any step of the algorithm, all colored vertices of B are colored in blue, and all vertices of R are colored in red. Indeed, if the coloring satisfies this property at a given step of the algorithm, we choose next a vertex v with nonnul color-degree. Because the graph is bipartite, all its neighbors are in the same set and have the same color: red if $v \in B$, or blue if $v \in R$. Vertex v, therefore, is colored in red if it is in R or in blue if it is in B. □

We exhibit a counterexample to show that *Dsatur* is not optimal on arbitrary graphs.

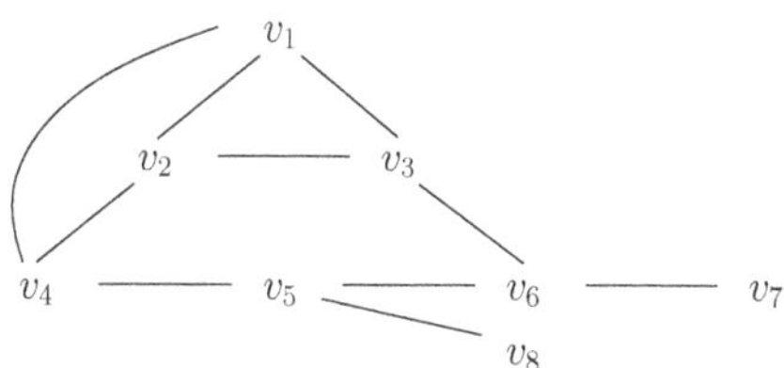

Dsatur can choose v_4 first because it has the maximum degree 3; it is colored with 1. Between the vertices with color-degree 1, the algorithm can (arbitrarily) choose v_5, which is colored with 2. Then the algorithm can choose to color v_6, using color 1. Then, v_1 is chosen between vertices of color-degree 1 and degree 3, and it is colored with 2. We finally need to use colors 3 and 4 for v_2 and v_3, while this graph could have been colored with only three colors (v_1, v_5, v_7 with color 1; v_2, v_6, v_8 with color 2; and v_3, v_4 with color 3).

To build this counterexample, we force *Dsatur* to make a wrong decision by coloring both v_4 and v_6 with color 1 and v_1 with color 2, which forces four colors because of v_2 and v_3. Note that it would be easy to build an example without any tie (thereby avoiding random choices) by increasing the degree of some vertices (for instance, in the example, v_7 and v_8 are there just to increase the degrees of v_5 and v_6).

The problem of coloring general graphs is NP-complete, as will be shown in Chapter 7. However, for a particular class of graphs, a smart greedy algorithm can return the optimal solution, as we detail below.

3.3.3 Coloring interval graphs

We focus now on interval graphs. Given a set of intervals, we define a graph whose vertices are intervals and whose edges connect intersecting intervals. The following example shows such a graph, obtained with a set of seven intervals.

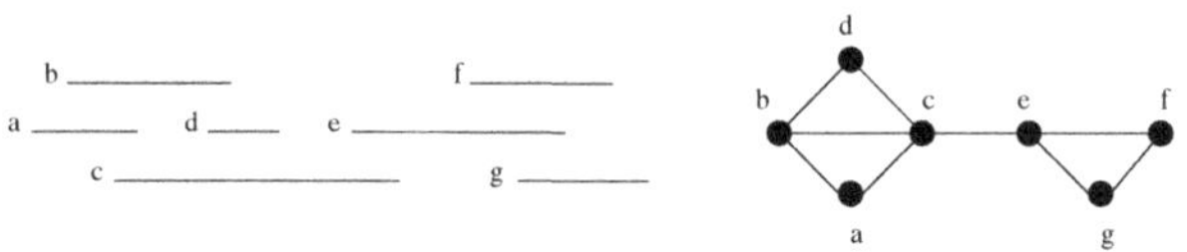

The problem of coloring such a graph is quite similar to the sports hall problem. Indeed, one can see each interval as representing an event, with its starting and ending times, and the color as representing a sports hall. Then, only compatible events will be colored with the same color, and we could use one sports hall per set of compatible events. If we minimize the number of colors, we minimize the number of sports halls that are needed to organize all events.

Graphs that are obtained from a set of intervals are called interval graphs. We define the following greedy algorithm: intervals (i.e., vertices) are sorted by nondecreasing starting times (or left extremity). In the example, the order is a, b, c, d, e, f, g. Then, the greedy coloring is done as before; for each chosen vertex, we color it with the smallest compatible color. On the example, we obtain the coloring $1, 2, 3, 1, 1, 2, 3$, which is optimal, as the graph contains a cycle of length 3.

We prove now that this greedy algorithm is optimal for any interval graph. Let G be such a graph, and let d_v be the starting time of interval v corresponding to vertex v. We execute the greedy algorithm; it uses k colors. If vertex v receives color k, then this means that $k-1$ intervals that start no later than d_v intersect this interval and had all been colored with colors 1 to $k-1$; otherwise, v would be colored with a color $c \leqslant k-1$. All of these intervals are thus intersecting because they all contain the point d_v; therefore, graph G contains a clique of size k. Since all vertices of a clique must be colored with distinct colors, we cannot color the graph with fewer than k colors. The greedy algorithm, therefore, is optimal.

Once again, we point out that the order chosen by the greedy algorithm is vital because we could force the greedy algorithm to make a wrong decision, even on a bipartite graph as below, if we would not proceed from left to right. We could first color a, then d, leading to the use of three colors instead of two.

3.4 Theory of matroids

In this section are elementary results on matroids, a framework that allows us to guarantee the optimality of a generic greedy algorithm in some situations. Unfortunately, it is not easy to characterize which problems can be captured as matroid instances. Still, the theory is beautiful, and we outline its main ideas.

Matroids. The term *matroid* was introduced in 1935 by H. Whitney [107], while working on the linear independence of the vector columns of a matrix. We define it below and illustrate the concept through a canonical example.

DEFINITION 3.1. $(S, \mathcal{I})$ is a matroid if S is a set of n elements, and $\mathcal{I}$ is a collection of subsets of S, with the following properties:

i. $X \in \mathcal{I} \Rightarrow (\forall\, Y \subset X,\ Y \in \mathcal{I})$ (*hereditary property*), and
ii. $(A, B \in \mathcal{I},\ |A| < |B|) \Rightarrow \exists x \in B \backslash A$ s.t. $A \cup \{x\} \in \mathcal{I}$ (*exchange property*).
If $X \in \mathcal{I}$, X is said to be an *independent set.*

Readers familiar with linear algebra will immediately see that linearly independent subsets of a given vector set form a matroid. The canonical computer science example follows.

Example of matroid: Forests of a graph. Let $G = (V, E)$ be a (nondirected) graph. We define a matroid with $S = E$ (the elements are the edges of the graph), and $\mathcal{I} = \{A \subset E \mid A \text{ has no cycle}\}$. Therefore, a set of edges is an independent set if and only if this set of edges is a forest of the graph, i.e., a set of trees (a tree is a connected graph with no cycle). We check that this matroid satisfies both properties.

(i) The hereditary property. It is pretty obvious that a subset of a forest is a forest; if we remove edges from a forest, we cannot create a cycle, thus we still have a forest.

(ii) The exchange property. Let A and B be two forests of G (i.e., $A, B \in \mathcal{I}$) such that $|A| < |B|$. $|A|$ is the number of edges in forest A, and every vertex is part of a tree (an isolated vertex with no edges is a tree made of a single vertex). Then A (resp. B) contains $|V| - |A|$ (resp. $|V| - |B|$) trees. Indeed, each time an edge is added to the independent set, two trees are connected, therefore decrementing the number of trees by one. Thus, B contains fewer trees than A, and there exists a tree T of B that is not included in a tree of A, i.e., two vertices u and v of tree T are not in the same tree of A. On the path from u to v in T, there are two vertices, connected by an edge (x, y), that are not in the same tree of A. Then, if we add this edge to the forest A, we still have a forest, i.e., $A \cup \{(x, y)\} \in \mathcal{I}$, which concludes the proof.

DEFINITION 3.2. Let $F \in \mathcal{I}$; $x \notin F$ is an *extension* of F if $F \cup \{x\} \in \mathcal{I}$, i.e., $F \cup \{x\}$ is an independent set. An independent set is *maximal* if it has no extensions.

In our running example, any edge connecting two distinct trees of a forest is an extension. A forest is maximal if adding any edge to it would create a cycle. A maximal independent set in the example of the forest is a spanning tree (or spanning forest if G is not connected).

LEMMA 3.1. *All maximal independent sets are of same cardinal.*

Proof. If this lemma were not true, we could find an extension to the independent set of smaller cardinal thanks to the exchange property, which would mean that it was not maximal. □

We introduce a last definition: We add weights to the elements of the matroid and, therefore, obtain a *weighted matroid.*

DEFINITION 3.3. In a *weighted matroid*, each element of S has a weight: $x \in S \mapsto w(x) \in \mathbb{N}$. The weight of a subset $X \subset S$ is defined as the sum of the weights of its elements: $w(X) = \sum_{x \in X} w(x)$.

Greedy algorithms on a weighted matroid. The problem is to find an independent set of maximum weight. The idea of the greedy algorithm is to sort elements of S by nonincreasing weights. We start with the empty set, which always is an independent set because of the hereditary property. Then, we add elements into this set, as long as we keep an independent set. This generic algorithm is formalized in Algorithm 3.1.

```
1 Sort elements of S = {s_1, ..., s_n} by nonincreasing weight:
  w(s_1) ⩾ w(s_2) ⩾ ··· ⩾ w(s_n)
2 A ← ∅
3 for i = 1 to n do
4     if A ∪ {s_i} ∈ I then
5         A ← A ∪ {s_i}
```

ALGORITHM 3.1: Independent set of maximum weight.

THEOREM 3.3. *Algorithm 3.1 returns an optimal solution to the problem of finding an independent set of maximum weight in the weighted matroid.*

Proof. Let s_k be the first independent element of S, i.e., the first index i of the algorithm such that $\{s_i\} \subset \mathcal{I}$. We first prove that there exists an optimal solution that contains s_k.

Let B be an optimal solution, i.e., an independent set of maximum weight. If $s_k \in B$, we are finished. Otherwise, let $A = \{s_k\} \in \mathcal{I}$. While $|B| > |A|$, we apply the exchange property to add an element of B to the independent set A. We obtain the independent set with $|B|$ elements, $A = \{s_k\} \cup B \backslash \{s_j\}$, where $\{s_j\}$ is the one element of B that has not been chosen for the extension (there is already element s_k in A, and at the end, $|A| = |B|$; therefore, all elements of B but one are extensions of A).

We now compare the weights. We have $w(A) = w(B) - w(s_j) + w(s_k)$. Moreover, $w(s_k) \geqslant w(s_j)$, because s_j is independent (by hereditary property), and $j > k$ (by definition of s_k). Finally, $w(A) \geqslant w(B)$, and since B is an optimal solution, $w(A) = w(B)$. The independent set A is of maximal weight, and it contains s_k, which proves the result.

To prove the theorem, we show by induction that the greedy algorithm returns the optimal solution; we restrict the search to a solution that contains s_k, and we start the reasoning again with $S' = S \backslash \{s_k\}$, and $\mathcal{I}' = \{X \subset S' \mid X \cup \{s_k\} \in \mathcal{I}\}$. □

Back to the running example. Theorem 3.3 proves the optimality of Kruskal's algorithm to build a minimum weight spanning tree [67]. Edges are sorted by nondecreasing weight, and we choose greedily the next edge that does not add a cycle when added to the current set of edges. Of course, we should discuss a suitable data structure so that we can easily check the condition "no cycle has been created." With a simple array, we can check the condition in $O(n^2)$, and it is possible to achieve a better complexity with other data structures [27]. In any case, the complexity of the greedy algorithm remains polynomial.

Example: A semimatching problem. In this very simple example, we are given a directed weighted graph. The problem is to find a maximum weight subset of the edges so that no two starting points are the same. A natural greedy algorithm would sort all edges according to their weight in nonincreasing order, then consider all edges in this order, selecting an edge (i, j) if and only if no edge (i, j') had been selected earlier. In fact, this greedy algorithm selects for every node the outgoing edge that has maximum weight; hence, it can be easily implemented in time $O(n+m)$, where n is the number of nodes and m the number of edges of the directed graph. While the optimality of this greedy algorithm is not difficult to prove directly, we prove it using matroid theory.

The problem can be cast in terms of a matrix W with nonnegative entries, with the goal to select a set of entries whose sum is maximal, subject to the constraint that no two entries are from the same row of the matrix. There are n rows in W, one per node in the graph. Let W_{ij} be the entry in row i and column j of the matrix W, and let $x_{ij} \in \{0, 1\}$ be the indicator of whether W_{ij} is selected. We aim at maximizing $\sum_{i,j} W_{ij}x_{ij}$ subject to the set of constraints $\sum_j x_{ij} \leqslant 1$ for each row i. The greedy algorithm chooses entries one at a time in order of weight, largest first (and breaking ties arbitrarily), rejecting an entry only if an entry in the same row has already been chosen. Here is an example, where chosen entries are underlined:

$$W = \begin{pmatrix} \underline{12} & 7 & 10 & 11 \\ 8 & 6 & 4 & \underline{16} \\ 3 & \underline{5} & 2 & 1 \\ 14 & 13 & 9 & \underline{15} \end{pmatrix}.$$

To prove that the greedy algorithm is optimal, we exhibit the matroid; independent sets are sets of entries such that no two of them are from the same row of the matrix. We show that both properties hold. The hereditary property is obvious. Indeed, when removing entries from an independent set, we cannot create a row with two entries or more. The exchange property is not difficult either. Let A and B be two independent sets with $|A| < |B|$. There is at most one element per row in A and B, so there must be a row that contains an element of B and no element of A. Adding this element to A preserves its independence. This concludes the proof of optimality of the greedy algorithm.

As mentioned before, it is not easy to exhibit matroid structures for which interesting and efficient greedy algorithms can be derived. A more complicated example that involves scheduling tasks with deadlines is studied in Exercise 3.5. We refer the reader to [70, 94] for much more material on matroids and greedoids.

3.5 Exercises

Exercise 3.1: Interval cover (solution p. 68)

We are given a set $X = \{x_1, \ldots, x_n\}$ of n points on a line.

1. Design a greedy algorithm that determines the smallest set of closed intervals of length 1 that contains all the points.

2. Prove the optimality of the algorithm and give its complexity.

3. Could you use the theory of matroids to prove the optimality of the algorithm?

Exercise 3.2: Memory usage (solution p. 69)

Given a memory of size L, we want to store a set of n files $P = (P_1, \ldots, P_n)$. File P_i $(1 \leqslant i \leqslant n)$ is of size a_i, where a_i is an integer. If $\sum_{i=1}^{n} a_i > L$, we cannot store all files. We need to select a subset $Q \subseteq P$ of files to store, such that $\sum_{P_i \in Q} a_i \leqslant L$. We sort the files P_i by nondecreasing sizes ($a_1 \leqslant \cdots \leqslant a_n$).

1. Write a greedy algorithm that maximizes the number of files in Q. The output must be a Boolean table S such that $S[i] = 1$ if $P_i \in Q$, and $S[i] = 0$ otherwise. What is the complexity of this algorithm in number of comparisons and number of arithmetic operations?

2. Prove that this strategy always returns a maximal subset Q. We define the utilization ratio as $\frac{\sum_{P_i \in Q} a_i}{L}$. How small can it be with our strategy?

3. We now want to maximize the utilization ratio, i.e., fill the memory as much as possible. Design a greedy algorithm for this new objective function.

4. Is the latter greedy algorithm optimal? How small can the utilization ratio be with this algorithm? Prove the result.

Exercise 3.3: Scheduling dependent tasks on several machines

(solution p. 71)

Let $G = (V, E)$ be a directed acyclic graph (DAG). Here G is a task graph. In other words, each node $v \in V$ represents a task, and each edge $e \in E$ represents a precedence constraint, i.e., if $e = (v_1, v_2) \in E$, then the execution of v_2 cannot start before the end of the execution of v_1. We need to schedule the tasks on an unlimited number of processors. Moreover, the execution time of task $v \in V$ is $w(v)$. The problem is to find a valid schedule, i.e., a start time $\sigma(v)$ for each task v such that no precedence constraints are violated, and that minimizes the total execution time. The reader may refer to Section 6.4.4, p. 140, for more background on scheduling.

1. Define formally (by induction) the top level $tl(v)$ of a task $v \in V$, which is the earliest possible starting time of task v.

2. Propose a greedy schedule of the tasks, based on the top levels, and prove its optimality. This schedule is called σ_{free}.

3. We define the bottom level $bl(v)$ of a task as the largest weight of a path from v to an output task, i.e., a task with no successor. The weight of the path includes the weight of v. Define bottom levels formally, and propose a schedule of the tasks, based on the bottom levels, that is called σ_{late}.

4. Show that any optimal schedule σ satisfies:

$$\forall v \in V,\ \sigma_{free}(v) \leqslant \sigma(v) \leqslant \sigma_{late}(v).$$

5. Give an example of a DAG that has at least three different optimal schedules.

Exercise 3.4: Scheduling independent tasks with priorities (solution p. 72)

We need to schedule n independent tasks, T_1, T_2, $\ldots$, T_n, on a single processor. Each task T_i has an execution time w_i and a priority p_i. Because we execute the tasks sequentially on a single processor and as we target an optimal schedule, we can focus on schedules that execute tasks as soon as possible. A schedule is then fully defined by the order followed to execute the tasks. In other words, here, a schedule of tasks $T_1, \ldots, T_n$ is a permutation $T_{\sigma(1)}$, $T_{\sigma(2)}$, $\ldots$, $T_{\sigma(n)}$, specifying the order in which tasks are executed. We assume that the first task to be executed is processed from time 0 on. The cost of a schedule is defined as $\sum_{i=1}^{n} p_i C_i$, where C_i is the completion time of task T_i, i.e., the date at which its processing was completed. We look for a schedule that minimizes this cost.

1. Consider any schedule and two tasks T_i and T_j that are executed consecutively under this schedule. Which task should be executed first in order to minimize the cost?

2. Design an optimal greedy algorithm. What is its complexity?

Exercise 3.5: Scheduling independent tasks with deadlines (solution p. 73)

The goal here is to exhibit a matroid to prove the optimality of a greedy algorithm. We need to schedule n independent tasks, T_1, T_2, $\ldots$, T_n, on a single processor. Each task T_i is executed in one time unit, but it has a deadline d_i that should not be exceeded. If a task does not complete its execution before its deadline, there is a cost w_i to pay. The objective here is to find a schedule that minimizes the sum of the costs of the tasks that are completed after their deadlines. A schedule, in this exercise, will be a function, $\sigma : \mathcal{T} \rightarrow \mathbb{N}$, that associates with each task its execution time, such that two tasks cannot be scheduled at the same time, i.e., for all $1 \leqslant i, j \leqslant n$, $\sigma(T_i) \neq \sigma(T_j)$. The first task can be executed at time 0.

We say that a task is *on time* if it finishes its execution before its deadline, and that it is *late* otherwise. Note that minimizing the cost of late tasks is equivalent to maximizing the cost of on-time tasks. A *canonical* schedule is such that (i) on-time tasks are scheduled before late tasks, and (ii) on-time tasks are ordered by nondecreasing deadlines.

1. Prove that there is always an optimal schedule that is canonical, i.e., we can restrict the search to canonical schedules.

2. Design a greedy scheduling algorithm to solve the problem. What is its complexity?

3. Illustrate the greedy algorithm on the following example with seven tasks; the tasks are sorted by nonincreasing w_i ($w_i = 8-i$, for $1 \leqslant i \leqslant 7$), and their deadlines are as follows: $d_1 = 4$, $d_2 = 2$, $d_3 = 4$, $d_4 = 3$, $d_5 = 1$, $d_6 = 4$, and $d_7 = 6$.

4. Prove the optimality of the algorithm by exhibiting a matroid.

While Exercises 3.4 and 3.5 deal with simple uniprocessor scheduling problems for which the greedy algorithm is optimal, there are many more complex scheduling problems [21]. These include, for instance, scheduling problems with tasks with different execution times, several machines, precedence constraints between tasks, and so on. More scheduling problems are described in Section 6.4.4, p. 140.

Exercise 3.6: Edge matroids (solution p. 74)

This exercise aims at illustrating the matroid theory. The goal here is to exhibit a weighted matroid, design the corresponding greedy algorithm, and prove its optimality.

This exercise is a generalization of the semimatching algorithm presented in Section 3.4. We are given a directed graph $G = (V, E)$ whose edges have integer weights. Let $w(e)$ be the weight of edge $e \in E$. We also are given a constraint $f(u) \geqslant 0$ on the out-degree of each node $u \in V$. The goal is to find a subset of edges of maximal weight and whose out-degree at any node satisfy the constraint. We see that if $f(u) = 1$ for all nodes, we retrieve the semimatching algorithm.

1. Define independent sets and prove you have a matroid.

2. What is the cardinal of maximal independent sets?

3. What is the complexity of the (optimal) greedy algorithm?

Exercise 3.7: Huffman code (solution p. 75)

Let Σ be a finite alphabet with at least two elements. A binary code is an injective application from Σ to the set of finite suites of 0 and 1 (i.e., a binary word, also called *code word*). The code can be naturally extended by concatenation to a mapping defined on the set Σ^* of words using the alphabet Σ. A code is said to be *of fixed length* if all the letters in Σ are coded by binary words of same size. A code is said to be a *prefix* code if no code word is a prefix of another code word. Given the code of a word in Σ^*, the decoding operation consists of finding the original word.

1. Prove that the decoding operation has a unique solution, both for a code of fixed length and for a prefix code.

2. Represent a prefix code by a binary tree, where leaves are the letters of the alphabet Σ.

3. Consider a text in which each letter $c \in \Sigma$ appears with a frequency $f(c) \neq 0$. With each prefix code of this text, represented by a tree T, is associated a cost, defined by $B(T) = \sum_{c \in \Sigma} f(c) \times l_T(c)$, where $l_T(c)$ is the size of the code word of c. If $f(c)$ is exactly the number of occurrences of c in the text, then $B(T)$ is the number of bits in the encoded text. A prefix code T is *optimal* if, for this text, $B(T)$ is minimum. Prove that for any optimal prefix code there is a corresponding binary tree with $|\Sigma|$ leaves and $|\Sigma| - 1$ internal nodes.

4. Prove that there is an optimal prefix code such that two letters of smallest frequencies are siblings in the tree (i.e., their code words have the same size and differ only by the last bit).

   Hint: Prove also that these two letters are leaves of maximal depths.

5. Given x and y, two letters of smallest frequencies, we consider the alphabet $\Sigma' = (\Sigma \setminus \{x, y\}) \cup \{z\}$, where z is a new letter with frequency $f(z) = f(x) + f(y)$. Let T' be the tree of an optimal code for Σ'. Prove that the tree T obtained from T' by replacing the leaf associated with z with an internal node with two leaves x and y is an optimal code for Σ.

6. Using both previous questions, design an algorithm that returns an optimal code, and give its complexity. Illustrate the algorithm on the following problem instance: $\Sigma = \{a, b, c, d, e, g\}$, $f(a) = 45$, $f(b) = 13$, $f(c) = 12$, $f(d) = 16$, $f(e) = 9$, and $f(g) = 5$.

3.6 Solutions to exercises

Solution to Exercise 3.1: Interval cover

1. Algorithm 3.2 is a greedy algorithm to solve the interval cover problem. It builds a maximal length interval starting at the first point on the line, removes all points included in that interval, and then iterates.

2. Sorting the points costs $O(n \log(n))$ when the execution of the while loop costs $O(n)$. Therefore, the algorithm runs in $O(n \log(n))$.

   We prove the optimality of Algorithm 3.2 by induction on the number n of points. If $n = 1$, there is a single point, the algorithm returns a single interval and thus is optimal. Now, assume we have proved the

```
1 Sort the x_i in nondecreasing order
2 X ← {x_1, ..., x_n}
3 I ← ∅
4 while X ≠ ∅ do
5 |   x_k ← min(X)
6 |   I ← I ∪ {[x_k, x_k + 1]}
7 |_  X ← X\[x_k, x_k + 1]
8 return I
```

ALGORITHM 3.2: Interval cover.

optimality of Algorithm 3.2 for any set containing at most n points, and consider an instance X including $n+1$ points. Let I_{opt} be an optimal cover: $I_{opt} = \{[a_1, a_1+1], \ldots, [a_p, a_p+1]\}$ with $a_1 < \cdots < a_p$. Let I_{greedy} be the interval cover built by Algorithm 3.2: $I_{greedy} = \{[g_1, g_1+1], \ldots, [g_m, g_m+1]\}$ with $g_1 < \cdots < g_m$. We need to show that $m = p$. We let $I = \Big(I_{opt} \backslash \{[a_1, a_1+1]\}\Big) \cup \{[g_1, g_1+1]\}$. Because $g_1 = x_1$, I is also an interval cover of X. Indeed, because I_{opt} is an interval cover, $a_1 \leqslant x_1$ and thus $a_1 + 1 \leqslant x_1 + 1 = g_1 + 1$. Therefore, $X \cap [a_1, a_1+1] \subseteq X \cap [g_1, g_1+1]$. In other words, $I_{opt} \backslash \{[a_1, a_1+1]\}$ is an interval cover of $X \backslash [g_1, g_1+1]$. The induction hypothesis tells us that Algorithm 3.2 builds an optimal interval cover of $X \backslash [g_1, g_1+1]$ because this set contains at most n points. This optimal solution is exactly $\{[g_2, g_2+1], \ldots, [g_m, g_m+1]\}$. Because $I_{opt} \backslash \{[a_1, a_1+1]\}$ is an interval cover of $X \backslash [g_1, g_1+1]$ with $p-1$ intervals, we have $p - 1 \geqslant m - 1$ and thus $p \geqslant m$. However, I_{opt} is, by definition, an optimal interval cover of X and contains p intervals, while the cover I_{greedy} contains m intervals. Therefore, $p \leqslant m$ and, thus, $p = m$. Algorithm 3.2 is optimal on our instance containing $n+1$ points.

3. The obvious idea would be to consider sets of nonoverlapping intervals. Elements in this set would satisfy the hereditary property. However, they would not satisfy the exchange property. We do not know how to prove the optimality of Algorithm 3.2 using the theory of matroids.

Solution to Exercise 3.2: Memory usage

1. The P_is are sorted in nondecreasing size: $a_1 \leqslant \cdots \leqslant a_n$. Algorithm 3.3 is a greedy algorithm that iteratively tries to add the remaining smallest file to the set of already picked files. Each time we try to add a file, we need to compare the new total size with L to check whether the file

will fit. We, therefore, need n comparisons in the worst case, and $n-1$ additions.

```
1 for i = 1 to n do S[i] ← 0
2 size ← a_1
3 i ← 1
4 while size ⩽ L and i ⩽ n do
5     S[i] ← 1
6     i ← i + 1
7     if i ⩽ n then size ← size + a_i
8 return S
```

ALGORITHM 3.3: Largest number of files that fit in a space of size L.

2. Let G be the solution built by Algorithm 3.3. Let Q be an optimal solution that has the largest number of files in common with G. We compare Q and G. If G contains as many files as Q, G is optimal. Otherwise, $|G| < |Q|$ and, therefore, Q contains strictly more files than G. Then let P_k be any file of $Q \setminus G$, i.e., any file of Q not belonging to G. One can easily see that G is made up of the $|G|$ smallest files. Therefore, the size of P_k is larger than or equal to the size of any file of G. Let P_j be the smallest file of $G \setminus Q$, i.e., the smallest file of G that is not a file of Q. Such a P_j exists because otherwise we would have $G \subset Q$. Indeed, when considering file P_k, Algorithm 3.3 should have added P_k to its current solution as $(G \cup \{P_k\}) \subset Q$, and thus $G \cup \{P_k\}$ fits in a memory of size L. This contradicts the definition of P_k. Therefore, P_j does exist. Then, let $Q' = (Q \setminus \{P_k\}) \cup \{P_j\}$. Q' contains as many files as Q and is also a valid solution because we have shown that $a_k \geqslant a_j$. Then, Q' is an optimal solution that has one more file in common with G than Q. This contradicts the definition of Q.

   If we make no assumptions on the a_is, they can be all larger than L, in which case the utilization ratio is null. The interesting case is obviously when, for any $i \in [1; n]$, $a_i \leqslant L$. In that case, the worst utilization ratio achieved by Algorithm 3.3 is $\frac{1}{L}$ as the a_is are integers by hypothesis (this is achieved, for instance, when $n = 2$, $a_1 = 1$, and $a_2 = L$).

3. This algorithm is the same as the previous one, except that we consider the files in the reverse order.

4. Let us consider the following instance containing three files: $a_1 = \frac{L}{2} + 1$ and $a_2 = a_3 = \frac{L}{2}$ (where L is even). The greedy algorithm will return

the solution $\{P_1\}$ whose utilization ratio is $\frac{1}{2} + \frac{1}{L}$, while the optimal solution is $\{P_2, P_3\}$ whose utilization ratio is 1.

The minimum utilization ratio is $\frac{1}{2} + \frac{1}{L}$ if no a_i is greater than L (it is equal to 0 otherwise). Indeed, as the sum of the sizes of all files is strictly greater than L, the sum of the sizes of the solution of the greedy algorithm is strictly greater than $\frac{L}{2}$ and thus greater than or equal to $1 + \frac{L}{2}$. Indeed, either $L \geqslant a_n > L/2$, or there exists an index k such that $\sum_{i=k}^{n} a_i \leqslant \frac{L}{2}$ and $a_{i-1} + \sum_{i=k}^{n} a_i > \frac{L}{2}$ (as the files are sorted by nondecreasing sizes). Furthermore, a sum of file sizes of $\frac{L}{2} + 1$ is reached on the counterexample to optimality.

Remark. The reader may be frustrated that we are not building any polynomial-time optimal algorithm for this problem. The reason will later become clear; in Exercise 7.20 (p. 155) we will show the NP-completeness of the problem 2-PARTITION (all these notions will be defined in Chapter 6). The reader will then be able to show the NP-completeness of the current problem when $L = \frac{1}{2} \sum_{i=1}^{n} a_i$.

Solution to Exercise 3.3: Scheduling dependent tasks on several machines

1. The earliest start time of a task is the earliest time when all the tasks that it depends upon are completed. Therefore, $tl(v) = 0$ if v does not depend on any task, i.e., if we have $\{(x, y) \in E \mid y = v\} = \emptyset$. Otherwise,

$$tl(v) = \max_{(u,v) \in E} (tl(u) + w(u)).$$

2. We just let $\sigma_{free}(v) = tl(v)$. By definition of top levels, all precedence constraints are satisfied by this schedule. The optimality also comes from the definition of the top level of a task; it is the earliest time at which the execution of the task can start in any schedule.

3. Bottom levels are defined recursively, the same way top levels are defined:

$$bl(v) = w(v) + \max_{(v,u) \in E} bl(u).$$

The meaning of the bottom level is that the earliest time at which the schedule can complete is a time $bl(v)$ after the beginning of the execution of v. Let $MS(\sigma_{free}, +\infty)$ be the execution time of the σ_{free} schedule. We define σ_{late} as follows:

$$\sigma_{late}(v) = MS(\sigma_{free}, +\infty) - bl(v).$$

While σ_{free} is an as-soon-as possible (ASAP) schedule, σ_{late} is an as-late-as-possible (ALAP) schedule. One can easily check that σ_{late} is a valid schedule, i.e., that it satisfies all the precedence constraints. Furthermore, this is also an optimal schedule.

4. These inequalities follow directly from the definitions of the top level and of the bottom level. Indeed, by definition of the top level, the earliest time at which the execution of a task v can start under any schedule is $\sigma_{free}(v)$. By definition of the bottom level, the execution of a task v must start no later than at time $\mathcal{M} - bl(v)$ for the whole execution to be able to complete by time $\mathcal{M}$.

5. Figure 3.4 presents a task graph that admits (at least) three distinct optimal schedules (all tasks have unit execution time in this example). Under the σ_{free} schedule: At date 0 one executes A and B; at date 1, C, D, and E; at date 2, F and G; and, finally, at date 3, H. Under the σ_{late} schedule: At date 0 one executes B; at date 1, A and D; at date 2, C, F, and E; and, finally, at date 3, H and G. We also can build a schedule that is a mix of the two previous ones: At date 0 one executes A and B; at date 1, C, D, and E; at date 2, F; and, finally, at date 3, G and H.

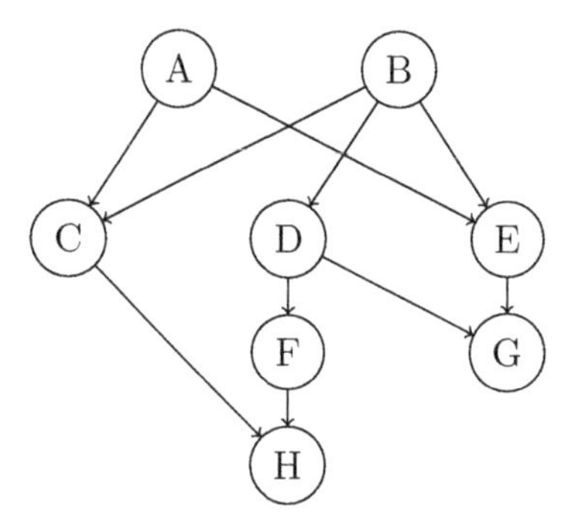

FIGURE 3.4: Task graph admitting three different optimal schedules.

Solution to Exercise 3.4: Scheduling independent tasks with priorities

1. In an optimal schedule, tasks are executed as soon as possible. The processing of the $(i+1)$-th task starts as soon as the processing of the i-th completes. Therefore, under the schedule $T_{\sigma(1)}, T_{\sigma(2)}, \ldots, T_{\sigma(n)}$, the processing of the i-th task is completed at time $C_{\sigma(i)} = \sum_{j=1}^{i} w_{\sigma(j)}$.

   We consider two schedules whose only difference is the order in which two consecutive tasks, T_i and T_j, are executed. In the first schedule, T_i is executed right before T_j; in the second schedule, T_i is executed right after T_j. The costs of the two schedules are identical except for the terms corresponding to tasks T_i and T_j. Let w be the sum of the execution times of the tasks executed before T_i in the first schedule (or

T_j in the second). Then there exists a constant C such that:

- The cost of the first schedule is equal to:
  $C + (w + w_i)p_i + (w + w_i + w_j)p_j$.
- The cost of the second schedule is equal to:
  $C + (w + w_j)p_j + (w + w_j + w_i)p_i$.

Then the cost of the first schedule is not greater than that of the second schedule if and only if:

$$C + (w + w_i)p_i + (w + w_i + w_j)p_j \leqslant C + (w + w_j)p_j + (w + w_j + w_i)p_i$$

that is, if and only if $w_i p_j \leqslant w_j p_i$.

2. In an optimal solution, tasks must be scheduled by nondecreasing values of the ratio $\frac{w_i}{p_i}$ (this is known as "Smith's ratio rule"). An optimal greedy algorithm first schedules one of the tasks that minimizes the ratio $\frac{w_i}{p_i}$. Such an algorithm has a complexity in $O(n \log n)$ because tasks must be sorted.

Solution to Exercise 3.5: Scheduling independent tasks with deadlines

1. Canonical schedules are not a restriction for this problem: (i) If a late task were executed before an on-time task, we could exchange them without impacting the total cost to pay. (ii) If we had two tasks on time, such that $\sigma(T_i) < \sigma(T_j)$ and $d_i > d_j$, we could exchange them; therefore, T_j would stay on time because it would be executed earlier, and T_i would also stay on time because T_j initially was on time and $d_i > d_j$. The total cost would be exactly the same.

2. The greedy algorithm sorts tasks by nonincreasing cost w_i. This algorithm tries successively to add each task to the schedule. A task is kept if it can be executed on time with a canonical schedule. The rationale is that this algorithm selects the most costly tasks and decreases the cost of late tasks as much as possible.

   Note that the complexity of the greedy algorithm is in $O(n^2)$; the number of steps is n, and it takes a time $O(n)$ to check that all tasks in the current set can be executed on time.

3. In the following, a canonical schedule is written as an ordered list of on-time task indices; for instance, the schedule $\{3, 1\}$ means that $\sigma(T_3) = 0$ and $\sigma(T_1) = 1$, while the other tasks are late.

   If we apply the algorithm on the example, we obtain successively the following schedules: $\{1\}$; $\{2, 1\}$; $\{2, 1, 3\}$; $\{2, 4, 1, 3\}$. However, we cannot add task T_5 because in the canonical order $\{5, 2, 4, 1, 3\}$ task T_3 is late. We cannot add task T_6 either, but $\{2, 4, 1, 3, 7\}$ is a set of on-time tasks. This is actually the optimal solution, with a cost of 5.

4. To prove the optimality of this greedy algorithm, we exhibit a matroid.

   We define the matroid $(S, \mathcal{I})$, where S is the set of tasks and where a subset of tasks is an independent set if and only if these tasks can all be executed on time.

   Before proceeding with the proof that $(S, \mathcal{I})$ is a matroid, we establish the following property. Let A be a set of tasks and $N_t(A)$ be the number of tasks of A with a deadline less than or equal to t. Then, the three following propositions are equivalent:

   (a) A is an independent set;
   (b) For any $t \in [1; n]$, $N_t(A) \leqslant t$;
   (c) If all tasks of A are executed with a canonical schedule, there are no late tasks.

   We first prove that (a) implies (b). If A is an independent set but $N_t(A) > t$, then there is at least one task that will be executed after t, which would mean that A is not independent. It is straightforward to prove that (b) implies (c); finally, (c) implies (a) by definition of the independent sets. Therefore, the equivalence is true.

   We now prove that $(S, \mathcal{I})$ is a matroid and, therefore, that the greedy algorithm is optimal.

   The hereditary property is obvious. Any subset of an independent set is an independent set, i.e., we can always execute on time any subset of a set of tasks that can all be executed on time.

   For the exchange property, we consider two independent sets, A and B ($A, B \in \mathcal{I}$), such that $|A| < |B|$. We need to find a task $T_i \in B$ such that $A \cup \{T_i\}$ is an independent set. For $t = 0$, we have $N_0(A) = N_0(B) = 0$. Let $m = \max_{1 \leqslant i \leqslant n} d_i$. Then, for $t = m$, $N_m(A) = |A| < |B| = N_m(B)$. We search for the largest value of t, $0 \leqslant t \leqslant m$, such that $N_t(A) \geqslant N_t(B)$. Then, $N_{t+1}(B) > N_{t+1}(A)$, and there are more tasks of deadline equal to $t+1 \leqslant m$ in B than in A. We choose $T_i \in B \setminus A$ with a deadline $d_i = t + 1$, and then $A \cup \{T_i\}$ is an independent set.

Solution to Exercise 3.6: Edge matroids

1. Define the independent sets as all subsets of edges whose out-degree does not exceed $f(u)$ at each node u. For an independent set F, define $d_F(u)$ the number of out-edges at u that belong to F. By definition of independent sets, we have $d_F(u) \leqslant f(u)$ for each node u.

   The hereditary property is obvious because removing some edge whose source is a node u will decrease $d_F(u)$ and further relax the constraint on $f(u)$. As for the exchange property, consider two independent sets of edges A and B, with $|A| < |B|$. Because there are fewer edges in A, there must exist a node u that is the source of fewer edges in A than in B, i.e., such that $d_A(u) < d_B(u)$. Let e be an edge in $B \setminus A$ whose source is u. We can safely add e to A without violating the degree constraint $f(u)$ because $d_A(u) + 1 \leqslant d_B(u) \leqslant f(u)$. Hence, $A \cup \{e\}$ is independent.

2. Let $d(u)$ be the out-degree of node u. Clearly, all maximal independent sets have the same cardinal $\sum_{u \in V} \min(f(u), d(u))$.

3. Because we have exhibited a weighted matroid, we have to use only Algorithm 3.1 and Theorem 3.3 (p. 62) to define an optimal greedy algorithm and prove its optimality. In other words, to build an optimal solution, all we have to do is sort the edges by nonincreasing weights and add the edges greedily, with an edge being added to the current solution if and only if, after addition, the new solution is still an independent set.

   Checking whether an edge can be added can be done in constant time, so the complexity of the greedy algorithm is dominated by the time needed to sort the edges by nonincreasing weights, which requires $O(|E| \log |E|)$ steps.

Solution to Exercise 3.7: Huffman code

1. The decoding operation for a fixed-length code is obviously unique; otherwise, two different letters would have to share the same code word. Let us assume that there exists a prefix code whose decoding does not always lead to a unique solution. Then, there exists a text x that can be decomposed in $x = ua$ and $x = vb$ where u and v are two different code words. Then, either u is a prefix of v or v is a prefix of u. Because the code is a prefix code, then $u = v$, and we have a contradiction.

2. A prefix tree can be represented by a binary tree as follows: A 0 is associated with each left branch and a 1 with each right branch. The code of a letter is obtained by reading the labels associated with the branches visited on the path from the root to the leaf associated with the letter. Figure 3.5 presents a prefix code and its associated tree representation.

3. Each optimal prefix code corresponds to a binary tree where each internal node has two children. Indeed, if an internal node had only a single child, we could "raise" the subtree rooted at the single child by deleting this child (this process is illustrated in Figure 3.6). This would decrease $l_T(c)$ or leave it unchanged for any letter c: Either the letter c belonged to the raised subtree and then $l'_T(c) = l_T(c) - 1$, or it did not belong to it and then $l'_T(c) = l_T(c)$. By decreasing $l_T(c)$, or leaving it unchanged, we negate the optimality assumption, hence reaching a contradiction.

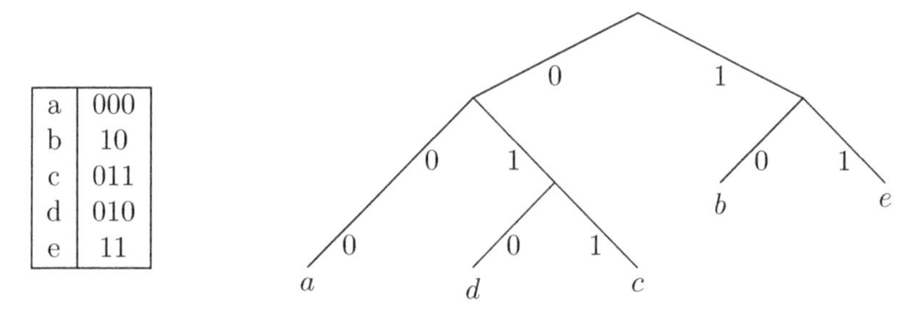

a	000
b	10
c	011
d	010
e	11

FIGURE 3.5: A prefix code and its binary tree representation.

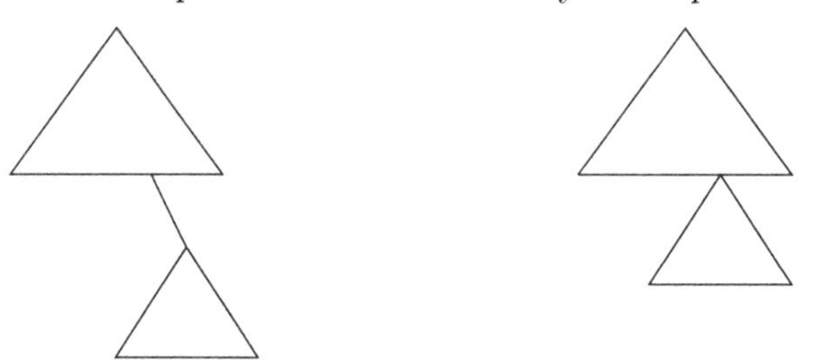

FIGURE 3.6: Prefix code optimization by the deletion of an internal node with a single child.

Such a tree contains $|\Sigma|$ leaves as it codes for $|\Sigma|$ letters. We show by induction that it contains $|\Sigma| - 1$ internal nodes.

- A tree with two leaves contains one internal node.
- If the tree is made of two subtrees with i and j leaves, then the two subtrees have $i - 1$ and $j - 1$ internal nodes, respectively, because of the induction hypothesis. The whole tree then has $(i-1)+(j-1)+1 = i+j-1$ internal nodes, and the property is satisfied.

4. To prove the desired property, along with the property proposed in the hint, we consider the binary tree T representing an optimal code. Any tree T' corresponding to a prefix code thus satisfies $B(T) \leq B(T')$. Let x and y be two letters of smallest frequencies. Let a and b be two letters that are siblings at the maximal depth in the tree T. From T we build a new tree T'' by exchanging x and a, and y and b. This transformation is illustrated in Figure 3.7.

   To prove that T'' also corresponds to an optimal code, we show that $B(T) \geqslant B(T'')$.

$$\begin{aligned} B(T) - B(T'') = {} & f(a)l_T(a) + f(b)l_T(b) + f(x)l_T(x) + f(y)l_T(y) \\ & -f(a)l_{T''}(a) - f(b)l_{T''}(b) - f(x)l_{T''}(x) - f(y)l_{T''}(y). \end{aligned}$$

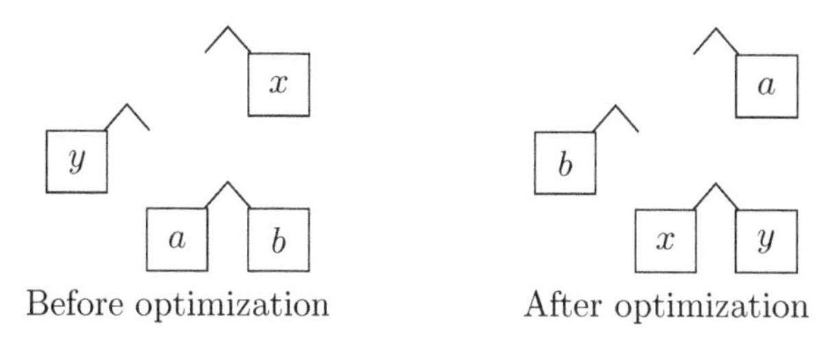

FIGURE 3.7: Tree optimization by pairing the two letters of lowest frequency.

By definition of the transformation, we have the following equalities: $l_{T''}(a) = l_T(x)$, $l_{T''}(b) = l_T(y)$, $l_{T''}(x) = l_T(a)$, and $l_{T''}(y) = l_T(b)$. Therefore,

$$B(T)-B(T'') = (f(a) - f(x)(l_T(a)-l_T(x))+(f(b)-f(y))(l_T(b)-l_T(y)).$$

Because x and y are two letters of smallest frequencies, $f(a) \geq f(x)$ and $f(b) \geq f(y)$. Because a and b are two letters at the maximal depth, $l_T(a) \geq l_T(x)$ and $l_T(b) \geq l_T(y)$. Therefore, $B(T) - B(T'') \geq 0$ and, thus, T'' also corresponds to an optimal code. T'' corresponds to an optimal code where two letters of smallest frequencies are siblings at the maximal depth in the tree. This property will define our *greedy choice*.

5. We consider $\Sigma' = \Sigma \setminus \{x, y\} + \{z\}$, where z is a new letter of frequency $f(z) = f(x) + f(y)$. Let T' be the tree of an optimal code for Σ'. Let T be the tree obtained from T' by replacing the leaf z with an internal node whose children are the leaves x and y. We want to show that T is optimal for Σ.

   Let T'' be an optimal tree for Σ. Using the property established at the previous question, we can assume that the leaves x and y are siblings in T''. We then build from T'' a tree T''' by replacing the node parent of the leaves x and y with the leaf z. We then have $B(T''') = B(T'') - f(x)l_T(x) - f(y)l_T(y) + (f(x)+f(y))(l_T(x)-1) = B(T'') - f(x) - f(y)$. Similarly, $B(T') = B(T) - f(x) - f(y)$. Because T' is optimal for Σ, $B(T') \leq B(T''')$. Therefore, $B(T) - f(x) - f(y) \leq B(T'') - f(x) - f(y)$ and thus $B(T) \leq B(T'')$. As T'' is by definition optimal for Σ, so is T.

6. Algorithm 3.4 builds the Huffman code for an alphabet Σ and its associated frequency function f. For a binary heap,

   - inserting an element costs $O(\log(n))$;
   - finding an element with minimal key costs $O(1)$;
   - extracting an element with minimal key costs $O(\log(n))$.

   Therefore, the complexity of any iteration of the loop is $O(\log(n))$. The overall complexity is $O(n \log(n))$ because there are $n-1$ iterations of the loop and because the initial construction of the heap costs $O(n \log(n))$.

```
1  F ← build_binary_heap(Σ, f)
2  n ← |Σ|
3  for i = 1 to n − 1 do
4      z ←allocate_node()
5      x ←extract_min(F)
6      y ←extract_min(F)
7      z(left) ← x
8      z(right) ← y
9      f(z) ← f(x) + f(y)
10     Insert(F, z, f(z))
11 return extract_min(F)
```

ALGORITHM 3.4: Building a Huffman code.

Figure 3.8 presents the tree associated with the Huffman code for the example. In this tree, nodes are labeled by frequencies, and edges by the associated code digit.

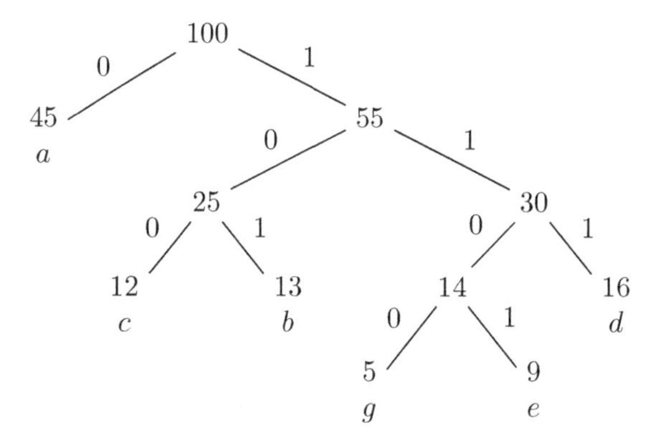

FIGURE 3.8: Tree of the Huffman code for the instance: $\Sigma = \{a, b, c, d, e, g\}$, $f(a) = 45$, $f(b) = 13$, $f(c) = 12$, $f(d) = 16$, $f(e) = 9$, and $f(g) = 5$.

3.7 Bibliographical notes

Section 3.1 is an extension of an example from the book by Cormen, Leiserson, Rivest, and Stein [27]. Section 3.3 is inspired by the nice book on graph algorithms by West [106]. Exercise 3.5 (scheduling independent tasks with deadlines) comes from the book by Cormen, Leiserson, Rivest, and Stein [27].

Chapter 4

Dynamic programming

In this chapter, we focus on how to find optimal dynamic-programming algorithms. Particular attention is paid to problem size in order to avoid exponential-cost algorithms. The chapter is illustrated with two classical examples: the coin changing problem and the knapsack problem. These techniques are then further illustrated with a set of exercises in Section 4.4, with solutions found in Section 4.5.

4.1 The coin changing problem

The problem is the following: If we want to make change for S cents, and we have an infinite supply of each coin in the set $Coins = \{v_1, v_2, \dots, v_n\}$, where v_i is the value of the i-th coin, what is the minimum number of coins required to reach the value S?

Greedy algorithm. We propose a greedy algorithm to solve the problem. First, we sort coins by nonincreasing values, then for each coin value we take as many coins as possible. The algorithm is formalized as Algorithm 4.1.

```
1 Sort elements of Coins = {v_1, ..., v_n} by nonincreasing values:
  v_1 ≥ v_2 ≥ ··· ≥ v_n
2 R ← S        { R is the remaining sum to reach; it is initially S }
3 for i = 1 to n do
4   | c_i = ⌊R / v_i⌋       { c_i is the number of coins of value v_i that are taken }
5   | R ← R − c_i × v_i     { R is updated }
```

ALGORITHM 4.1: Greedy algorithm for the coin changing problem.

We first assume that $Coins = \{10, 5, 2, 1\}$ (a typical European set of coins). In this case, we can prove that Algorithm 4.1 is optimal:

- An optimal solution returns, at most, one coin of value 5 (if there are two, it is better to use one single coin of value 10).
- An optimal solution returns, at most, one coin of value 1 (otherwise, we can use a coin of value 2).
- An optimal solution returns, at most, two coins of value 2 (otherwise, to obtain $6 = 2 + 2 + 2$, we would rather use $6 = 5 + 1$: one coin 5 and one coin 1).

Therefore, in the optimal solution, there cannot be more than four coins that are not of value 10, and $5 + 2 + 2 + 1 = 10$, so if there are four such coins, we would rather use a coin of 10. Thus, the optimal solution uses, at most, three coins that are not of value 10, and their total is at most 9. We can then conclude that the optimal number of coins of value 10 is $\lfloor \frac{S}{10} \rfloor$, which is the number selected by the greedy algorithm. It is then easy to conclude that the greedy algorithm always selects the optimal number of coins of each value.

Note, however, that the greedy algorithm is not optimal for any set of coins. For instance, if $Coins = \{6, 4, 1\}$ and $S = 8$, the greedy algorithm requires three coins $8 = 6 + 1 + 1$, while the optimal solution requires two coins of value 4. Still, U.S. readers will be pleased to know that the greedy algorithm is optimal for the set $Coins = \{25, 10, 5, 1\}$. The proof follows an ad hoc case analysis very similar to that conducted for European coins. Because the greedy algorithm is not always optimal, we explore another idea to solve the problem.

An optimal algorithm. The problem is to find the minimum number of coins required to reach sum S, with coins of value $\{v_1, \ldots, v_n\}$, which we denote as $z(S, n)$. Because the greedy algorithm may fail, we try to solve more subproblems so that we do not take a bad greedy choice as we did in the previous example. We also allow ourselves to come back to a choice already made and try another set of coins.

We investigate a way to solve the problem that is in appearance more complex than the initial problem. In other words, we artificially ask for more than requested and aim at finding $z(T, i)$, the minimum number of coins required to reach sum $T \leqslant S$ with the first i coins, i.e., coins selected from the subset $\{v_1, \ldots, v_i\}$ (where $0 \leqslant i \leqslant n$). Instead of computing only $z(S, n)$, the original problem, we compute $S \times n$ values $z(T, i)$. But now, we have a recurrence relation to compute $z(T, i)$:

$$z(T, i) = \min \begin{cases} z(T, i-1) & i\text{-th coin not used;} \\ z(T - v_i, i) + 1 & i\text{-th coin used (at least) once.} \end{cases}$$

The recurrence must be properly initialized; values of i and T are decreasing, so we consider the cases $i = 0$ and $T \leqslant 0$:

- $z(T, 0) = +\infty$ for $T > 0$: There are no more coins; therefore, we cannot reach the sum $T > 0$, and this solution cannot be correct.
- $z(0, i) = 0$: We do not need any coin to reach the sum $T = 0$.

- $z(T, i) = +\infty$ for $T < 0$: We have exceeded the sum; this solution cannot be correct.

Thanks to the recurrence relation and the initialization conditions, we are now able to compute $z(S, n)$ and to solve the original problem. This kind of algorithm is called a *dynamic-programming* algorithm.

If the recurrence is applied without memoizing which values have already been computed, using a recursive algorithm, there will be an exponential number of computations. Note that the word *memoization* comes from "memo": the idea consists of memoizing the values so that we can look them up later.

However, we need to compute only $S \times n$ values of the function $z(T, i)$ ($1 \leqslant T \leqslant S$ and $1 \leqslant i \leqslant n$). This can be done either recursively, by memoizing the values that have already been computed, or iteratively, with, for instance, a loop with increasing i and then a loop with increasing T, so that we always have the values required to compute $z(T, i)$, i.e., $z(T, i-1)$ and $z(T - v_i, i)$, as shown in Algorithm 4.2. The precedence constraints are shown in Figure 4.1, and they are always enforced with this algorithm (for details about precedence constraints, see Section 6.4.4, p. 140). Note that we ensure that we never call the function with $T < 0$, and, therefore, we do not need the third initialization condition.

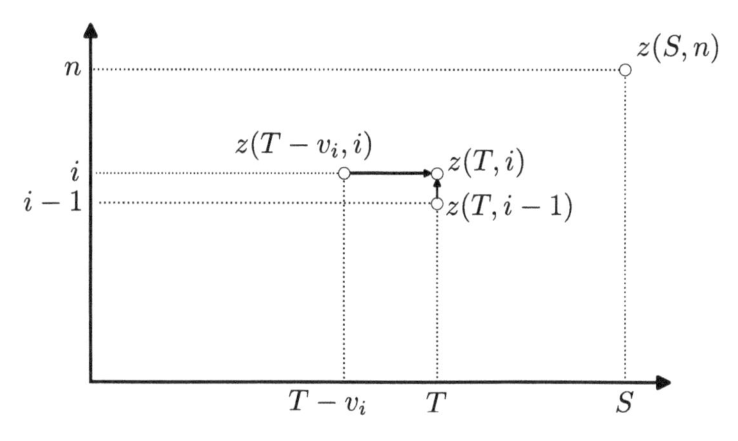

FIGURE 4.1: Precedence constraints for the coin changing dynamic-programming algorithm.

The complexity of the dynamic-programming algorithm is $O(n \times S)$, while the greedy algorithm has a complexity in $O(n \log n)$ (the execution is linear, but sorting the coins requires a time in $O(n \log n)$).

Finally, note that characterizing the set of coins for which the greedy algorithm is optimal is still an open problem. It is easy to find sets that work. For instance, coins $\{1, B, B^2, B^3, \ldots\}$ with $B \geqslant 2$. However, the general case seems tricky. There are several variants of the coin changing problem and

```
1 for T = 1 to S do
2 └ z(T,0) ← +∞        { Initialization: case i = 0 }
3 for i = 0 to n do
4 └ z(0,i) ← 0         { Initialization: case T = 0 }
5 for i = 1 to n do
6 │ for T = 1 to S do
7 │ │ z(T,i) ← z(T,i−1)
  │ │     { z(T,i−1) computed at previous iteration, or case i = 0 }
8 │ │ if T − v_i ⩾ 0 then
9 │ │ │ z(T,i) ← min(z(T,i), z(T − v_i, i))
  └ └ └     { z(T − v_i, i) computed earlier in this loop, or case T = 0 }
```

ALGORITHM 4.2: Dynamic-programming algorithm for the coin changing problem.

many dynamic-programming algorithms to solve them. The interested reader may refer to the following papers: [84, 97]. We move in the next section to another classical problem: the knapsack problem.

4.2 The knapsack problem

We have a set of items, each with a weight and a value, and we want to determine the items to include in the collection so that the total weight does not exceed a given limit, and the total value is as large as possible. Formally, there are n items $I_1, \ldots, I_n$, and item I_i has a weight w_i and a value c_i ($1 \leqslant i \leqslant n$). We are also given a maximum total weight W. The goal is to find a subset K of $\{1, \ldots, n\}$ that maximizes $\sum_{i \in K} c_i$, under the constraint $\sum_{i \in K} w_i \leqslant W$. The analogy with the problem of packing the best items for a well-deserved vacation should be clear.

Greedy algorithm. Here again, we start by designing a greedy algorithm to solve the problem. The idea consists of selecting first those items that have a good value per unit of weight, $\frac{c_i}{w_i}$. Therefore, we sort items by nonincreasing $\frac{c_i}{w_i}$, and then we greedily add them in the knapsack as long as the total weight is not exceeded.

However, the algorithm is not optimal because items are not divisible. We cannot take only a fraction of an item, i.e., either we take it or we discard it. A counterexample for the greedy algorithm can be designed as follows, with three items. The first item with the greatest ratio c_1/w_1 is such that it fills up the knapsack by itself (no other item can fit in the knapsack once I_1 has been chosen, i.e., $w_1 + w_i > W$, for $i \geqslant 2$). Then, two more items are such that $w_2 + w_3 \leqslant W$ (they fit together in the knapsack), and $c_2 + c_3 > c_1$ (they have more value than the first item alone). If we are able to construct such an example, the greedy algorithm chooses the first item, while a better solution consists of choosing items 2 and 3. A possible set of items is the following, with $W = 10$: $(w_1 = 6, w_2 = 5, w_3 = 5)$ and $(c_1 = 7, c_2 = 5, c_3 = 5)$.

If we consider the problem of the fractional knapsack, in which it is possible to take only a fraction of an object, then the greedy algorithm is optimal. In the example, it would take the whole item 1 and then a fraction (4/5) of item 2 to fill the remaining space in the knapsack. The value would then be $c_1 + \frac{4}{5}c_2 = 7 + 4 = 11$, which is optimal. It is easy to prove the optimality of the greedy algorithm in this case. If an optimal solution is not making the greedy choice, we can always exchange a fraction of item of the optimal solution with the fraction of item of better value per weight unit that was not greedily chosen, and the total value can only increase.

Dynamic-programming algorithm. We come back to the integer knapsack problem, and since the greedy algorithm is not optimal, we try to solve a more complex problem, as in the coin changing problem, in order to be able to establish a recurrence. The two parameters are the total weight and the number of items considered. We want to compute $C(v, i)$, which is the maximum value that can be obtained when filling up a knapsack of maximum total weight v, using only some of the first i items $\{I_1, \ldots, I_i\}$. The original problem is the value $C(W, n)$: The knapsack is of maximum total weight W, and we have the n items at our disposal.

To write the recurrence, we have two choices: (1) Either we have chosen the last object, or (2) we have not, therefore leading to:

$$C(v,i) = \max \begin{cases} C(v, i-1) & \text{last object not chosen;} \\ C(v - w_i, i-1) + c_i & \text{last object chosen;} \end{cases}$$

with the initialization conditions:

- $C(v, i) = 0$ for $v = 0$ or $i = 0$;
- $C(v, i) = -\infty$ if $v < 0$ (capacity exceeded).

The optimal solutions of all subproblems that we solve allow us to compute the optimal solution of the original problem. Similarly to the coin changing problem, we need to respect the precedence constraints of the computations carefully, and we never want to compute twice the same value of the function $C(v, i)$. The algorithm is formalized in Algorithm 4.3. The precedence constraints are shown in Figure 4.2. Because the computation is done row by row, these constraints are always respected.

The complexity of the greedy algorithm is in $O(n \log n)$ because the n items must be sorted. However, the complexity of the dynamic-programming algo-

1 **for** $i = 0$ to n **do**
2 $\quad C(0, i) \leftarrow 0$ { *Initialization: case* $v = 0$ }
3 **for** $v = 1$ to W **do**
4 $\quad C(v, 0) \leftarrow 0$ { *Initialization: case* $i = 0$ }
5 **for** $i = 1$ to n **do**
6 $\quad$ **for** $v = 1$ to W **do**
7 $\quad\quad C(v, i) \leftarrow C(v, i - 1)$
8 $\quad\quad$ **if** $v - w_i \geqslant 0$ **then**
9 $\quad\quad\quad C(v, i) \leftarrow \max(C(v, i), C(v - w_i, i) + c_i)$

ALGORITHM 4.3: Dynamic-programming algorithm for the knapsack problem.

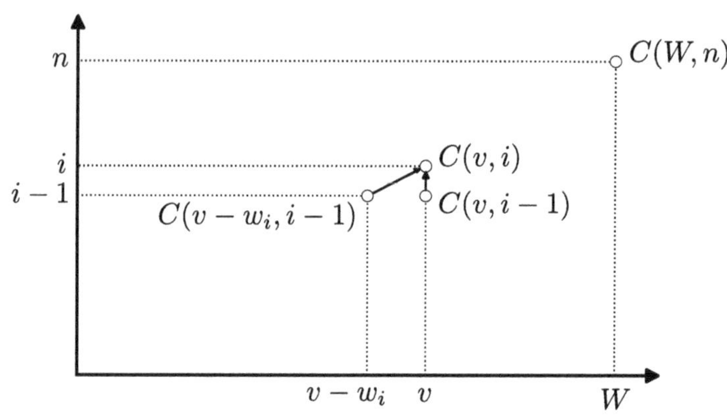

FIGURE 4.2: Precedence constraints for the knapsack dynamic-programming algorithm.

rithm is in $O(n \times W)$ because we need to compute $n \times W$ values of the function $C(v, i)$, and each computation takes constant time.

4.3 Designing dynamic-programming algorithms

In the previous two sections, we have given examples of dynamic-programming algorithms. The basic reasoning to obtain the optimal algorithm is similar in both cases:

1. Identify subproblems whose optimal solutions can be used to build an optimal solution to the original problem. Conversely, given an optimal

solution to the original problem, identify subparts of the solution that are optimal solutions for some subproblems. Usually, this step means that we identify a more complex problem derived from the original problem.

2. Write the recurrence.

3. Write the initial cases.

4. Write the algorithm, usually as an iterative algorithm, and take care to enforce precedence constraints (use a figure to check that these constraints are indeed satisfied). A recursive algorithm may be used, but it requires tests to avoid redundant computations.

5. Study the complexity of the algorithm (usually straightforward from the iterative version of the algorithm).

Such an algorithm is *bottom-up*; we need results of the multiple subproblems to make a choice and compute the optimal solution, while the greedy algorithms are *top-down*, making a local choice at each step.

With dynamic-programming algorithms, one must be particularly cautious about the size of the data. It is not unusual to write nonpolynomial dynamic-programming algorithms. For instance, in the knapsack problem, the cost of the dynamic-programming algorithm is $O(nW)$. However, data can be encoded in $\sum_{i=1}^{n} \log w_i + \sum_{i=1}^{n} \log c_i \leqslant n(\log W + \log C)$, which means that W is in fact exponential in the problem size. This important encoding issue is related to weak NP-completeness and pseudo-polynomial algorithms, which we come back to in Section 6.6, p. 145.

4.4 Exercises

Exercise 4.1: Matrix chains (solution p. 90)

Consider n matrices $A_1, \ldots, A_n$, where A_i is of size $P_{i-1} \times P_i$ $(1 \leqslant i \leqslant n)$. We want to compute $A_1 \times A_2 \times \cdots \times A_n$. The problem is to decide in which order the multiplications should be done and, therefore, to add parentheses to the expression, in order to minimize the number of operations. Note that it costs $P_a \times P_b \times P_c$ to multiply a matrix of size $P_a \times P_b$ by a matrix of size $P_b \times P_c$.

Propose a dynamic-programming algorithm to solve the problem and give its complexity. Be careful to define the initial conditions and the recurrence.

Exercise 4.2: The library (solution p. 91)

The library is planning to move. It has a collection of n books $b_1, b_2, \ldots, b_n$. Book b_i has a width w_i and a height h_i. The books are stored on identical shelves of width L. Each shelf is used to store a set of books of consecutive indices. In other words, for each shelf, there exist two indices i and j such that the shelf exactly includes the books b_i, b_{i+1}, $\ldots$, b_{j-1}, b_j.

1. We assume first that all heights are identical: $h_i = h$, for $1 \leqslant i \leqslant n$, and we want to minimize the number of shelves that are used. Propose a greedy algorithm to solve the problem and prove that it is optimal.

2. Now, books have different heights, but we can adjust the distance between two shelves. The new objective criteria is the total space usage, defined as the sum of the heights of the higher book on each shelf. Give an example where the greedy algorithm of the previous question is no longer optimal, design an optimal algorithm to solve this problem, and give its complexity.

3. We come back to the problem with identical heights. Now, we want to place the n books on k shelves of same length L, and the objective is to minimize L, while k is fixed. In other words, we need to partition the n books into k sets, where the width of the widest set is as small as possible. Design an algorithm to solve the problem, and give its complexity in terms of n and k.

Exercise 4.3: Polygon triangulation (solution p. 93)

We consider planar convex polygons. A triangulation of a polygon is a set of lines that do not intersect inside the polygon and that divide the polygon into triangles. Here, the triangulation lines all pass through polygon vertices.

Let $P = \langle v_0, \ldots, v_n \rangle$ be a convex polygon, where $v_0, \ldots, v_n$ are the polygon vertices numbered in the direct order, and let w be a weight function defined on the triangles formed by the sides and the lines drawn in P. For instance, $w(i, j, k)$ can be the perimeter of the triangle defined by the vertices v_i, v_j, and v_k. The problem is to find a triangulation that minimizes the sum of the weight of the triangles induced by the triangulation.

1. For $1 \leqslant i < j \leqslant n$, we define $t(i, j)$ as the weight of an optimal triangulation of the polygon $\langle v_{i-1}, \ldots, v_j \rangle$, with $t(i, i) = 0$ for $1 \leqslant i \leqslant n$. Express a recurrence to compute t, derive an algorithm to solve the problem, and give its complexity.

2. If the weight function can be anything, how many values do we need to know for the function to be defined on all polygon triangles? Compare with the complexity of the algorithm.

3. If the weight of a triangle is equal to its surface, what can you say about the algorithm that you have designed?

Exercise 4.4: Square of ones (solution p. 96)

Given a matrix A of size $n \times m$ with coefficients in $\{0, 1\}$, we want to find the maximum width K of a square of ones in A, as well as the coordinates (I, J) of the top left corner of such a square. In other words, for all i, j such that $I \leqslant i \leqslant I + K - 1$ and $J \leqslant j \leqslant J + K - 1$, we have $A[i, j] = 1$.

1. Design a dynamic-programming algorithm to solve this problem.

2. What is the complexity of your algorithm?

(*Hint:* Consider $t[i, j]$, the width of the biggest square of ones whose top left corner is (i, j).)

Exercise 4.5: The wind band (solution p. 98)

In a wind band, there are n musicians of size $t_1, t_2, \ldots, t_n$. For concerts, the orchestra has m suits ($m \geqslant n$) of size $u_1, u_2, \ldots, u_m$. Every year, some musicians leave the band and are replaced by new ones, and we need to give each musician a suit of appropriate size: $\alpha(i)$ is the index of the suit given to the musician of size t_i.

1. Yves, the drum player, believes that the objective is to minimize the average difference between the size of a musician and the size of his or her suit, i.e., minimize $\frac{1}{n} \sum_{i=1}^{n} |t_i - u_{\alpha(i)}|$. He proposes a greedy algorithm. We find i and j such that $|t_i - u_j|$ is minimum, we give the suit of size u_j to the musician of size t_i, and we iterate until everybody receives a suit. Is this algorithm optimal?

2. Anne, the horn player, believes that it is more fair to minimize the average square of differences: $\frac{1}{n} \sum_{i=1}^{n} (t_i - u_{\alpha(i)})^2$. Show in an example the advantage of this objective function, compared to Yves's. Is the greedy algorithm optimal for this objective function?

3. If there are as many suits as musicians (i.e., $n = m$), then design an optimal algorithm for Anne's objective function.

4. Design an optimal algorithm for the general case $m \geqslant n$ (and Anne's objective function).

Exercise 4.6: Ski rental (solution p. 98)

The problem is to distribute m pairs of skis of lengths $s_1, \ldots, s_m$ to n persons of size $h_1, \ldots, h_n$, all wanting to go skiing. We assume that there are enough skis in the rental shop for everybody (i.e., $m \geqslant n$). The allocation is defined by an injective function $f : \{1, \ldots, n\} \to \{1, \ldots, m\}$, and f is optimal when it minimizes $A(n, m) = \sum_{k=1}^{n} |s_{f(k)} - h_k|$.

1. Design an efficient algorithm that returns an optimal allocation of the skis.

   (*Hint:* Prove that the tallest person can be allocated the longest pair of skis used.)

2. What is the complexity of the algorithm? You should refine the analysis to guarantee that the algorithm is in $O(n \log n)$ if $m = n$.

3. Prove that we can obtain a better complexity when $n^2 = o(m)$.

   (*Hint:* Restrict to $O(n^2)$ pairs of skis.)

Exercise 4.7: Building set (solution p. 102)

We want to build a tower as high as possible from a set of bricks. We have n different types of bricks and as many bricks of each type as we want. The brick of type i is a parallelepiped of size $\{x_i, y_i, z_i\}$, and it can be oriented in any way, two dimensions being the base of the brick, and the third one being the height. When we build the tower, a brick can be placed on top of another only if the two dimensions of its base are *strictly* smaller than the dimensions of the brick on which we want to place it.

1. Design an optimal dynamic-programming algorithm to build a tower of maximum height.

2. What is the complexity of this algorithm?

4.5 Solutions to exercises

Solution to Exercise 4.1: Matrix chains

We want to compute $A_1 \times \cdots \times A_n$. Let us look for the optimal cost of computing the product $A_i \times \cdots \times A_j$. We denote this cost by $C(i, j)$. The optimal solution for the problem will be obtained for $i = 1$ and $j = n$.

We define $C(i, j)$ by induction. We partition in two the product of matrices $(A_i, \ldots, A_j)$ to indicate which two matrices were multiplied in the last matrix multiplication. In other words, if we cut $(A_i, \ldots, A_j)$ after the position k, this means that the last multiplication was between matrix $A_i \times A_{i+1} \times \cdots \times A_k$ and matrix $A_{k+1} \times \cdots \times A_j$. Let us assume that the optimal solution was to cut $(A_i, \ldots, A_j)$ after the matrix A_k. Then the optimal cost to compute

$A_i \times \cdots \times A_j$ is equal to the optimal cost of computing $A_i \times \cdots \times A_k$, plus the optimal cost of computing $A_{k+1} \times \cdots \times A_j$, plus the cost of computing the final matrix product: $(A_i \times A_{i+1} \times \cdots \times A_k) \times (A_{k+1} \times \cdots \times A_j)$. Therefore, we have the induction:

$$C(i,j) = \min_{k=i}^{j-1}\{C(i,k) + C(k+1,j) + (P_{i-1} \times P_k \times P_j)\}.$$

Thus, to compute $C(i,j)$, one needs to have already computed all the $C(i,k)$s and all the $C(k,j)$s, with $k \in [i; j-1]$. The initial conditions are $C(i,i) = 0$ and $C(i,i+1) = P_{i-1} \times P_i \times P_{i+1}$.

The dynamic-programming algorithm works as follows: First, it initializes all $C(i,i)$s and all $C(i,i+1)$s. Then, it computes all the values $C(i,i+2)$, then all the $C(i,i+3)$, and so on. Computing a single $C(i,j)$ costs $O(n)$ and, thus, computing all $C(i,i+s)$ costs $O(n^2)$. Therefore, the dynamic-programming algorithm runs in $O(n^3)$. To reconstruct the optimal solution from the output of the dynamic program, one needs only to memoize for any couple (i,j) the place $k(i,j)$ of the cut that minimizes $C(i,j)$.

Solution to Exercise 4.2: The library

1. The greedy algorithm stores as many books as possible on the first shelf, starting with book b_1 and finishing with some book b_{k_1} (included). It then stores as many books as possible on the second shelf, starting with book b_{k_1+1}, and so on.

   To prove the optimality of this greedy algorithm, we consider any instance and we compare the solution of the greedy algorithm on that instance, denoted G, to an optimal solution, denoted O. Let i be the rank of the first shelf for which the two solutions differ. Remember that books are stored by consecutive indices. Therefore, by definition of i, the first book on shelf i is the same under both solutions, say b_k. By definition of the greedy algorithm and of i, G contains strictly more books on shelf i than O. We build a new solution O' as follows: The first shelves, up to shelf i included, contain the same books as for the solution G; the remaining shelves have the same composition as for O, except that we have discarded the books already put on shelf i. This new solution has the same number of shelves as O and, thus, is optimal. The first $i+1$ shelves have the same composition under G and O'. By iterating this process, since there are at most n shelves in an optimal solution, we prove that G is an optimal solution.

2. In this question, books have different heights. For the counterexample to the optimality of the previous greedy algorithm, we take an instance with three books of same width equal to 1 ($w_1 = w_2 = w_3 = 1$) and where at most two books can fit on a shelf ($L = 2$). We pick for the heights: $h_1 = 1$, $h_2 = 2$, and $h_3 = 3$. The greedy algorithm puts the

first two books on the first shelf, for a height of 2, and the last one alone on the second shelf, for a height of 3. The total cost of this solution is thus $2+3=5$. An optimal solution is to put the first book alone on the first shelf, for a height of 1, and the last two books on the second shelf for a height of 3. The total cost of this solution is $1+3=4$.

To find the principle needed to design a dynamic-programming algorithm, let us consider an optimal solution for storing books b_1 through b_n. Let b_k be the last book on the first shelf under this optimal solution. Then, the way the books b_{k+1} through b_n are stored is an optimal solution for the problem where the entire collection of books is b_{k+1}, b_{k+2}, ..., b_n. Then, the cost of the whole solution is the cost of the book organization on the first shelf, that is, the maximal height of a book on the first shelf, plus the cost of the reminder of the solution. Therefore, to design an optimal solution, the "only" remaining problem is to find which book is the last one on the first shelf, knowing that the sum of widths of the books on the first shelf cannot be greater than L. Letting $C(i)$ be the cost of storing optimally books b_i through b_n, we have:

$$C(i) = \min_{\substack{i \leqslant k \leqslant n \\ \sum_{j=i}^{k} w_j \leqslant L}} \left(\max\{h_i, \ldots, h_k\} + C(k+1) \right).$$

To compute $C(i)$, we must already know the values of all $C(j)$s for $j > i$. Algorithm 4.4 is a dynamic-programming algorithm to compute these values $C(i)$s. In this algorithm, *LastBook*[i] is the index of the last book on the shelf starting with book i.

The complexity of Algorithm 4.4 is $O(n^2)$.

3. Once again, to find the principle needed to design a dynamic-programming algorithm, consider an optimal solution for storing books b_1 through b_n on k shelves. Let b_i be the last book on the first shelf in this optimal solution. Then, without loss of generality, we can assume that the distribution of the books b_{i+1} through b_n is an optimal solution for the problem where the entire collection of books is b_{i+1}, b_{i+2}, ..., b_n, and where it should be stored on $k-1$ shelves. Indeed, if this book distribution were not optimal for that problem, we could replace it with an optimal solution to find another optimal solution of the original problem (all n books on k shelves). To see that the distribution of the books b_{i+1} through b_n on the last $k-1$ shelves may not be optimal, let us consider the following example with four books ($n=4$) and three shelves ($k=3$): $w_1 = 3$, $w_2 = 2$, $w_3 = 1$, and $w_4 = 1$. L cannot be smaller than the width of the largest book. Therefore, the solution with b_1 on the first shelf, b_2 and b_3 on the second one, and b_4 on the last one is optimal because it achieves $L=3$. Nevertheless, the optimal distribution of the last three books on two shelves is b_2 on the first shelf and b_3 and b_4 on the last one, for a width of 2.

```
1  C(n) ← h_n
2  for i = n − 1 downto 1 do
3      ShelfWidth ← w_i
4      ShelfHeight ← h_i
5      C(i) ← h_i + C(i + 1)
6      LastBook[i] ← i
7      for j = i + 1 to n do
8          ShelfWidth ← ShelfWidth + w_j
9          if h_j > ShelfHeight then ShelfHeight ← h_j
10         if ShelfWidth ⩽ L and ShelfHeight + C(j + 1) ⩽ C(i) then
11             C(i) ← ShelfHeight + C(j + 1)
12             LastBook[i] ← j
```

ALGORITHM 4.4: Organization of books on library shelves so as to minimize the overall storage space.

Let $M_{i,j}$ be the cost of storing books b_i through b_n on j shelves. Following the previous analysis, we have:

$$\begin{cases} \forall i \in [1;n], \forall j \in [2;k], \quad M_{i,j} = \min_{i \leqslant k \leqslant n} \max \left\{ \sum_{l=i}^{k} w_l, M_{k+1,j-1} \right\} \\ \forall i \in [1;n], \quad M_{i,1} = \sum_{k=i}^{n} w_k \\ \forall j \in [1;k], \quad M_{n+1,j} = 0. \end{cases}$$

Algorithm 4.5 is a dynamic-programming algorithm implementing the above recursive computation and whose complexity is $O(k \times n^2)$.

Solution to Exercise 4.3: Polygon triangulation

1. To find a recursion on t, we just need to remark that the line segment $[v_{i-1}, v_j]$ is one side of a triangle in any triangulation considered here. The only vertex of that triangle that is not yet defined is one of the vertices of the polygon $\langle v_{i-1}, \dots, v_j \rangle$. Therefore, we need to look among the vertices v_i, ..., v_{j-1} for the vertex v_k defining the best triangulation. This is illustrated in Figure 4.3. The weight of an optimal triangulation is then equal to the weight of the triangle defined by v_{i-1}, v_k, and v_j, plus the weight of an optimal triangulation of $\langle v_{i-1}, v_i, \dots, v_k \rangle$, plus the weight of an optimal triangulation of $\langle v_k, v_{k+1}, \dots, v_j \rangle$. Therefore,

$$t(i,j) = \min_{i \leqslant k \leqslant j-1} \left(w(i-1,k,j) + t(i,k) + t(k+1,j) \right).$$

```
1  M_{n,1} = w_n
2  for i = n − 1 downto 1 do M_{i,1} = w_i + M_{i+1,1}
3  for j = 2 to k do
4      for i = n − 1 downto 1 do
5          L ← M_{i,j−1}    /* The optimal is to leave the last shelf empty */
6          LastBook[j] ← n + 1
7          ShelfWidth ← 0
8          for m = i to n − 1 do
9              ShelfWidth ← ShelfWidth + w_m
10             if max{ShelfWidth, M_{m+1,j−1}} < L then
11                 L ← max{ShelfWidth, M_{m+1,j−1}}
12                 LastBook[j] ← m
```

ALGORITHM 4.5: Organization of books on k library shelves so as to minimize the width of the widest shelf.

(We remark that this expression is valid when $j = i + 1$ thanks to the convention $t(i, i) = 0$.)

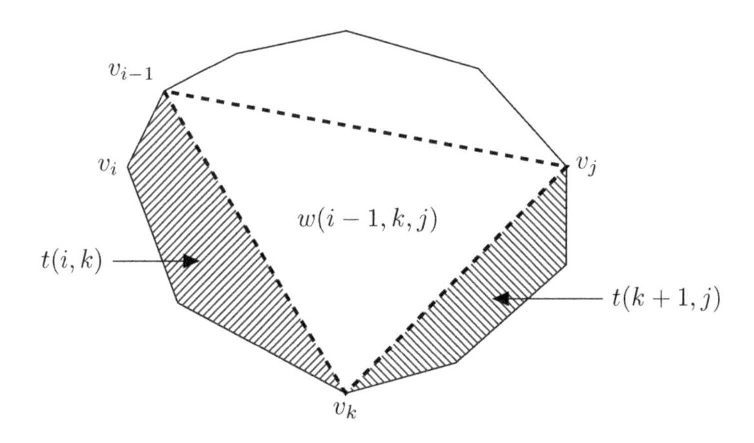

FIGURE 4.3: Recursive polygon triangulation.

We are now ready to write a dynamic-programming algorithm to compute $t(1, n)$ from the values $t(k, j)$ where $i + 1 \leqslant k \leqslant j$ and from the values $t(i, k)$ where $i \leqslant k \leqslant j - 1$. In other words, if we represent the values of $t(i, j)$ to be computed with an (i, j)-diagram as the one presented in Figure 4.4, $t(1, n)$ is computed from the values that are "below" and "to the right" of the point $(1, n)$. Therefore, to compute $t(1, n)$, we must compute the values $t(i, j)$ for i going from 1 to n and with $j \geqslant i$, and this should be done by increasing values of $(j - i)$. Algorithm 4.6 performs such a computation.

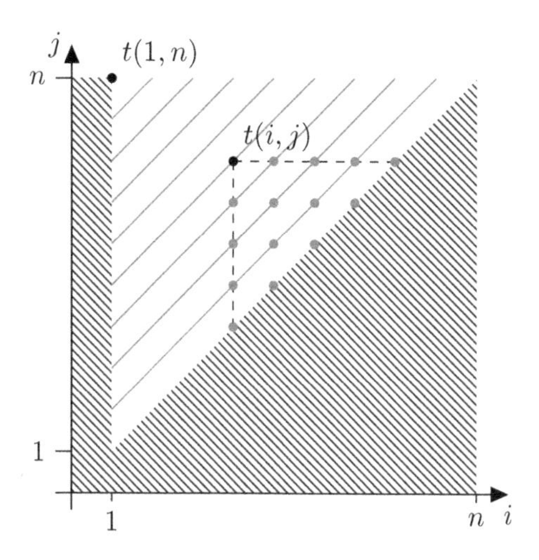

FIGURE 4.4: (i, j)-diagram of the values to be computed by the dynamic-programming algorithm.

To compute $t(i, j)$, Algorithm 4.6 performs $2(j - i)$ additions and $j - i - 1$ comparisons. Overall, the number of additions performed is

$$A_n = \sum_{d=1}^{n-1} ((n - d) \times (2d))$$

where $(n - d)$ is the number of values $t(i, j)$ to be computed on the diagonal $j - i = d$ and where $2d$ is the number of additions needed in the computation of $t(i, j)$ where $j - i = d$. We obtain:

$$A_n = 2n \left(\sum_{d=1}^{n-1} d \right) - 2 \left(\sum_{d=1}^{n-1} d^2 \right) = \frac{(n-1) \times n \times (n+1)}{3} = \Theta(n^3).$$

To compute the number T_n of tests needed, one just has to remark that when computing $t(i, j)$, the number of tests is equal to half the number of additions, minus one. Then, $T_n = A_n/2 - C_n$ where C_n is the total number of values $t[i, j]$ that are computed. Therefore:

$$T_n = \frac{1}{2} A_n - \sum_{d=1}^{n-1} (n - d) = \frac{n^3 - 3n^2 + 2n}{6} = \Theta(n^3).$$

The overall complexity of Algorithm 4.6 is, therefore, $\Theta(n^3)$.

```
1 for i = 0 to n do
2   t(i,i) ← 0
3 for d = 1 to n − 1 do
4   for i = 1 to n − d do
5     t(i,i+d) ← min_{i⩽k⩽i+d−1} (t(i,k)+t(k+1,i+d)+w(i−1,k,i+d))
6 return t(1,n)
```

ALGORITHM 4.6: Dynamic-programming algorithm to compute an optimal polygon triangulation.

2. In the general case, to define fully the function w, one needs one value for each possible triangle, that is, $\binom{n+1}{3} = \Theta(n^3)$. Each of these values must be read at least once to define an optimal triangulation. Therefore, any algorithm must have a complexity of at least $\Omega(n^3)$, and Algorithm 4.6 is optimal (at least for the order of magnitude of its complexity).

3. When the weight of a triangle is equal to its area, all triangulations have the same weight, which is the area of the polygon. Algorithm 4.6 is not well suited to that case. We can replace it with Algorithm 4.7. This algorithm does not perform a single test and only $n - 3$ additions (far fewer than the $\Theta(n^3)$ of Algorithm 4.6).

```
1 Let the triangulation be defined by the line segments (v_0, v_i), for
  2 ⩽ i ⩽ n − 1
2 return Sum of the weights of the triangles
```

ALGORITHM 4.7: Program to compute an arbitrary polygon triangulation.

Solution to Exercise 4.4: Square of ones

1. Let $C_{i,j}$ denote the largest square of ones of matrix A whose top left corner is the element $A_{i,j}$. Let $t(i,j)$ be the width of $C_{i,j}$. If $A_{i,j} = 0$, then $t(i,j) = 0$. Otherwise, since $C_{i,j}$ has size $t(i,j)$, then all elements $A_{k,l}$ with $i \leqslant k \leqslant i + t(i,j) - 1$ and $j \leqslant l \leqslant j + t(i,j) - 1$ should be equal to one. Therefore, for $C_{i,j}$ to have size $t(i,j)$, then:

- $C_{i,j+1}$ must have a size at least $t(i,j) - 1$;
- $C_{i+1,j}$ must have a size at least $t(i,j) - 1$;
- $C_{i+1,j+1}$ must have a size at least $t(i,j) - 1$.

Therefore, we have the following recursive formula for any $i \in [1; n]$ and $j \in [1; m]$:

$$\begin{cases} t(i,j) = 0 \text{ if } A_{i,j} = 0 \\ t(i,j) = \min\big(t(i,j+1), t(i+1,j), t(i+1,j+1)\big) + 1 \text{ otherwise} \end{cases}$$

with the notation extension $t(n+1,j) = t(i,n+1) = 0$. Algorithm 4.8 implements this recursion.

```
1  for i = 1 to n do
2      t(i, m) ← A_{i,m}
3  for j = 1 to m − 1 do
4      t(n, j) ← A_{n,j}
5  for i = n − 1 downto 1 do
6      for j = m − 1 downto 1 do
7          if A_{i,j} = 0 then
8              t(i, j) ← 0
9          else
10             t(i, j) ← 1 + min(t(i, j + 1), t(i + 1, j), t(i + 1, j + 1))
11 K ← 0
12 for i = 1 to n do
13     for j = 1 to m do
14         if t(i, j) > K then
15             K ← t(i, j)
16             I ← i
17             J ← j
18 return (K, I, J)
```

ALGORITHM 4.8: Dynamic-programming algorithm to compute the largest square of ones in a matrix with values 0 and 1.

2. The algorithm has a complexity of $O(n \times m)$.

Solution to Exercise 4.5: The wind band

1. We consider the following instance with two musicians and two suits: $t_1 = 1$, $t_2 = 4$, $u_1 = 3$, and $u_2 = 6$. The greedy algorithm produces the solution $\alpha(2) = 1$ and $\alpha(1) = 2$ whose score is $(1+5)/2 = 3$. The optimal solution is $\alpha(1) = 1$ and $\alpha(2) = 2$ whose score is $(2+2)/2 = 2$.

2. This objective function has a strong negative impact on the large differences in sizes. We consider the same instance as in the previous question. The greedy algoritm produces the same solution whose score is now $(1+5^2)/2 = 13$. The optimal solution is the same as the previous one, and its score is now $(2^2 + 2^2)/2 = 4$.

3. We assume that musicians and suits are sorted by nondecreasing sizes. Suppose that in a solution there are two musicians i and j, with $i < j$, such that $\alpha(i) > \alpha(j)$. Let S be the score of that solution and let S' be the score of the solution where we have exchanged the suits allocated to those two musicians.

$$\begin{aligned} S' - S &= \left((t_i - u_{\alpha(j)})^2 + (t_j - u_{\alpha(i)})^2\right) - \left((t_i - u_{\alpha(i)})^2 + (t_j - u_{\alpha(j)})^2\right) \\ &= 2(u_{\alpha(i)} - u_{\alpha(j)})(t_i - t_j) \,. \end{aligned}$$

   By hypothesis, because $i < j$, $t_i \leqslant t_j$, and because $\alpha(i) > \alpha(j)$, $u_{\alpha(i)} \geqslant u_{\alpha(j)}$. Therefore, $S' - S \leqslant 0$, and the new solution is better than the original one. In other words, in an optimal solution, a taller musician must receive a larger suit.

   An optimal algorithm then sorts musicians and suits and assigns the i-th suit to the i-th musician. Its complexity is $O(n \log(n))$.

4. From what we have proved at the previous question, the largest of all suits either is assigned to the tallest of the musicians or is not used. Let us denote by $M_{i,j}$ the cost of the optimal solution when assigning to the j smallest musicians suits among the i smallest ones. From what precedes, either the largest suit is not used and $M_{i,j} = M_{i,j-1}$, or the largest suit is assigned to the tallest musician and $M_{i,j} = M_{i-1,j-1} + (u_i - t_j)^2$. This gives us the following recursive definition of the optimal solution:

$$M_{i,j} = \min\left\{M_{i-1,j}, M_{i-1,j-1} + (u_i - t_j)^2\right\}.$$

   Algorithm 4.9 computes this solution following a dynamic-programming approach.

Solution to Exercise 4.6: Ski rental

1. In order to be able to follow the hint easily, we assume that the skiers are sorted by nondecreasing heights and that the pairs of skis are sorted by nondecreasing lengths. We want to express the optimal solution through

```
1 M_{0,0} ← 0
2 for i = 1 to n do
3     M_{i,i} ← M_{i-1,i-1} + (u_i − t_i)^2
4     for j = i + 1 to m do
5         if M_{i,j-1} < M_{i-1,j-1} + (u_i − t_j)^2 then
6             M_{i,j} ← M_{i,j-1}
7         else
8             M_{i,j} ← M_{i-1,j-1} + (u_i − t_j)^2
9             A_i ← j        /* We record the assignment */
```

ALGORITHM 4.9: Assignment of suits to musicians.

some recursion. Therefore, we consider the restriction of our problem to the first i skiers and the first j pairs of skis. We assume that the property proposed by the hint is true. In other words, we restrict our search space by assuming that the longest pair of skis used is assigned to the tallest person. Then we focus on the j-th pair of skis, that is, on the pair of longest skis. There are two cases:

(a) The optimal solution for the first i skiers and first j ski pairs does not use the j-th pair of skis. Then, $A(i,j) = A(i,j-1)$.

(b) The optimal solution for the first i skiers and first j ski pairs uses the j-th pair of skis. Then, according to the hint, that pair is used for the tallest person, that is, the i-th one. Then, $A(i,j) = A(i-1,j-1) + |s_j - h_i|$.

Gathering the two cases, we obtain:

$$A(i,j) = \min\{A(i,j-1), A(i-1,j-1) + |s_j - h_i|\}. \quad (4.1)$$

However, to achieve this result, we have assumed that the property proposed in the hint is true. Therefore, we have to establish it.

We assume that there is an instance, and an optimal ski allocation function f, such that, in this instance, f does not allocate the longest pair of skis used to the tallest person. We then build a new ski assignment, f', whose cost is not greater than the one of f, and that satisfies the desired property. As previously, we assume that the skiers are sorted by nondecreasing heights and that the pairs of skis are sorted by nondecreasing lengths. Without loss of generality, we can assume that the longest pair of skis used is the j-th one. Indeed, if this is not the case, and if the longest pair of skis used is the k-th one, we remove from the previous instance all the pairs of skis except the first k ones to obtain an example with the desired property. Let p, $p \in [1;n]$, be the person assigned the j-th pair of skis: $f(p) = j$. Then, we build a new allocation function f' identical to f except that we swap the assignments of the persons p and i: $f'(p) = f(i)$, $f'(i) = f(p) = j$, and $f'(k) = f(k)$ for any $k \in [1; n] \setminus \{i, p\}$. When evaluating the objective function, the only terms that change are the ones involving i and p. Therefore, we focus on the terms:

$$A = |s_j - h_p| + |s_{f(i)} - h_i| \quad \text{and} \quad B = |s_{f(i)} - h_p| + |s_j - h_i| \,.$$

We will have established the desired property if we succeed in proving that we always have $A \geqslant B$. We must consider all the different relative orderings of $s_{f(i)}$, s_j, h_p, and h_i.

- $h_p \leqslant h_i \leqslant s_{f(i)} \leqslant s_j$:
  $A - B = (s_j - h_p - h_i + s_{f(i)}) - (-h_p + s_{f(i)} + s_j - h_i)$
  $= 0$
- $h_p \leqslant s_{f(i)} \leqslant h_i \leqslant s_j$:
  $A - B = (s_j - h_p + h_i - s_{f(i)}) - (-h_p + s_{f(i)} + s_j - h_i)$
  $= 2(h_i - s_{f(i)}) \geqslant 0$
- $s_{f(i)} \leqslant h_p \leqslant h_i \leqslant s_j$:
  $A - B = (s_j - h_p + h_i - s_{f(i)}) - (h_p - s_{f(i)} + s_j - h_i)$
  $= 2(h_i - h_p) \geqslant 0$
- $s_{f(i)} \leqslant h_p \leqslant s_j \leqslant h_i$:
  $A - B = (s_j - h_p + h_i - s_{f(i)}) - (h_p - s_{f(i)} - s_j + h_i)$
  $= 2(s_j - h_p) \geqslant 0$
- $s_{f(i)} \leqslant s_j \leqslant h_p \leqslant h_i$:
  $A - B = (h_p - s_j + h_i - s_{f(i)}) - (h_p - s_{f(i)} - s_j + h_i)$
  $= 0$

Therefore, in all cases, we have $B \geqslant A$, and f' defines a solution whose quality is at least as good as that of f. We can then safely use equation (4.1) to compute an optimal solution. This is what Algorithm 4.10 does (with the convention that $A(0, j) = 0$ for any $j \in [0; m]$), where $s(i, j)$ records which pair of skis is assigned to skier i when the choice is made among the first j pairs of skis. The bounds of the loop at Step 4 may surprise the reader at first sight. These bounds state only that:

- There should be at least $i - 1$ pairs of skis available for the first $i - 1$ skiers (and we know then that none of the first $i - 1$ pairs of skis are going to be assigned to one of the last $n - (i - 1)$ skiers).
- There should be at least $n - i$ pairs of skis available for the last $n - i$ skiers (and we know then that none of the last $n - i$ pairs of skis are going to be assigned to one of the first i skiers).

2. We remark that for the iteration i of the outermost loop, we compute $A(i, i)$ and then $(m - n + i) - (i + 1) + 1 = m - n$ iterations of the loop

```
1  for i = 1 to n do
2  |  A(i,i) ← A(i−1,i−1) + |s_i − h_i|
3  |  s(i,i) ← i
4  |  for j = i+1 to m−n+i do
5  |  |  if A(i−1,j−1) + |s_i − h_j| < A(i,j−1) then
6  |  |  |  A(i,j) ← A(i−1,j−1) + |s_i − h_j|
7  |  |  |  s(i,j) ← j
8  |  |  else
9  |  |  |  A(i,j) ← A(i,j−1)
10 |  |  |  s(i,j) ← s(i,j−1)
```

ALGORITHM 4.10: Ski allocation algorithm.

at Step 4. Hence, the complexity of Algorithm 4.10 is

$$\sum_{i=1}^{n}(1+m-n) = n(m-n+1).$$

When taking into account the presorting of the pairs of skis and of the persons, we obtain an overall complexity in $O(m\log(m) + n(m-n))$ because $n \leqslant m$. When $n = m$ the complexity is just $O(m\log(m))$, that is the complexity of the initial sorts.

3. We are dealing here with a very large choice of skis. We first remark that we can restrict the algorithm, for each skier i, to look for a solution among its n most suitable pairs of skis, a set we denote S_i. We use a subset of size n, instead of just 1, to take into account potential conflicts between skiers. The pairs of skis defined this way give rise to a set $S = \bigcup_{i=1}^{n} S_i$ containing at most n^2 pairs of skis (some of the S_i sets may not be distinct and thus share some pairs of skis). We then apply Algorithm 4.10 to the set S. We compute the set S in time $O(n(\log(m)+n))$: For each of the n persons, we find the best pair of skis with a binary search and then visit a neighborhood of size $O(n)$ around this best pair to find the n best pairs. The overall complexity is then:

$$O(n\log(n) + m\log(m) + n(\log(m)+n) + n^2\log(n^2) + n(n^2-n)) \\ = O(n^3 + m\log(m))$$

when taking into account the original sorts. This complexity is lower than the original one as $n^3 = o(n \times m)$ because we have assumed that $n^2 = o(m)$. However, we still have to prove that we obtain an optimal solution when we restrict the search to the set S.

We remark that, by construction, for any pair of skis $j \notin S$ and for any skier i, there exist at least n pairs of skis $s_{j_1}, \ldots, s_{j_n}$ in S_i such that $|s_j - h_i| \geqslant |s_{j_k} - h_i|$ $(1 \leqslant k \leqslant n)$ because we have taken pairs of skis *around* the optimal pair of skis. Therefore, if there exists an optimal assignment that uses the pair of skis j for a skier i (along, potentially, with other pairs of skis not in S), we can replace j with a pair from S_i. This is always possible because the other skiers use at most $n-1$ pairs of skis and because S_i contains n pairs of skis. Furthermore, as we have just seen, such an exchange does not increase the objective function. Hence, the new solution is also optimal.

Solution to Exercise 4.7: Building set

1. Given any brick $\{x_i,\ y_i,\ z_i\}$, there are a priori up to six ways to put it on top of the tower: There are three choices for the base and then two orientations (without loss of generality, we assume bricks are always laid out so that their faces are parallel). The base is then defined by a length and a width, where the length is greater than the width. One can remark that if it is valid to put the new brick on the one at the top of the tower so that the length of the new brick is parallel to the width of the one at the top, then it is also valid to put the new brick on the top of the tower with its width parallel to the width of the brick at the top. We, therefore, can assume, without loss of generality, that bricks are put on top of each other so that their widths are parallel. This reduces to three the number of ways a brick can be put on top of the tower.

   In the remainder of this solution, the brick (L_i, l_i, h_i) designs a parallelepiped of size $\{L_i, l_i, h_i\}$ that can be laid down only so that its base is $L_i \times l_i$, its length L_i, its width l_i (with, thus, $L_i \geqslant l_i$), and its height h_i. Therefore, from the original set of n parallelepipeds $\{\{x_i, y_i, z_i\}_{1 \leqslant i \leqslant n}\}$, we build a set of $3n$ bricks $\cup_{1 \leqslant i \leqslant n}\{(z_i, y_i, x_i), (z_i, x_i, y_i), (y_i, x_i, z_i)\}$ (with the assumption that, for any i, $x_i \leqslant y_i \leqslant z_i$). With these new notations, one can put the brick (L_i, l_i, h_i) on top of the brick (L_j, l_j, h_j) if and only if $L_i < L_j$ and $l_i < l_j$.

   For any value of i, $1 \leqslant i \leqslant 3n$, let H_i be the maximum height of a tower whose top brick is the i-th brick. Then, for any i, $1 \leqslant i \leqslant 3n$, we can write the induction:

$$H_i = h_i + \max\{0,\ \max\{H_j \mid 1 \leqslant j \leqslant 3n,\quad L_j > L_i,\ l_j > l_i\}\}.$$

   (We introduced the "maximum with zero" term in case there does not exist a single brick on which the brick i can be put.) In order to be able to compute such an induction, the idea is to sort the $3n$ bricks by nonincreasing values of L_i. The induction can then be rewritten:

$$H_i = h_i + \max\{0,\ \max\{H_j \mid 1 \leqslant j \leqslant i-1,\quad L_j > L_i,\ l_j > l_i\}\}.$$

We then have a natural ordering to compute the H_is. The optimal solution is then defined by the maximum of the H_is. Algorithm 4.11 implements this computation (under the assumption that, for any i, $x_i \leqslant y_i \leqslant z_i$).

```
1  list ← ∅
2  for i = 1 to n do
3  └ list ← list ∪ {(z_i, y_i, x_i), (z_i, x_i, y_i), (y_i, x_i, z_i)}
4  Sort list = {(L_i, l_i, h_i) | i ∈ {1, ..., 3n}} by nonincreasing L_i
5  H_1 ← h_1
6  for i = 2 to 3n do
7  │ 𝓗 ← {H_j | L_j > L_i, l_j > l_i, 1 ⩽ j < i}
8  │ if 𝓗 = ∅ then
9  │ └ H_i ← h_i
10 │ else
11 │ └ H_i ← h_i + max{h | h ∈ 𝓗}
   └
12 H ← max_{1⩽i⩽3n} H_i
13 return H
```

ALGORITHM 4.11: Tallest tower constructed from a building set.

2. Building the set of the $3n$ bricks and sorting it can be done in $O(n\log(n))$. Then, to compute the value of H_i, one must compare L_i to L_j and l_i to l_j for any $j \in [1, i-1]$, which costs $2(i-2)$ comparisons. Then, the maximum must be taken over $i-1$ values, which costs $i-2$ additional comparisons. The computation of all the H_is thus has a quadratic cost. The final determination of the optimal solution is the search for the maximum other $3n$ values, which costs $3n-1$. Therefore, the overall complexity of Algorithm 4.11 is $O(n^2)$.

4.6 Bibliographical notes

Exercise 4.1 (matrix chains) comes from the book by Cormen, Leiserson, Rivest, and Stein [27]. Exercise 4.3 (polygon triangulation) is inspired from the book by Goodrich and Tamassia [43].

Chapter 5

Amortized analysis

In this chapter, we briefly discuss amortized analysis, the goal of which is to average the cost of n successive operations. This should not be confused with the average cost of an operation. We first describe the three classical methods with examples (Section 5.1) and then proceed with exercises in Section 5.2, with solutions in Section 5.3.

5.1 Methods for amortized analysis

First, we introduce two examples to illustrate the methods used to conduct an amortized analysis. Then, we present the three classical methods: aggregate analysis, the accounting method, and the potential method.

5.1.1 Running examples

The first example is a k-bit counter that we want to increment. Initially, the counter has a value of 0, and each operation increments it. Formally, this counter is represented by an array A of k bits, where $A[i]$ is the $(i+1)$-th bit, for $0 \leqslant i \leqslant k-1$. A number x represented by this counter is such that $x = \sum_{i=0}^{k-1} A[i].2^i$. For instance, if $k = 6$ and if we perform $n = 4$ operations, we obtain the following sequence:

$$\begin{array}{c} 0\,0\,0\,0\,0\,0 \\ 0\,0\,0\,0\,0\,1 \\ 0\,0\,0\,0\,1\,0 \\ 0\,0\,0\,0\,1\,1 \\ 0\,0\,0\,1\,0\,0 \end{array}$$

The cost of an increment is defined as the number of bits that should be modified. This cost is not constant for each value of the counter; it is equal to the number of successive 1s at the right of the counter, plus 1 (switching the first 0 to 1).

The second example consists of inserting n elements in a table, dynamically, starting from an empty table. We insert a new element directly in the table if there is space, with a cost 1. Otherwise, we create a new table that has twice the size of the original table (or a table of size 1 for the first insertion); we copy the content of the original table and insert the new element. The cost is then the size of the original table plus 1. Note that the table is always at least half full (an empty table is considered full), so even if the cost may be high for some operations, we then have free space for the next operations.

For both examples, the amortized analysis consists of asking the following question: What is the cost of n successive operations?

5.1.2 Aggregate analysis

The goal of this method is to show that the cost of n successive operations can be bounded by $T(n)$. Therefore, in the worst case, the cost per operation on average, i.e., the amortized cost per operation, is bounded by $T(n)/n$.

For the k-bit counter, it is obvious that the cost of n successive increment operations is bounded by nk. However, this upper bound can be improved. Indeed, the right-most bit flips each time, the second one flips every second time, and so on. Therefore, the cost of n operations is at most $n+\frac{n}{2}+\frac{n}{4}+\cdots \leqslant 2n$, regardless of the value of k. This leads to an amortized cost per operation of 2.

For the table insertion, for any integer $k \geqslant 0$, the cost of the (2^k+1)-th insertion is $c(i) = 2^k + 1$, i.e., the size of the table is doubled. Otherwise, the cost is $c(i) = 1$ (including the cost of the first insertion, $c(1)$). Therefore, we have:

$$\sum_{i=1}^{n} c(n) \leqslant n + \sum_{k=0}^{\lfloor \log_2(n) \rfloor} 2^k \leqslant 3n$$

and an amortized cost of 3.

5.1.3 Accounting method

The principle of this method is to pay in advance for costly operations that may happen afterwards, hence, keeping a constant cost per operation. One has to guarantee that at the time of operation i, one has enough credit (including advance payment and the payment for the operation) to cover the cost of the operation.

For the k-bit counter, each time we flip a bit from 0 to 1, we decide to pay 2 euros[1]: 1 euro for the flip and another one so that we will be able to flip back the bit from 1 to 0 without having to pay. For this example, since at each increment there is only one bit to flip from 0 to 1, the cost is 2 at each

[1] Yes, \$2 would be okay, too.

increment, and, hence, an upper bound of $2n$ for n operations (note that we may have paid for some operations that have not been done yet).

For the table insertion, we decide to pay €3 at each insertion: €1 is used to pay for the insertion, a second one will be used to pay for the transfer of the element when a new table will be required, and a third one is assigned to an element in the first half of the table that also will need to be transferred later when the table is full. Therefore, each time the size of the table is doubling, we can transfer all elements at no cost. This leads to an upper bound of $3n$.

5.1.4 Potential method

This last method consists of representing the prepaid work of the accounting method by a potential that can be used to pay for future operations. The prepaid work of the accounting method is no longer associated with objects but rather with the data structure itself. We define a potential function, which associates to each data structure a potential. This potential function should always be greater or equal to the potential function of the data structure before the first operation, so that there is always enough potential to pay for an operation. We introduce the following notations:

- Φ_0 is the potential before the first operation.
- $\Phi_i \geqslant \Phi_0$ is the potential of the data structure after i operations.
- c_i is the cost of operation i.
- $\hat{c}_i = c_i + \Phi_i - \Phi_{i-1}$ is the amortized cost of operation i. A costly operation may have a small amortized cost if the potential function has decreased with operation i, i.e., $\Phi_i - \Phi_{i-1} < 0$.

Therefore, the amortized cost of a sequence of n operations can be computed as:

$$\sum_{i=1}^{n} \hat{c}_i = \sum_{i=1}^{n} c_i + \Phi_n - \Phi_0.$$

Because $\Phi_n - \Phi_0 \geqslant 0$, the total amortized cost $\sum_{i=1}^{n} \hat{c}_i$ gives an upper bound on the total actual cost $\sum_{i=1}^{n} c_i$. Note that we often define Φ so that $\Phi_0 = 0$ and $\Phi_i \geqslant 0$, for convenience.

For the k-bit counter, Φ_i is the number of bits that are at value 1 after operation i. This number is always positive or null, and it is initially null. Let $t(i)$ be the number of right-most successive 1s just before operation i. The potential after operation i is, therefore, $\Phi_i = \Phi_{i-1} - t(i) + 1$ because $t(i)$ 1s have been reset to 0, and one 0 has taken the value 1. Moreover, the cost of operation i is $c_i = t(i) + 1$:

$$\begin{array}{ll} \cdots 0\, 1 \cdots 1 & \Phi_{i-1} \\ \cdots 1\, \underbrace{0 \cdots 0}_{t(i)} & \Phi_i = \Phi_{i-1} - t(i) + 1. \end{array}$$

Therefore, the amortized cost of operation i is $\hat{c}_i = c_i + \Phi_i - \Phi_{i-1} = t(i) + 1 + (-t(i) + 1) = 2$.

For the table insertion, the potential can be seen as the richness of the table; a table is rich when it is full. The table potential equals twice the number of elements in the table minus the size of the table. Because the table is always at least half full, this value cannot be negative. Formally, let num_i be the number of elements after i operations, and let $size_i$ be the size of the table after i operations. Initially, $num_0 = size_0 = \Phi_0 = 0$, and the potential function is expressed as $\Phi_i = 2num_i - size_i \geqslant 0$. Because we perform only insertions, $num_i = num_{i-1} + 1 = i$.

If the size of the table remains identical after operation i, we have $size_i = size_{i-1}$ and $c_i = 1$. Therefore, $\hat{c}_i = c_i + \Phi_i - \Phi_{i-1} = 1 + 2 = 3$. However, if $size_i = 2size_{i-1}$, this means that the cost of the operation was $c_i = num_i$ and that the table was full after operation $i - 1$, i.e., $size_{i-1} = num_{i-1}$, and, therefore, $\hat{c}_i = c_i + \Phi_i - \Phi_{i-1} = num_i + 2 - size_{i-1} = 3$.

5.2 Exercises

Exercise 5.1: Binary counter (solution p. 112)

Consider the running example of the k-bit counter introduced in Section 5.1.1 with an *increment* function.

1. Show that if we had also a *decrement* function on the counter, then a sequence of n operations could have a cost in $\Theta(nk)$.

2. We keep only the *increment* function, and we add a *reset* operation that resets the counter to its initial value 0. Show how to implement this counter as a table of bits so that any sequence of n operations (increment or reset) takes a time $O(n)$ on a counter with initial value 0, with the accounting method.

   (*Hint:* Keep a pointer to the highest-order 1.)

Exercise 5.2: Inserting and deleting (solution p. 113)

Consider the running example of table insertion introduced in Section 5.1.1. We consider now that it is also possible to delete elements from the table.

1. If we double the size of the table when it is full, and we halve the size of the table when it is less than half empty, what would be the amortized cost?

2. Propose an implementation of the insert and delete functions with a constant amortized cost. Apply the accounting method and then the potential method to compute the amortized cost.

Exercise 5.3: Stack (solution p. 114)

We consider a stack with the following operations: *push*(S, x) pushes object x onto stack S, *pop*(S) pops the top of the stack and returns the popped object (it returns an error if the stack is empty), and, finally, *multipop*(S, k) removes the k top objects of the stack, and the entire stack if it contains fewer than k objects (it tests at each step whether the stack is empty). Initially, the stack is empty.

1. What is the time complexity of each of these operations? Use the aggregate analysis to obtain the amortized cost of a sequence of n operations.

2. Use the accounting method to analyze the amortized cost.

3. Use the potential method to analyze the amortized cost.

4. Propose an implementation of a first-in first-out queue with two stacks, such that adding an element in the queue and removing an element from the queue both have an amortized cost of $O(1)$.

Exercise 5.4: Deleting half the elements (solution p. 115)

We want to implement a data structure S with real numbers, with the following operations: *insert*(S, x) inserts the object x in S, and *delete*(S) removes the $\lceil |S|/2 \rceil$ largest elements of S. Propose an implementation such that the amortized cost of both operations is cost.

(*Hint:* You can find in linear time the median of a list, see Section 9.3 of [27].)

Exercise 5.5: Searching and inserting (solution p. 116)

We consider a data structure for n elements. Let $k = \lceil \log(n + 1) \rceil$, and let $(n_{k-1}, n_{k-2}, \ldots, n_0)$ be the binary representation of n. The data structure consists of k sorted arrays $A_0, A_1, \ldots, A_{k-1}$, and the size of A_i is 2^i for $0 \leqslant i \leqslant k - 1$. The array A_i is full if $n_i = 1$, and empty otherwise, so that the total number of elements is $n = \sum_{i=0}^{k-1} n_i 2^i$. Note that each individual array is sorted, but there is no particular relationship between elements of two different arrays.

1. Propose a *search* operation for this data structure (find if an element is in the data structure), and analyze its worst-case running time.

2. Propose an *insert* operation for this data structure (insert a new element in the data structure), and analyze its worst-case and amortized running times.

3. Discuss how to implement a *delete* operation.

4. Compare the costs achieved by this data structure with the costs of searching and inserting in a sorted array of size n.

Exercise 5.6: Splay trees (solution p. 117)

The problem is to perform a sequence of m access operations on a set of n elements that are totally ordered. The elements are represented as a binary search tree: There is one element per node, and for any node x, all the elements in the left subtree of x are smaller that x, while all the elements in the right subtree of x are greater than x. The operation $access(i)$ is then in $O(d)$, where d is the depth of node x containing element i. In order to reduce the total access cost in a sequence of n accesses, we aim at moving frequently accessed elements toward the root. Therefore, each time any element x is accessed, we use the *splaying* heuristic. We repeat the following splaying steps until x is the root of the tree (see Figure 5.1).

- **zig**: If $p(x)$, the parent of x, is the tree root, rotate the edge joining x with $p(x)$ (this case is terminal).
- **zig-zig**: If $p(x)$ is not the root and x and $p(x)$ are both left or both right children, rotate the edge joining $p(x)$ with $p(p(x))$, and then rotate the edge joining x with $p(x)$.
- **zig-zag**: Otherwise, rotate first the edge joining x with $p(x)$ and then the edge joining x with the new $p(x)$ (that was initially $p(p(x))$).

1. Apply the splaying heuristic on node a of the tree below:

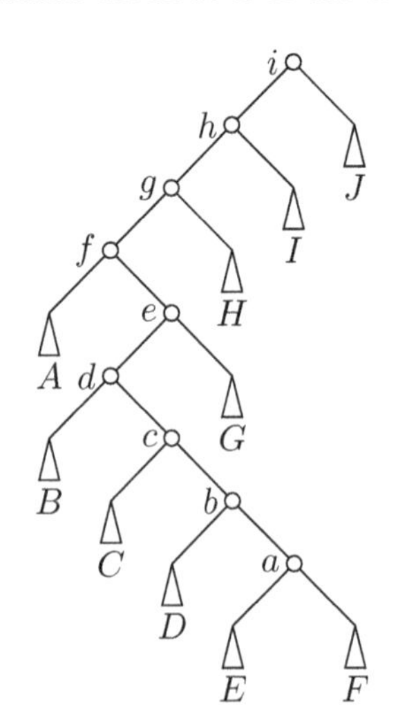

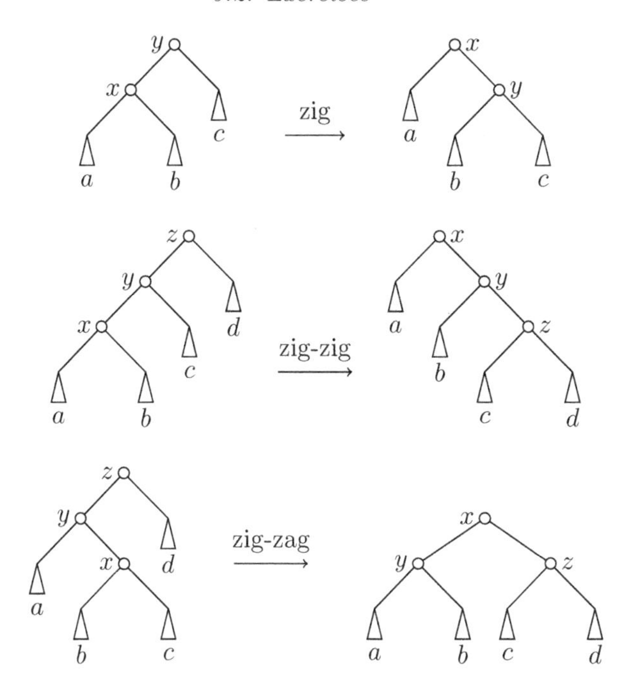

FIGURE 5.1: The different splaying steps, where x is the accessed element.

2. What is the time complexity of the splaying heuristic, in terms of number of rotations?

3. To analyze the amortized complexity of splaying, we use a potential function defined as follows: We assume that each element i has a positive weight $w(i)$, whose value is arbitrary but fixed. The size $s(x)$ of a node x is the sum of the weights of all elements in the subtree rooted in x, and the rank of x is $r(x) = \log s(x)$. The potential of a tree is the sum of the ranks of all its nodes. The cost of an operation is the number of rotations, but we still charge 1 if there is no rotation.

   Let $r(x)$ (resp. $r'(x)$) be the rank of x after (resp. before) the operation. Show that the amortized cost of a zig is at most $1+3(r'(x)-r(x))$ and the amortized cost of a zig-zig or a zig-zag is at most $3(r'(x)-r(x))$. (Note that if $a, b > 0$, $a+b \leqslant 1$, then $\log(a)+\log(b) \leqslant -2$.)

4. Deduce that the amortized time to splay a tree with root t at a node x is at most $3(r(t)-r(x))+1 = O(\log(s(t)/s(x)))$. Note that this is true for any positive weights.

5. Prove that the total access time is $O((m+n)\log n + m)$ (recall that n is the number of elements in the tree, and m is the number of accesses). Hint: Assign a weight to each element.

6. For any element i, $q(i)$ is the access frequency of i, i.e., the total number of times i is accessed (within the m accesses). Show that if every element is accessed at least once, then the total access time is $O\left(m + \sum_{i=1}^{n} q(i) \log\left(\frac{m}{q(i)}\right)\right)$. Hint: Assign a weight to each element.

Exercise 5.7: Half perimeter of a polygon (solution p. 119)

We consider a polygon with n vertices, numbered in the clockwise order from 0 to $n-1$. The edge from i to $i+1 \mod n$, for $0 \leqslant i < n$, has a length a_i.

1. We aim at finding the two vertices i and j that minimize the absolute value of the difference between the two portions of perimeters that they define, i.e., that minimize (the sums are modulo n) $\left|\sum_{l=i}^{j-1} a_l - \sum_{l=j}^{i-1} a_l\right|$.

   (a) Design a naive algorithm and give its complexity.

   (b) Design a linear-time algorithm.

2. Find in linear time the three vertices i, j, and k that minimize the difference between the larger *third* and the smaller *third* portions of the perimeter that they define, i.e.,
$\max\left(\sum_{l=i}^{j-1} a_l, \sum_{l=j}^{k-1} a_l, \sum_{l=k}^{i-1} a_l\right) - \min\left(\sum_{l=i}^{j-1} a_l, \sum_{l=j}^{k-1} a_l, \sum_{l=k}^{i-1} a_l\right)$.

5.3 Solutions to exercises

Solution to Exercise 5.1: Binary counter

1. After $2^{k-1} - 1$ operations, the value of the counter is $0\ 1\ 1\ \cdots\ 1$. If we perform a sequence of operations (increment, decrement), the counter will alternate between $1\ 0\ 0\ \cdots\ 0$ and $0\ 1\ 1\ \cdots\ 1$, hence, having a cost k for each operation, and for n operations, a cost in $\Theta(nk)$.

2. We introduce a new variable, max_A, that contains the index of the highest-order 1 in the counter A. Initially, $max_A = -1$ because there are only 0s in the counter. This value is updated at each operation, see Algorithms 5.1 and 5.2, where $|A| = k$.

```
1 i ← 0
2 while i < |A| and A[i] = 1 do
3 |   A[i] ← 0
4 |_  i ← i + 1
5 if i < |A| then
6 |   A[i] ← 1
7 |   if i > max_A then max_A ← i
8 |_  else max_A ← −1
```

ALGORITHM 5.1: Increment.

```
1 for i ← 0 to max_A do
2 |_  A[i] ← 0
3 max_A ← −1
```

ALGORITHM 5.2: Reset.

We use the accounting method, with a cost 4 for an increment operation, and a cost 1 for a reset, assuming that it costs 1 to flip a bit and also 1 to update max_A.

For an increment operation, similarly to the counter with increment only, we pay 1 for the bit that is changed to 1 (line 6 of Algorithm 5.1), and we give 1 to the bit so that it can be flipped back to 0, either with a reset operation or with another increment operation. We give an extra 1 to the bit, in case it is examined by a reset operation while it was already flipped back to 0, in the *for* loop of Algorithm 5.2. We also need to pay for the update of max_A, hence, having an extra cost of 1 for increment (and thus a total cost of 4). For the reset, we give a cost of 1 to update max_A. All operations were already prepaid because we explore only bits that have been paid for with increment operations (we stop at bit max_A). The amortized cost for each operation is bounded by 4, and, therefore, a sequence of n operations has a cost bounded by $O(n)$ in the worst case.

Solution to Exercise 5.2: Inserting and deleting

1. With this solution, the amortized cost is not constant. Once the table has a size n with $n-1$ elements, if the sequence of operations consists of inserting two elements then deleting two elements, and so on, we alternate between a table of size n and a table of size $2n$, paying a

cost n for every other operation.

2. The idea is to double the size of the table if it is full (as before) and to halve the size of the table when only a quarter of the table is full.

   Accounting method: We charge 3 euros for an insertion as before and only 2 euros for a deletion. Indeed, we use 1 to delete the element, and we store 1 in the emptied slot that will be used to move elements when the table size is halved. This way, all operations can be paid in constant time.

   Potential method: The potential function is now $\Phi_i = 2num_i - size_i$ if the table is at least half full, i.e., $2num_i \geqslant size_i$. Otherwise, $\Phi_i = size_i/2 - num_i$. In both cases, $\Phi_i \geqslant 0$. We need to compute the amortized cost in all cases. The i-th operation may be an insertion or a deletion, and depending on the ratio of the number of elements in the table to the table size, the potential differs.

   Let us start with the case in which the i-th operation is an insertion: $num_i = num_{i-1} + 1$.

   - If $num_{i-1} \geqslant size_{i-1}/2$, then $num_i \geqslant size_i/2$, and we have exactly the same potential function as in the case without deletion. The analysis is identical to that of Section 5.1.4, and $\hat{c}_i = 3$.
   - Assume now that $num_{i-1} < size_{i-1}/2$. Then, after an insertion, we have $size_i = size_{i-1}$ and $c_i = 1$. If $num_i < size_i/2$, then $\hat{c}_i = 1 + (size_i/2 - num_i) - (size_{i-1}/2 - num_{i-1}) = 0$. Otherwise, $\hat{c}_i = 1 + (2num_i - size_i) - (size_{i-1}/2 - num_{i-1}) = 3 + 3num_{i-1} - \frac{3}{2}size_{i-1} < 3$, since $num_{i-1} < size_{i-1}/2$.

   Now, consider that the i-th operation is a deletion: $num_i = num_{i-1} - 1$.

   - If $num_{i-1} \geqslant size_{i-1}/2$, then, after a deletion, we have $size_i = size_{i-1}$, and $c_i = 1$. Either $num_i \geqslant size_i/2$, and then $\hat{c}_i = 1 + (2num_i - size_i) - (2num_{i-1} - size_{i-1}) = -1$. Otherwise, $\hat{c}_i = 1 + (size_i/2 - num_i) - (2num_{i-1} - size_{i-1}) = 2 - 3num_{i-1} + \frac{3}{2}size_{i-1} \leqslant 2$.
   - If $num_{i-1} < size_{i-1}/2$, then the size of the table may remain the same: $size_i = size_{i-1}$ and $c_i = 1$. In this case, $\hat{c}_i = 1 + (size_i/2 - num_i) - (size_{i-1}/2 - num_{i-1}) = 2$. Otherwise, the size of the table is halved, and we have $size_{i-1}/4 = num_{i-1}$, $size_i = size_{i-1}/2$ and $c_i = num_{i-1}$. Therefore, $\hat{c}_i = num_{i-1} + (size_i/2 - num_i) - (size_{i-1}/2 + num_{i-1}) = 1 + num_{i-1} - size_{i-1}/4 = 1$.

   In all cases, $\hat{c}_i \leqslant 3$ and, therefore, the amortized cost of each operation is bounded above by a constant.

Solution to Exercise 5.3: Stack

1. The *push* and *pop* operations are in $O(1)$, while *multipop* is in $O(\min(|S|, k))$.

Each object can be popped from the stack at most once for each *push* of the same object. Therefore, there cannot be more calls to *pop* than the number of calls to *push*, even if we count the calls to *pop* from within the *multipop* function. In a sequence of n operations, there are at most n calls to *push* and, hence, no more than n calls to *pop*, and it takes a time in $O(n)$. The amortized cost of each operation, therefore, is in $O(1)$.

2. For the accounting method, we associate a cost with each operation. The amortized cost of *push* is 2, and the two other operations, *pop* and *multipop*, have an amortized cost 0. Indeed, the cost of the *pop* operations is prepaid when pushing an object at the top of the stack. We then pay 1 to push the object, and we keep 1 to pay later the *pop* of this object. For a sequence of n operations, the total amortized cost, therefore, is in $O(n)$.

3. We define the potential function Φ_i of the stack as the number of objects in the stack after i operations. Therefore, $\Phi_i \geqslant 0$ for all i. Initially, $\Phi_0 = 0$ since the stack is empty. If operation i is a *push*, the amortized cost is $\hat{c}_i = c_i + (\Phi_i - \Phi_{i-1}) = 1 + 1 = 2$. If operation i is a *pop*, the amortized cost is $\hat{c}_i = c_i + (\Phi_i - \Phi_{i-1}) = 1 - 1 = 0$. Finally, if operation i is *multipop*(S, k), the cost of this operation is $c_i = \min(|S|, k)$, and we have removed exactly c_i objects from the stack during the multipop. Therefore, the amortized cost is $\hat{c}_i = c_i + (\Phi_i - \Phi_{i-1}) = c_i - c_i = 0$. We have obtained the same amortized cost as with the accounting method, and they are all in $O(1)$.

4. To implement a queue, we use two stacks: S_{enter} and S_{exit}. When an object is added to the queue, we push it into stack S_{enter}. To remove an object from the queue, we pop an object from stack S_{exit}. If S_{exit} is empty, we pop objects from S_{enter} and push them back in S_{exit}.

   We use the accounting method, with a cost 3 for adding an object in the queue and a cost 1 for removing an object from the queue. Indeed, once an object is added in the queue, it may require a cost of 3. We pay 1 to push the object in S_{enter}, and we associate 2 with the object so that we can pay the transfer into S_{exit} (one pop and one push). When removing an object from the queue, we need to pay only 1 for one pop because the cost of the transfers from S_{enter} to S_{exit} has already been prepaid.

Solution to Exercise 5.4: Deleting half the elements

S is implemented with an unsorted list, so that the insertion takes a time $O(1)$. For the *delete* operation, it can be done in $O(|S|)$ in the worst case. We find the median of S in $O(|S|)$, and then we go through the list and remove $\lceil |S|/2 \rceil$

elements that are greater than or equal to the median. This second step is also done in $O(|S|)$. Let a be a constant such that this *delete* operation takes a time at most $a|S|$.

We use the potential analysis to find the amortized cost of each operation. The potential function is $\Phi_i = an_i$, where n_i is the size of the table before operation i.

If operation i is an insertion, then $c_i = 1$ and $n_i = n_{i-1} + 1$. Therefore, $\hat{c}_i = 1 + (\Phi_i - \Phi_{i-1}) = 1 + a$, which is a contant. If i is a *delete*, then $c_i \leqslant an_i$ and $n_i \leqslant n_{i-1}/2$. Therefore, $\hat{c}_i \leqslant an_i + a(n_{i-1}/2 - n_{i-1}) \leqslant 0$. Therefore, the amortized cost of both operations is, at most, $a + 1$.

Solution to Exercise 5.5: Searching and inserting

1. The search is done by repeatedly searching in all arrays. The search in an array of size m is done by binary search, in $\log(m)$, since the arrays are sorted. In the worst case, all arrays are full and the search takes $\Theta(\log(2^i))$ for array A_i, for $0 \leqslant i \leqslant k-1$. Therefore, we obtain a time of:

$$\begin{aligned} T(n) &= \Theta(\log(2^{k-1}) + \log(2^{k-2}) + \cdots + \log(2^0)) \\ &= \Theta((k-1) + (k-2) + \cdots + 1 + 0) = \Theta(k(k-1)/2) \end{aligned}$$

   and, hence, a worst-case running time in $\Theta(\log^2(n))$.

2. For the insertion, we create a new sorted array, A, with only the element to be inserted. If A_0 is empty, we replace A_0 with A. Otherwise, we merge A_0 and A, thus obtaining an array of size 2. If A_1 is empty, we replace it with A; otherwise we merge again until we find an empty array. The merge of two arrays of size 2^i is done in the worst case in time $\Theta(2 \times 2^i) = \Theta(2^{i+1})$, and it may happen that the $k-2$ first arrays are full, hence, leading to a worst-case running time in $\Theta(n)$:

$$\begin{aligned} T(n) &= \Theta(2^1 + 2^2 + \cdots + 2^{k-1}) \\ &= \Theta(2^k) = \Theta(n). \end{aligned}$$

   To obtain the amortized cost, we compute the cost of a sequence of n insertions, starting from an empty data structure. The analysis is similar to the analysis for the binary counter, and we use the aggregate analysis. Let r be the position of the right-most 0 in the binary representation of n. For $j < r$, $n_j = 1$. The cost of the $(n+1)$-th insertion is $\sum_{j=0}^{r-1} 2^{j+1} = O(2^r)$.

   Since we have $r = 0$ for every other operation, $r = 1$ for every fourth operation, and so on, there are at most $\lceil n/2^r \rceil$ insert operations for each value of r, and we can bound the cost of n insert operations by:

$$O\left(\sum_{r=0}^{\lceil\log(n+1)\rceil}\left(\left\lceil\frac{n}{2^r}\right\rceil\right)2^r\right) = O(n\log(n)).$$

Finally, the amortized cost of an insert operation is in $O(\log(n))$.

3. To delete an element x, we apply the following procedure:
   - Find the smallest j such that A_j is full.
   - Find the array A_i containing x.
   - Delete x from A_i and replace it with the smallest element of A_j (this element should be inserted in the array so that it remains sorted).
   - Now, A_j has only $2^j - 1$ elements, and we successfully place the first element of A_j in A_0, the next two elements in A_1 (in order), and so on, until the last 2^{j-1} elements in A_{j-1}. All these arrays are initially empty by definition, and we fill all of them since $1 + 2 + .. + 2^{j-1} = 2^j - 1$. The new arrays are already sorted because we add elements in order.

4. In a sorted array of size n, the search is done by binary search in $O(\log(n))$, and the insertion is done in $O(n)$. With the new structure, the amortized cost of the insertions is down to $O(\log(n))$, with only a small increase in the search time ($O(\log^2(n))$).

Solution to Exercise 5.6: Splay trees

1. The solution is depicted in the figure below:

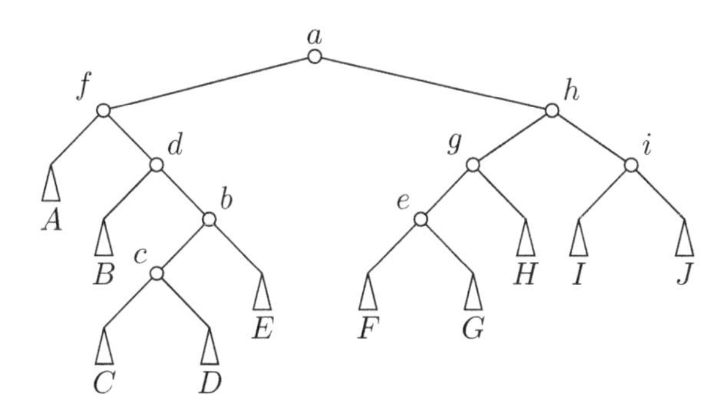

2. The splaying heuristic on a node x at depth d takes a time $\Theta(d)$, i.e., a time proportional to the time needed to access the element at node x.

3. The amortized cost is $\hat{c} = c + \Phi' - \Phi$, where c is the cost of the execution, Φ the potential before the operation, and Φ' the potential after the operation. Before the operation, $y = p(x)$ and $z = p(y)$ (if it exists). We consider the three different operations:
   - **zig:** If the operation is a zig, then only one rotation is done ($c = 1$). Only nodes x and y can change rank, and, furthermore, $r'(x) \geqslant r(x)$ and $r(y) \geqslant r'(y)$. Finally, the amortized cost is

$$\begin{aligned}\hat{c} &= 1 + r'(x) + r'(y) - r(x) - r(y)\\ &\leqslant 1 + r'(x) - r(x)\\ &\leqslant 1 + 3(r'(x) - r(x)).\end{aligned}$$

   - **zig-zig:** In this case, there are two rotations ($c = 2$), and only x, y, and z can change rank. Furthermore, $r'(x) = r(z)$, $r'(x) \geqslant r'(y)$, and $r(y) \geqslant r(x)$. Finally,

$$\begin{aligned}\hat{c} &= 2 + r'(x) + r'(y) + r'(z) - r(x) - r(y) - r(z)\\ &= 2 + r'(y) + r'(z) - r(x) - r(y)\\ &\leqslant 2 + r'(x) + r'(z) - 2r(x).\end{aligned}$$

     To prove that this last sum is at most $3(r'(x) - r(x))$, we need to have $2r'(x) - r(x) - r'(z) \geqslant 2$. Because, for any element t, $r(t) = \log s(t)$, $r(x) + r'(z) - 2r'(x) = \log(s(x)/s'(x)) + \log(s'(z)/s'(x))$. Because the operation is a zig-zig, $s(x) + s'(z) \leqslant s'(x)$. Moreover, the sizes are all positive. Therefore, $0 \leqslant s(x)/s'(x) + s'(z)/s'(x) \leqslant 1$. For any $a \in]0, 1[$ and $b \in [0, 1 - a[$, $\log(a) + \log(1 - a - b) \leqslant \log(a) + \log(1 - a) \leqslant 2\log(1/2) = -2$ (proved by differentiating or by using the concavity of the log function). Therefore, $r(x) + r'(z) - 2r'(x) \leqslant -2$, which allows us to conclude that the amortized cost is at most $3(r'(x) - r(x))$.
   - **zig-zag:** In this case, there are two rotations ($c = 2$), and only x, y, and z can change rank. Furthermore, as in the zig-zig case, $r'(x) = r(z)$ and $r(x) \leqslant r(y)$. Finally,

$$\begin{aligned}\hat{c} &= 2 + r'(x) + r'(y) + r'(z) - r(x) - r(y) - r(z)\\ &\leqslant 2 + r'(y) + r'(z) - 2r(x).\end{aligned}$$

     Because the operation is a zig-zag, $s'(y) + s'(z) \leqslant s'(x)$. Therefore, following the study of the zig-zig case, we establish that $r'(y) + r'(z) - 2r'(x) = \log(s'(y)/s'(x)) + \log(r'(z)/r'(x)) \leqslant -2$. Therefore, $\hat{c} \leqslant 2(r'(x) - r(x))$.

4. If there are no rotations, then the bound is immediate because $t = x$ and the cost is 1 in this case. If there are rotations, we total the amortized costs at each step. First, we remark that only the last rotation can be a zig. Therefore, all intermediate terms in the sum are cancelled. In the

worst case, the last operation is a zig, and, therefore, the total amortized cost is at most $3(r'(x) - r(x)) + 1$, where $r(x)$ is the initial rank of x, and $r'(x)$ is the final rank of x, i.e., it is equal to $r(t)$.

5. We assign a weight of $1/n$ to each element. Therefore, $W = \sum_{i=1}^{n} w(i) = 1$, and the size of any node x is such that $1/n \leqslant s(x) \leqslant 1$. Thus, for any element, the amortized cost of access is at most $3\log(s'(x)/s(x)) + 1 \leqslant 3\log n + 1$, that is, $3m\log n + m$ for the m accesses. Moreover, the size of a node varies between $1/n$ and 1, and its rank, therefore, varies between $-\log(n)$ and 0. The decrease in potential over the sequence of m accesses is, therefore, at most $O(n\log(n))$. Therefore, the total access cost is at most $O((m+n)\log n + m)$ (the amortized cost plus the potential difference).

6. The reasoning is similar to the previous question. This time, we assign a weight $q(i)/m$ to element i. Once again, $W = 1$. The amortized cost to access element i is now $O(\log(m/q(i)))$, and it is accessed $q(i)$ times. The amortized cost for the m accesses is, therefore, $O(m + \sum_{i=1}^{n} q(i)\log(m/q(i)))$.

   Moreover, the decrease in potential over the sequence of m accesses is at most $\sum_{i=1}^{n} \log(m/q(i))$ because the rank of each element may only decrease by $\log(m/q(i))$. Because $q(i) \geqslant 1$ for all i (we assume that each element is accessed at least once), this is also in $O(\sum_{i=1}^{n} q(i)\log(m/q(i)))$, hence the result.

Solution to Exercise 5.7: Half perimeter of a polygon

1. (a) The obvious naive solution is to build all pairs (i, j) of endpoints and, for each pair, to compute the difference between the two portions of perimeters. This is what Algorithm 5.3 does.

   Algorithm 5.4 presents a slightly refined exhaustive search that runs in $O(n^2)$. First, Algorithm 5.4 computes the polygon perimeter (steps 1 through 3). Then, for any possible starting point i (loop at step 5), the algorithm identifies in linear time the first vertex j whose distance from i is at least equal to half the perimeter (steps 6 through 10). Then, the only two candidates to partition the polygon into halves are the vertices j (steps 11 through 13) and $j - 1$ (steps 14 through 18).

   (b) Algorithm 5.5 is a linear-time algorithm to solve the half-perimeter problem. The quadratic cost of Algorithm 5.4 comes from the linear-time computation for each vertex i of the first vertex j whose distance from i is at least equal to half the perimeter. In Algorithm 5.5, to compute the vertex "j" corresponding to vertex $i+1$, we do not restart from scratch as in Algorithm 5.4. On the contrary, we restart from the memorized vertex "j" for vertex i. As the

```
1 diff_min ← +∞
2 for i = 0 to n − 1 do
3     for j = i + 1 to i − 1 (increments are modulo n) do
4         diff ← | (Σ_{l=i}^{j−1} a_l) − (Σ_{l=j}^{i−1} a_l) |
5         if diff < diff_min then
6             diff_min ← diff
7             i_min ← i
8             j_min ← j
9 return (i_min, j_min, diff_min)
```

ALGORITHM 5.3: Naive algorithm in $O(n^3)$ to divide the perimeter of a polygon into halves.

starting vertex i visits all the vertices in the polygon one by one, it corresponding vertex j is never more than one perimeter away and, thus, the vertex j visits at most two times each vertex of the perimeter. Therefore, the computation of all the "j" vertices is done in linear time. In the worst case, however, the computation of one vertex "j" can be $\Omega(n)$. Only a global analysis, amortized over all vertices (aggregate analysis), leads to the proof that Algorithm 5.5 is indeed a linear-time algorithm.

Then, the only differences between Algorithms 5.5 and 5.4 are due to Algorithm 5.5 taking care not to overwrite data relative to vertex j (initializations are done outside of the loop and temporary variables are used at step 15) and to update the distance from the starting vertex to the candidate j vertex at each iteration of the loop (step 19).

2. Among the three sums $\sum_{l=i}^{j-1} a_l$, $\sum_{l=j}^{k-1} a_l$, and $\sum_{l=k}^{i-1} a_l$ defined by the three indices i, j, and k, at least one is greater than or equal to $P/3$, and at least one is smaller than or equal to $P/3$, where P denotes the perimeter. In other words, either two of the sums are greater than or equal to $P/3$, or two of them are smaller than or equal to $P/3$. To solve the problem in linear time, we design a linear-time algorithm for each of these two cases and take the best solution.

We first consider the case where two of the sums are greater than or equal to $P/3$. Without loss of generality, we can assume that these are the first and the second sums. For each possible value of i, j is then the first vertex after i such that the first sum is greater than or equal to $P/3$, and k is then the first vertex after j such that the second sum is

```
1  P ← 0
2  for i = 0 to n − 1 do
3  └ P ← P + a_i        { computation of the perimeter }
4  diff_min ← P
5  for i = 0 to n − 1 do
6  │  distance ← 0
7  │  j ← i
8  │  while 2 × distance < P do
9  │  │  distance ← distance + a_j
10 │  └  j ← (j + 1) mod n
11 │  diff ← |2 × distance − P|
12 │  if diff < diff_min then
13 │  └  i_min ← i; j_min ← j; diff_min ← diff
14 │  pred ← (j − 1) mod n
15 │  distance ← distance − a_pred
16 │  diff ← |2 × distance − P|
17 │  if diff < diff_min then
18 └  └  i_min ← i; j_min ← pred; diff_min ← diff
19 return (i_min, j_min, diff_min)
```

ALGORITHM 5.4: Algorithm in $O(n^2)$ to divide the perimeter of a polygon into halves.

greater than or equal to $P/3$. As the first two sums must be greater than or equal to $P/3$, when the first vertex moves forward on the perimeter from i to $i+1$, both vertices j and k either move forward or remain at the same position. Therefore, we can use the same amortized analysis argument as in the previous question to prove that the algorithm built this way runs in linear time.

The case where two sums are smaller than or equal to $P/3$ is similar. Without loss of generality, we assume that these are the first and the second sums. For each possible value of i, j is then the last vertex after i such that the first sum is smaller than or equal to $P/3$, and k is then the last vertex after j such that the second sum is smaller than or equal to $P/3$.

```
1  P ← 0
2  for i = 0 to n − 1 do
3  └ P ← P + a_i     { computation of the perimeter }
4  diff_min ← P
5  j ← 0
6  distance ← 0
7  for i = 0 to n − 1 do
8  │  while 2 × distance < P do
9  │  │  distance ← distance + a_j
10 │  └  j ← (j + 1) mod n
11 │  diff ← |2 × distance − P|
12 │  if diff < diff_min then
13 │  └  i_min ← i; j_min ← j; diff_min ← diff
14 │  pred ← (j − 1) mod n
15 │  pred_distance ← distance − a_pred
16 │  diff ← |2 × pred_distance − P|
17 │  if diff < diff_min then
18 │  └  i_min ← i, j_min ← pred, diff_min ← diff
19 └  distance ← distance − a_i
20 return (i_min, j_min, diff_min)
```

ALGORITHM 5.5: Linear-time algorithm to divide the perimeter of a polygon in two halves.

5.4 Bibliographical notes

Section 5.1 is inspired from the book by Cormen, Leiserson, Rivest, and Stein [27]. Several exercises also come from this book: Exercise 5.1 (binary counter), Exercise 5.2 (inserting and deleting), Exercise 5.3 (stack), Exercise 5.4 (deleting half the elements), and Exercise 5.5 (searching and inserting). Exercise 5.6 (splay trees) comes from [99]. Finally, Exercise 5.7 (half perimeter of a polygon) is an extension of an exercise from the book by Dijkstra [31].

Part II

NP-completeness and beyond

Chapter 6

NP-completeness

In this chapter, we introduce the complexity classes that are of paramount importance for algorithm designers: P, NP, and NPC. We take a strictly practical approach and determinedly skip the detour through Turing machines. In other words, we limit ourselves to NP-completeness, explaining its importance and detailing how to prove that a problem is NP-complete.

After introducing our approach in Section 6.1, we define the complexity classes P and NP in Section 6.2. NP-complete problems are introduced in Section 6.3, along with the practical reasoning to prove that a problem is NP-complete. Several examples are provided in Section 6.4. We discuss subtleties in problem definitions in Section 6.5 and strong NP-completeness in Section 6.6. Finally, we make our conclusions in Section 6.7.

6.1 A practical approach to complexity theory

This chapter introduces the key complexity classes that algorithm designers are confronted with: P, which stands for *Polynomial*, and NP, which stands for *Nondeterministic Polynomial*. In fact, we depart from the original definition of the class NP and use the (equivalent) characterization *Polynomial with Certificate*. Within the NP class, we focus on the subclass NPC of *NP-complete* problems.

When writing this chapter, we faced a cruel dilemma. Either we use a formal approach, which requires an introduction to Turing machines, explain their characteristics, and classify the languages that they can recognize, or we use a practical approach that completely skips the detour through the theoretical computer science framework and defines complexity classes out of nowhere (almost!). We firmly believe that there is no trade-off in between, and that a comprehensive exposure does require Turing machines. However, given the main objectives of this book, we chose the latter approach. The price to pay is that the reader will have to take for granted a key result, namely Cook's theorem [25], which we will state without proof. Cook's theorem provides the first NP-complete problem, and we will have to trust him on this. However, the main advantage is that we can concentrate on the art of the algorithm designer, namely *polynomial reduction*.

First, why Turing machines? To assess the complexity of a problem, we need to define its size and the number of time steps required to solve it. But what is appropriate within a time step? A formal answer relies on Turing machines. The size of a problem is the number of consecutive positions used to store its data on the (infinite) ribbon of the machine. The number of time steps is the number of moves before the Turing machine terminates the execution of its program, given the data initially stored on its ribbon. Instead, in the practical approach, we simply define the size of a problem as the number of memory locations, or bits, that are needed to store its data, and we define a time step as the maximum time needed to execute an elementary operation. Here, an elementary operation is defined as any *reasonable* computation. And, the trouble begins. Fetching the values of two memory locations, adding them and storing them back into some memory location, is that an elementary operation? Yes—well, provided that the access to the memory locations takes constant time, which may require that the total memory is bounded, or at least that two different memory locations used to solve the problem are not too far apart in storage. We are not far from moving the head of the Turing machine from one position to another! Similarly, adding two bits or two bytes or two double-precision floating point numbers (64 bits) is indeed an elementary operation, but adding two integers of unbounded length is not. In fact, an elementary operation is anything that can be done in polynomial time by a Turing machine, but this statement is helpful mostly to those who are familiar with Turing machines. Here is an example of an operation that is not reasonable. If we have two prime numbers p and q of r bits, we can compute their product $n = p \times q$ in $O(r^2)$, but given n, we cannot find p and q in time polynomial in r.

We refer the reader interested in the formal approach to some excellent books. The big classic is the book by Garey and Johnson [38] with a comprehensive treatment of NP-completeness. A very intuitive proof of Cook's theorem is given by Wilf [108]. More on complexity theory is provided by Papadimitriou [82].

6.2 Problem classes

In this section, we first emphasize the importance of polynomials in the theory. Then, we discuss how to define the problem size and how to encode data. This is illustrated through classical examples; integers are coded in a logarithmic size, but we should be careful if objects must be enumerated (set of nodes in a graph, list of tasks, etc).

6.2.1 Problems in P

The following remark, admittedly simple, is fundamental: *The composition of two polynomials is a polynomial.* Thanks to this observation, key values (time, size) can be defined up to a polynomial factor. From the point of view of complexity classes, values like n, n^3, or $n^{27} + 17n^5 + 42$ are totally equivalent; all these values are polynomial in n. Hence, there is no difference if an elementary operation of the algorithm would cost n^3 or $n^{27}+17n^5+42$ time steps of a Turing machine; as long as there is a polynomial number of such operations, the total number of time steps for the Turing machine remains polynomial.

The theory deals with decision problems, with a yes/no answer, rather than with optimization problems (this is related to languages that are accepted by Turing machines). A decision problem is in the complexity class P if it can be solved in *polynomial time.* Owing to the previous remark, we do not need to specify the degree of the polynomial, which is not relevant as far as theory is concerned (we come back to this last point below). Hence, the key for understanding this class P is the notion of "polynomial time." As mentioned before, one must decide what can be done within one unit of time. One usually assumes that one can add, multiply, or access memory in constant time, but the multiplication of large numbers (respectively, memory accesses) can depend on the size of the numbers (respectively, of the memory).

From an algorithmic point of view, we usually suppose that we can add, multiply, access memory within one unit of time, as long as numbers and memory size are bounded, which seems reasonable. These operations are then of polynomial time, and thus this model is polynomial with respect to the theoretical one with the Turing machine, as long as we are careful when dealing with nonbounded integers.

Also, the resolution time must be a polynomial of the *data size*, so one needs to define this "data size" carefully. This data size can strongly depend on the way an instance is encoded. Intuitively, integers can be coded in binary, therefore requiring a logarithmic size rather than a linear one (when encoded in unary). The encoding with any other basis $b \neq 2$ has the same size as the binary encoding, up to a constant factor ($\log_2(n)/\log_b(n) = 1/\log_b(2)$). However, some integers describing a problem instance should not be encoded in binary when they code objects to be enumerated. Otherwise, some "elementary" operations would have a cost exponential in the data size. We illustrate this by detailing two problem examples.

Example: 2-partition

DEFINITION 6.1 (2-PARTITION). Given n positive integers $a_1, \ldots, a_n$, is there a subset I of $\{1, \ldots, n\}$ such that $\sum_{i \in I} a_i = \sum_{i \notin I} a_i$?

The input data of a problem instance is a set of n integers. In theory, these n integers could be encoded either in unary or in binary. However, by convention, in complexity theory, any integer appearing in the coding of an instance must be encoded in binary. The only exception is for data whose encoding in unary would not change the overall data size of the instance (i.e., if the new data size remains polynomial in the original data size). For instance, for 2-PARTITION, the choice of encoding for the value n itself does not matter because encoding n integers requires a data size of at least one per integer and thus of at least n. Therefore, for the sake of simplicity, one usually encodes n in unary. Then, with the mandatory binary encoding of the n integers, the data size is $\sum_{1 \leqslant i \leqslant n} \log(a_i)$. With a unary encoding of the integers, the data size of an instance would have been $\sum_{1 \leqslant i \leqslant n} a_i$.

The choice of the encoding is vital for such a problem. Indeed, one can find an algorithm whose time is polynomial in $n \times \sum_{1 \leqslant i \leqslant n} a_i$. We design a simple dynamic-programming algorithm; we solve the problems $c(i, T)$, where $c(i, T)$ equals true if there is a subset of $\{a_1, \ldots, a_i\}$ of sum T (and false otherwise), for $1 \leqslant i \leqslant n$ and $0 \leqslant T \leqslant S = \sum_{1 \leqslant i \leqslant n} a_i$. The solution to the original problem is $c(n, \frac{S}{2})$. The recurrence relation is $c(i, T) = c(i-1, T-a_i) \vee c(i-1, T)$. This algorithm is in $O(nS)$. Therefore, this algorithm runs in a time that would be polynomial in the size of the data if we had allowed the integers to be coded in unary. However, the algorithm running time is exponential in the data size when integers are coded in binary, as mandated. Such an algorithm is said to be *pseudopolynomial*.

No one knows an algorithm that is polynomial in the data size (i.e., in $O(n \log(S))$), so the question whether the 2-PARTITION problem is or isn't in P is left open.

Example: Bipartite graphs

DEFINITION 6.2 (BIPARTITE). Given a graph $G = (V, E)$, is G a bipartite graph?

This is a decision problem; the answer must be yes or no. The input data of a problem instance is a graph (V, E), where V is the set of vertices and E the set of edges. The size of the data depends on how the graph is stored (or encoded). The graph consists of $|V| = n$ vertices. One usually codes n in unary rather than in binary. Independent vertices are vertices that are not endpoints of any edge. Independent vertices play a trivial role with respect to the problem, and they can therefore be safely discounted. Then, each vertex is the endpoint of at least one edge, $|E| \geqslant n/2$, and encoding n in binary does not change the data size. Therefore, for the sake of simplicity, in any graph problem the number of vertices is always encoded in unary for the same reason.

Then, the identifier of a vertex can be encoded in binary, thus in $\log(n)$ for one vertex leading to a total of $n \log(n)$, which is still polynomial in n. The number of edges is also polynomial in n because there are at most n^2 edges. Then, altogether, the total size of the problem data is a polynomial in n, where n is the number of vertices of the graph. When designing graph algorithms, one often denotes the size of data of a graph as $|V| + |E|$ (strictly speaking, it should be $|V| + |E| \log(|V|)$, but each expression is polynomial in the other one). This allows us to refine the cost study of the algorithms, in particular when $|E| \ll |V|^2$. However, $|V| + |E|$ is still polynomial in n, so this refinement does not alter the problem classification.

Now, given a graph, in order to answer the question (yes or no), we need to perform a number of operations that is polynomial in n (greedy graph coloring). This problem is, therefore, in the complexity class P because it can be solved in a time that is polynomial in the data size.

6.2.2 Problems in NP

To define the complexity class NP, we need to define the *certificate* of a problem, which is (an encoding of) a solution to the problem.

Problem solution: Certificate

Back to the 2-PARTITION problem, if we are given a subset $I \subseteq \{1, \ldots, n\}$, we can check in polynomial time (even in linear time) whether $\sum_{i \in I} a_i = \sum_{i \notin I} a_i$, and, therefore, we can answer whether the problem has a positive answer in polynomial time. Moreover, the size of the certificate I is $O(n \log(n))$, which is polynomial in the problem size (the certificate contains $O(n)$ identifiers, each coded on $O(\log(n))$ bits).

Another way to provide a solution to 2-PARTITION would be to give the certificate $\{a_i\}_{i \in I}$, but if the a_is are coded in unary in the certificate, it is of exponential size. However, if the a_is are coded in binary, then the certificate has polynomial size and it is perfectly acceptable; a certificate is valid if it is polynomial in the problem size.

For the bipartite graph problem, the certificate would be the set of indices of vertices of one of the two subsets of the graph, whose size is polynomial in the problem size. Given this set, it is then easy to check that it is a correct solution by looking at each edge of the graph, which takes a polynomial time.

(See also Section 6.4.4 on scheduling problems for an illustration of the care that must be taken to design a certificate of polynomial size.)

Definition of NP

We are now ready to define the problem class NP. This is the class of decision problems for which we can verify a certificate in a time that is polynomial in the problem size. By *verify*, we mean *check that the certificate is indeed a solution*, i.e., that the answer to the problem is *yes*. Both previous examples are, therefore, in NP because, if we are given a certificate of polynomial size, we can check in polynomial time whether it is a solution to the problem.

We make a short digression to explain that NP stands for *Nondeterministic Polynomial*, for reference to nondeterministic Turing machines that were originally used to define the class. As already mentioned, we define NP as *Polynomial with Certificate* in this book, and we ignore equivalent characterizations of the NP class, either older (nondeterministic Turing machines) or newer (the famous PCP theorem). (See [4] for more information.)

It is time to recapitulate; we have defined two classes of decision problems:

P: Given an instance I of the problem of size $|I|$ when encoded in binary, there is an algorithm whose running time is polynomial in $|I|$ and which reports whether the instance has a solution or not;

NP: Given an instance I of the problem of size $|I|$ when encoded in binary, and a certificate of size polynomial in $|I|$, there is an algorithm whose running time is polynomial in $|I|$ and which reports whether the certificate is indeed a solution to the instance.

We observe that P $\subseteq$ NP. If we can find a solution in polynomial time, then we can verify the solution in polynomial time, with an empty certificate. Most researchers believe that the inclusion is strict, i.e., P $\neq$ NP, because it should be easier to check whether a certificate is a solution to the problem than to find a solution to that problem. As you may have heard before, this question is open at the time of this writing.

We have already seen that BIPARTITE is in P; therefore, it is in NP. Also, 2-PARTITION is in NP, but we do not know whether it is in P or not. Below are a few more examples to illustrate the class NP.

Examples: Problems in NP

DEFINITION 6.3 (COLOR)**.** Given a graph $G = (V, E)$ and an integer k ($1 \leqslant k \leqslant |V|$), can we color G with at most k colors?

This is a graph coloring problem; two vertices connected with an edge cannot be assigned the same color. The size of the data is a polynomial in $|V| + \log(k)$. Indeed, we need to enumerate all vertices similarly to the problem bipartite, hence the term $|V|$, while the integer k is encoded in binary. Since $k \leqslant |V|$ (one never needs more colors than vertices), the size of the data is a polynomial in $|V|$. A certificate can be the list of the vertices together with their color, whose size is linear in the size of the problem instance. The verification would amount to checking that no two adjacent vertices are assigned the same color, and that no more than k colors are used in total, which can be done in linear time as well.

DEFINITION 6.4 (HC – Hamiltonian Cycle)**.** Given a graph $G = (V, E)$, is there a cycle that goes through each vertex once and only once?

Similarly to other graph problems, the size of the data is a polynomial in $|V|$. A certificate can be the ordered list of the vertices that constitute the cycle (with linear size again). As before, the verification is easy: Check that the cycle is built with existing edges in the graph, and that each vertex is visited once and only once.

DEFINITION 6.5 (TSP – Traveling Salesman Problem). Given a complete graph $G = (V, E)$, a cost function $w : E \rightarrow \mathbb{N}$ and an integer k, is there a cycle $\mathcal{C}$ going through each vertex once and only once, with $\sum_{e \in \mathcal{C}} w(e) \leqslant k$?

This classical traveling salesman problem is a weighted version of the HC problem. There are several variants of the problem with various constraints on the cost function w: The weights can be arbitrary, satisfy the triangular inequality, or correspond to the Euclidean distance. The variants do not change the problem complexity. The size of the data is a polynomial in $|V| + \sum_{e \in E} \log(w(e)) + \log(k)$. We need to enumerate vertices, and other integers are coded in binary. A certificate can be the ordered list of the vertices that constitute the cycle, and the verification is similar to that for the HC problem.

No one knows how to find a solution to these three problems in polynomial time.

Problems not in NP?

One rarely encounters a problem whose membership status, with respect to NP, is unknown. It is even rarer to come across a problem that is known not to belong to NP. These problems are usually not very interesting from an algorithmic point of view. They are, however, fundamental for the theory of complexity. We provide a few examples below.

Negation of TSP: Given a problem instance of TSP, is it true that there is no cycle in the graph of length $|V|/2$?

This problem is similar to TSP, but the question is asked in the reverse way. It is difficult to think of a certificate of polynomial size that would allow us to check in polynomial time that the answer to the question is yes. Whether this problem belongs to NP is an open question.

Square: Given n squares whose areas sum up to 1, can we partition the unit square into these n squares?

We are interested in this problem because it plays a prominent role in the case study of Chapter 14. Its complexity depends on the exact definition that is used. First, we give the variant of the problem that is used in Chapter 14; we are given n squares of size a_i, with $\sum_{1 \leqslant i \leqslant n} a_i^2 = 1$. The a_i are rational numbers, $a_i = b_i/c_i$, and the problem size is $\sum_{1 \leqslant i \leqslant n} \log(b_i) + \sum_{1 \leqslant i \leqslant n} \log(c_i)$. A certificate can be the position of each square, for $1 \leqslant i \leqslant n$, for instance the coordinates of its top left corner. This certificate is of polynomial size. We can then check in polynomial time whether it is a solution of the problem or not (however, writing such a verification procedure requires some care). Hence, this variant is in NP.

Another variant consists of having as input m_i squares of size a_i, for $1 \leqslant i \leqslant p$, with $n = \sum_{1 \leqslant i \leqslant p} m_i$ and $\sum_{1 \leqslant i \leqslant p} m_i a_i^2 = 1$. The size of the data is then $n + \sum_{1 \leqslant i \leqslant p} \log(m_i) + \sum_{1 \leqslant i \leqslant p} \log(b_i) + \sum_{1 \leqslant i \leqslant p} \log(c_i)$. We do not need to enumerate all squares but only the p basic squares, while the m_is can be coded in binary. Then, in a certificate of polynomial size, we cannot enumerate all the n squares to give their coordinates. There might exist a compact analytical formula that would characterize solutions (say, the j-th square of size a_i is placed at coordinates $f(i, j)$), but this is far from being obvious. We do not know whether this latter variant is in NP or not.

It is much harder to identify a problem that is known not to be in NP, at least without making any assumption like P $\neq$ NP. A (complicated) example is the problem of deciding whether two regular expressions represent different languages, where the expressions are limited to four operators: union, concatenation, the Kleene star (zero or more copies of an expression), and squaring (two copies of an expression). Any algorithm for this problem requires exponential space, hence, exponential verification time [77].

Another problem that is not in NP is the program termination problem, or halting problem (decide whether a program will terminate on a given input). However, this example is a little excessive because no algorithm can exist to solve it, regardless of its complexity [82].

6.3 NP-complete problems and reduction theory

As explained in the previous section, we do not know whether the inclusion P $\subseteq$ NP is strict or not. However, we are able to compare the complexity of problems in NP; Cook's idea was to prove that some problems of the NP class are at least as difficult as all other problems of the same class. These problems are called NP-complete and form the subclass NPC of the class NP. They are *the most difficult problems* of NP. If we are able to solve one NP-complete problem in polynomial time, then we will be able to solve all problems of NP in polynomial time, and we will have P = NP. The main objective of this section is to explain this line of reasoning in full detail and to explore some consequences.

We detail the theory of reduction, which aims at proving that a problem is *more difficult* than another one. However, if we want to prove that a problem is more difficult than any other one, we need to identify the first NP-complete problem, as explained in Section 6.3.2. Note that a set of NP-complete problems with the corresponding reductions is presented in Section 6.4.

6.3.1 Polynomial reduction

We start by explaining the mechanism of polynomial reduction, i.e., how to prove that a problem is more difficult than another. Consider two decision problems P_1 and P_2. How can we prove that P_1 is more difficult than P_2? We say that P_2 is polynomially reducible to P_1 and write $P_2 \xrightarrow{pr} P_1$ if, whenever we are given an instance $\mathcal{I}_2$ of problem P_2, we can convert it, with only a polynomial-time algorithm, into an instance $\mathcal{I}_1$ of P_1, in such a way that $\mathcal{I}_2$ has the answer "Yes" if and only if $\mathcal{I}_1$ has the answer "Yes."

Now, if P_2 is polynomially reducible to P_1, then P_1 must be more difficult than P_2 (or more precisely, at least as difficult as P_2). Indeed, if there exists a polynomial algorithm to solve P_1, then by applying the polynomial reduction, and because the composition of two polynomials is a polynomial, there exists a polynomial algorithm to solve P_2. Given an instance $\mathcal{I}_2$ of P_2, we can indeed convert it into instance $\mathcal{I}_1$ of P_1, and since there is an equivalence between solutions of $\mathcal{I}_1$ and $\mathcal{I}_2$, the polynomial algorithm for P_1 executed on instance $\mathcal{I}_1$ returns the solution for instance $\mathcal{I}_2$. Take the contrapositive of this statement. If there is no polynomial algorithm to solve P_2, then there is none to solve P_1 either, so P_1 is more difficult.

We point out that polynomial reduction is a transitive operation: If $P_3 \xrightarrow{pr} P_2$ and $P_2 \xrightarrow{pr} P_1$, then $P_3 \xrightarrow{pr} P_1$. Again, this is because the composition of two polynomials is a polynomial, nothing more.

Note also that it is mandatory to have the equivalence of solutions, i.e., if $\mathcal{I}_1$ has a solution then $\mathcal{I}_2$ has one, and if $\mathcal{I}_2$ has a solution then $\mathcal{I}_1$ has one. Otherwise, the polynomial reduction $P_2 \xrightarrow{pr} P_1$ would not imply that P_1 is more difficult than P_2.

6.3.2 Cook's theorem

The fundamental result of the P versus NP theory is Cook's theorem [25], which shows that the satisfiability problem SAT is the most difficult problem in NP. This means that all other problems in NP are polynomially reducible to SAT. We introduce SAT and give a brief intuitive sketch of Cook's proof.

DEFINITION 6.6 (SAT). Let F be a Boolean formula with n variables $x_1, \ldots, x_n$ and p clauses $C_1, \ldots, C_p$: $F = C_1 \wedge C_2 \wedge \cdots \wedge C_p$, with, for $1 \leqslant i \leqslant p$, $C_i = x^*_{i_1} \vee x^*_{i_2} \vee \cdots \vee x^*_{i_{f(i)}}$, $1 \leqslant i_k \leqslant n$ for $1 \leqslant k \leqslant f(i)$, and $x^* = x$ or $\overline{x}$. Does there exist an instantiation of the n variables such that F is true (i.e., C_i is true for $1 \leqslant i \leqslant p$)?

Clearly, SAT is in NPC, and a certificate can simply be the list of the instantiation of each variable (whether a given x_i is instantiated to true or false). However, it seems difficult to solve SAT without a certificate; because some clauses have x_i and other $\overline{x_i}$, we may have to try all 2^n possible instantiations to find one that satisfies the formula. In other words, SAT seems to be a hard problem indeed.

Cook's theorem states that all problems in NP are polynomially reducible to SAT. The main idea of the proof is the following: Consider any problem P in NP, and take an arbitrary instance I, together with its certificate C. The proof goes by simulating the execution of the Turing machine that accepts the couple (I, C) as input and outputs "Yes" after a polynomial number of steps.

Because Turing machines are simple, their behavior can be characterized by clauses linking a set of variables. We can define $x_{t,j,s}$ as a variable that is true if after t steps of computation, symbol s is in position j of the ribbon, and we can simulate the operation of the machine using these variables. There are many such variables, but only a polynomial number in $|I|$, and a polynomial number of clauses as well. A detailed, but easy-to-follow, proof is given by Wilf [108].

6.3.3 Growing the class NPC of NP-complete problems

Now that we have the first NP-complete problem handy, how can we find more? To prove that a problem, P_1, is in NPC, we merely have to prove that SAT is polynomially reducible to this problem. Indeed, by composition, all problems in NP are reducible to SAT, hence, to P_1. The reduction takes several steps:

1. Prove that $P_1 \in NP$: We must be able to build a certificate of polynomial size, and then, for any instance $\mathcal{I}_1$ of problem P_1, we must be able to check in polynomial time whether the certificate is a solution. Usually, this first step is easy, but it should not be forgotten.
2. Prove the completeness of P_1: We transform an arbitrary instance $\mathcal{I}$ of SAT into an instance $\mathcal{I}_1$ of P_1 in polynomial time, and such that:
   (a) the size of $\mathcal{I}_1$ is polynomial in the size of $\mathcal{I}$;
   (b) $\mathcal{I}_1$ has a solution $\Leftrightarrow \mathcal{I}$ has a solution.

Let us come back to the construction of instance $\mathcal{I}_1$. The construction should be done in polynomial time, but this is usually implicit because the size of $\mathcal{I}_1$ should be polynomial in the size of $\mathcal{I}$, and because we perform only "reasonable" operations.

Assume that we have polynomially reduced SAT to P_1. We now have two problems in NPC, namely, P_1 and SAT. If we want to extend the class to a third problem, $P_2 \in NP$, should we reduce SAT or P_1 to P_2? Of course, the answer is that either reduction works. Indeed, we have so far:

- $P_1 \xrightarrow{pr} SAT$ and $P_2 \xrightarrow{pr} SAT$ (both by Cook's theorem).
- $SAT \xrightarrow{pr} P_1$ (our previous reduction).

We can prove that $SAT \xrightarrow{pr} P_2$ either directly or via the reduction $P_1 \xrightarrow{pr} P_2$ because $SAT \xrightarrow{pr} P_1$ and because, as we have already stated, polynomial reduction is a transitive operation.

In other words, to show that some problem P_2 in NP is in NPC, we can pick any NP-complete problem P_1 in NPC and show that $P_1 \xrightarrow{pr} P_2$. This will show that P_2 is in NPC, and P_2 will itself become a candidate NP-complete problem to pick up for later reductions.

A decision problem is said to be NP-hard when it can be polynomially reduced from an NP-complete problem, but it is not known whether it belongs to NP.

6.3.4 Optimization problems versus decision problems

We have been focusing so far on decision problems, but, in many practical situations, we have to solve an optimization problem in which we want to maximize or minimize a given criterion. Optimization problems (also called search problems) are more complex than decision problems, but one can always restrict an optimization problem so that it becomes a decision problem.

For instance, the graph coloring problem is usually an optimization problem: What is the minimum number of colors required to color the graph? The restriction to the decision problem is the COLOR problem of Definition 6.3: Can we color the graph with at most k colors? If we can solve the optimization problem, we have immediately the solution to the decision problem, for any value of k. In this particular case, we also can go the other way round. If we are able to solve the decision problem, then we can find the answer to the optimization problem by performing a binary search on k $(1 \leqslant k \leqslant |V|)$ and computing the answer of the decision problem for each value of k. The binary search adds a factor $\log(|V|)$ to the algorithm complexity, so that if we had a polynomial algorithm, it remains polynomial. The two problems (optimization and decision) have the same complexity. In most cases, the optimization problem can be solved using a binary search as described above. However, this result is not always true; it can be difficult to find the answer to the optimization problem, even though we can solve the decision problem. In some extreme situations, there may be no solution to the optimization problem. For instance, there is no solution to the problem "Find the smallest rational number x such that $x^2 \geqslant 2$" because $\sqrt{2}$ is irrational, while it is easy to solve the decision problem in polynomial time: "Given a rational number x, do we have $x^2 \geqslant 2$?" (simply compute a square and compare it to 2).

Transforming a decision problem into an optimization problem may not be natural or even possible. However, we can always define the *associated* decision problem of an optimization problem: If the optimization problem aims at minimizing a value x with some constraints, the decision problem adds a value x_0 as an input to the problem, and the question is whether there is a solution achieving a value $x \leqslant x_0$.

A typical example is based on 2-PARTITION. Consider the scheduling problem with two processors, where we want to schedule n tasks of length a_i. Ideally, we want to 2-partition the tasks so that the execution finishes as soon as possible, but if this is not possible, we minimize the difference of finish times, which amounts to minimizing the global finish time. Formally, the objective is to find a subset I that minimizes $x = |\sum_{i \in I} a_i - \sum_{i \notin I} a_i|$. The associated decision problem with target value $x_0 = 0$ is exactly 2-PARTITION, that will be shown to be NP-complete (Exercise 7.20, p. 155). By misuse of language, we say that an optimization problem is NP-complete if the associated decision problem for some well-chosen target value is NP-complete. Hence, the scheduling problem with two processors as defined above is NP-complete.

6.4 Examples of NP-complete problems and reductions

At this point, we know that SAT is NP-complete. As already discussed, we proceed by reduction to increase the list of NP-complete problems. In this section, we show that 3-SAT, CLIQUE, and VERTEX-COVER are in NPC. We also give references for the NP-completeness of 2-PARTITION, HC (Hamiltonian Cycle), and then show that TSP (Traveling Salesman Problem) is NP-complete.

6.4.1 3-SAT

DEFINITION 6.7 (3-SAT). Let F be a Boolean formula with n variables $x_1, \ldots, x_n$ and p clauses $C_1, \ldots, C_p$: $F = C_1 \wedge C_2 \wedge \cdots \wedge C_p$, with, for $1 \leqslant i \leqslant p$, $C_i = x^*_{i_1} \vee x^*_{i_2} \vee x^*_{i_3}$, $1 \leqslant i_k \leqslant n$ for $1 \leqslant k \leqslant 3$, and $x^* = x$ or $\overline{x}$. Does there exist an instantiation of the variables such that F is true (i.e., C_i is true for $1 \leqslant i \leqslant p$)?

This problem is the restriction of SAT to the case where each clause consists of three variables, i.e., following the notations of Section 6.3.2, $f(i) = 3$ for $1 \leqslant i \leqslant p$. In fact, 3-SAT is so close to SAT that one might wonder why consider 3-SAT in addition to, or replacement of, SAT. The reason is that it is much easier to manipulate clauses with exactly three variables. Furthermore, proving the NP-completeness of 3-SAT is also a good exercise for our first reduction.

THEOREM 6.1. *3-SAT is NP-complete.*

Proof. This proof, as well as the next ones, follows the reduction method to prove that a problem is NP-complete.

First, we prove that 3-SAT is in NP. We can simply claim that it is in NP because it is a restriction of SAT, which itself is in NP. It also is easy to prove it directly. We consider an instance I of 3-SAT, which is of size $O(n + p)$. A certificate is a set of truth values, one for each variable. Therefore, it is of size $O(n)$, which is polynomial in the size of the instance. It is easy to check whether the certificate is a solution, and this takes a time $O(n + p)$. Altogether, 3-SAT is in NP.

To prove the completeness, we reduce an instance of SAT. So far, it is the only problem that we know to be NP-complete, thanks to Cook's theorem, so we have no choice.

Let $\mathcal{I}_1$ be an instance of SAT. First, we need to build an instance $\mathcal{I}_2$ of 3-SAT that will have a solution if and only if $\mathcal{I}_1$ has one. $\mathcal{I}_1$ consists of p clauses $C_1, \ldots, C_p$, of lengths $f(1), \ldots, f(p)$, and each clause is made of some of the n variables $x_1, \ldots, x_n$.

Instance $\mathcal{I}_2$ initially consists of the n variables $x_1, \ldots, x_n$. Then, we add to $\mathcal{I}_2$ variables and clauses corresponding to each clause C_i of $\mathcal{I}_1$. We build a set of clauses made of exactly three variables, and the goal is to have the equivalence between C_i and the constructed clauses. We consider various cases:

- If C_i has a single variable x, we add to instance $\mathcal{I}_2$ two new variables a_i and b_i and four clauses: $x \vee a_i \vee b_i$, $x \vee \overline{a_i} \vee b_i$, $x \vee a_i \vee \overline{b_i}$, and $x \vee \overline{a_i} \vee \overline{b_i}$.
- If C_i has two variables $x_1 \vee x_2$, we add to instance $\mathcal{I}_2$ one new variable c_i and two clauses: $x_1 \vee x_2 \vee c_i$ and $x_1 \vee x_2 \vee \overline{c_i}$.
- If C_i has three variables, we add it to $\mathcal{I}_2$.
- If C_i has k variables, with $k > 3$, $C_i = x_1 \vee x_2 \vee \cdots \vee x_k$, then we add $k-3$ new variables $z_1^i, z_2^i, \ldots, z_{k-3}^i$ and $k-2$ clauses: $x_1 \vee x_2 \vee z_1^i$, $x_3 \vee \overline{z_1^i} \vee z_2^i$, ..., $x_{k-2} \vee \overline{z_{k-4}^i} \vee z_{k-3}^i$, and $x_{k-1} \vee x_k \vee \overline{z_{k-3}^i}$.

Note that all clauses that are added to $\mathcal{I}_2$ are exactly made of three variables, and that the construction is done in polynomial time. Then, we must check the different points of the reduction.

First, note that $size(\mathcal{I}_2)$ is polynomial in $size(\mathcal{I}_1)$ (and even linear); indeed, $size(\mathcal{I}_2) = O(n + \sum_{i=1}^{p} f(i))$.

Then, we start with the easy side, which consists of proving that if $\mathcal{I}_1$ has a solution, then $\mathcal{I}_2$ has a solution. Let us assume that $\mathcal{I}_1$ has a solution. We have an instantiation of variables $x_1, \ldots, x_n$ such that C_i is true for $1 \leqslant i \leqslant p$. Then, a solution for $\mathcal{I}_2$ keeps the same values for the x_is, and set all a_j, b_j and c_j values to true. Therefore, if a clause with at most three variables is true in $\mathcal{I}_1$, all corresponding clauses in $\mathcal{I}_2$ are true. Consider now a clause C_i in $\mathcal{I}_1$ with $k > 3$ variables: $C_i = x_1 \vee x_2 \vee \cdots \vee x_k$. Let x_j be the first variable of the clause that is true. Then, for the solution of $\mathcal{I}_2$, we instantiate $z_1^i, \ldots, z_{j-2}^i$ to true and $z_{j-1}^i, \ldots, z_{k-3}^i$ to false. With this instantiation, all clauses of $\mathcal{I}_2$ are true, and thus $\mathcal{I}_1 \Rightarrow \mathcal{I}_2$.

For the other side, let us assume that $\mathcal{I}_2$ has a solution. We have an instantiation of all variables x_i, a_i, b_i, c_i, and z_j^i that is a solution of $\mathcal{I}_2$. Then, we prove that the same instantiation of $x_1, \ldots, x_n$ is a solution of the initial instance $\mathcal{I}_1$. First, for a clause with one or two variables, whatever the values of a_i, b_i, and c_i, we necessarily have x or $x_1 \vee x_2$ equal to true because we have added clauses constraining the extra variables. The clauses with three variables remain true since we have not modified them. Finally, let C_i be a clause of $\mathcal{I}_1$ with $k > 3$ variables, $C_i = x_1 \vee x_2 \vee \cdots \vee x_k$. We reason by contradiction. If this clause is false, then, necessarily, because of the first clause added to $\mathcal{I}_2$ when processing clause C_i, z_1^i must be true, and similarly we can prove that all z_j^i variables must be true. The contradiction arises for the last clause because it imposes that $\overline{z_{k-3}^i}$ should be true if x_{k-1} and x_k are both false. Therefore, by contradiction, at least one of the x_js must be true and the clause of $\mathcal{I}_1$ is true. We finally have $\mathcal{I}_2 \Rightarrow \mathcal{I}_1$, which concludes the proof. □

As a final remark, we point out that not all restrictions of a given NP-complete problem remain NP-complete. For instance, 2-SAT, the SAT problem where each clause contains exactly two variables, belongs to P. Several variants of 3-SAT are shown NP-complete in the exercises.

6.4.2 CLIQUE

We now consider a problem that is very different from SAT.

DEFINITION 6.8 (CLIQUE). Let $G = (V, E)$ be a graph and k be an integer such that $1 \leqslant k \leqslant |V|$. Does there exist a clique of size k (i.e., a complete subgraph of G with k vertices)?

This is a graph problem, and the size of the instance is polynomial in $|V|$ (recall that $|E| \leqslant |V|^2$, so we do not need to consider $|E|$ in the instance size).

THEOREM 6.2. *CLIQUE is NP-complete.*

Proof. First we prove that CLIQUE is in NP. The certificate is the list of vertices of a clique, and we can check in polynomial time (even quadratic time) whether it is a clique or not. For each vertex pair of the certificate, the edge between these vertices must be in E.

The completeness is obtained with a reduction from 3-SAT. We could do a reduction from SAT, but 3-SAT is more regular, so we give it preference for the reduction. Let $\mathcal{I}_1$ be an instance of 3-SAT with n variables and p clauses. Then we build an instance $\mathcal{I}_2$ of CLIQUE. We add three vertices to the graph for each clause (each vertex corresponds to one of the literals of the clause) and then we add an edge between two vertices if and only if (i) they are not part of the same clause and (ii) they are not antagonist (i.e., one corresponding to a variable x_i and the other to its negation $\overline{x_i}$). An example is shown in Figure 6.1, with the graph obtained for a formula with three clauses $C_1 \wedge C_2 \wedge C_3$, with $C_1 = x_1 \vee \overline{x_2} \vee \overline{x_3}$, $C_2 = \overline{x_1} \vee x_2 \vee x_3$, and $C_3 = x_1 \vee x_2 \vee x_3$.

Note that $\mathcal{I}_2$ is a graph with $3p$ vertices; the size of this instance, therefore, is polynomial in the size of $\mathcal{I}_1$. Moreover, we fix in instance $\mathcal{I}_2$ the integer k of the CLIQUE definition such that $k = p$. We are now ready to check the equivalence of the solutions.

Assume first that the instance $\mathcal{I}_1$ of 3-SAT has a solution. Then, we pick a vertex corresponding to a variable that is true in each clause, and it is easy to check that the subgraph made of these p vertices is a clique. Indeed, two of such vertices are not in the same clause, and they are not antagonistic; therefore, there is an edge between them.

On the other side, if there is a clique of size k in instance $\mathcal{I}_2$, then necessarily there is one vertex of the clique in each clause (otherwise, the two vertices

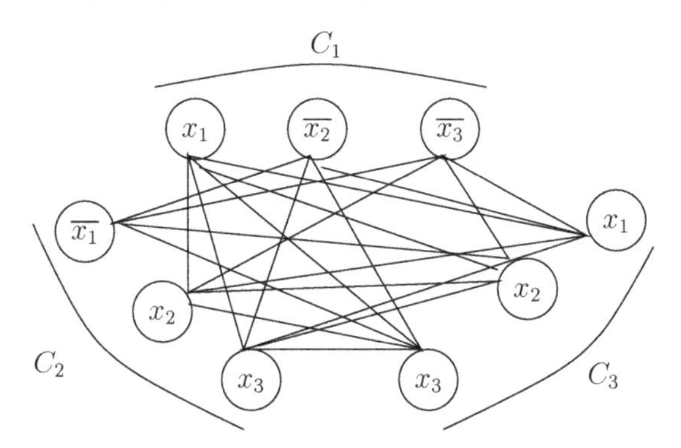

FIGURE 6.1: Example: Reduction of an instance of 3-SAT to an instance of CLIQUE.

within the same clause would not be connected). We choose these vertices to instantiate the variables, and we obtain a solution because we never make contradictory choices (because two antagonistic vertices cannot be part of the clique, there is no edge between them). This concludes the proof. □

We discuss variants of the CLIQUE problem in Section 6.5.

6.4.3 VERTEX-COVER

We continue to enrich the class NPC with another graph problem. We say that an edge $e = (u, v)$ is *covered* by its endpoints u and v.

DEFINITION 6.9 (VERTEX-COVER). Let $G = (V, E)$ be a graph and k be an integer such that $1 \leqslant k \leqslant |V|$. Do there exist k vertices $v_{i_1}, \ldots, v_{i_k}$ such that each edge $e \in E$ is covered by (at least) one of the v_{i_j}, for $1 \leqslant j \leqslant k$?

THEOREM 6.3. *VERTEX-COVER is NP-complete.*

Proof. It is easy to check that VERTEX-COVER is in NP. The certificate is a set of k vertices, $V_c \subseteq V$, and for each edge $(v_1, v_2) \in E$, we check whether $v_1 \in V_c$ or $v_2 \in V_c$. The verification is done in time $|E| \times k$, and, therefore, it is polynomial in the problem size.

This problem is once again a graph problem, so we choose to use a reduction from CLIQUE, which turns out to be straightforward. Let $\mathcal{I}_1$ be an instance of CLIQUE: It consists of a graph $G = (V, E)$ and an integer k. We consider the following instance $\mathcal{I}_2$ of VERTEX-COVER. The graph is $\overline{G} = (V, \overline{E})$, which is the complementary graph of G, i.e., an edge is in $\overline{G}$ if and only if it is not in G (see the example in Figure 6.2). Moreover, we set the size of the covering set to $|V| - k$.

If instance $\mathcal{I}_1$ has a solution, G has a clique of size k, and, therefore, the $|V| - k$ vertices that are not part of the clique form a covering set of $\overline{G}$. Reciprocally,

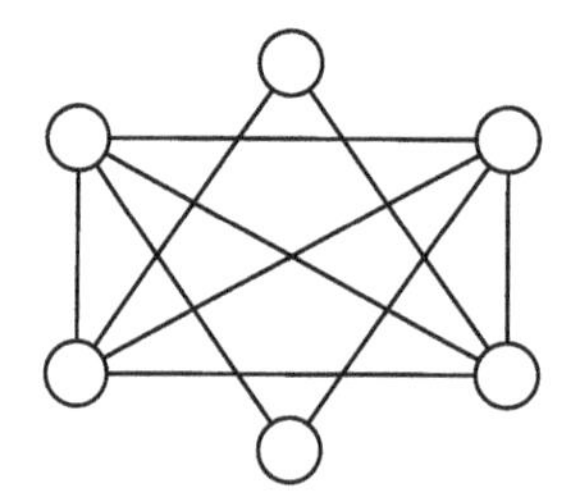
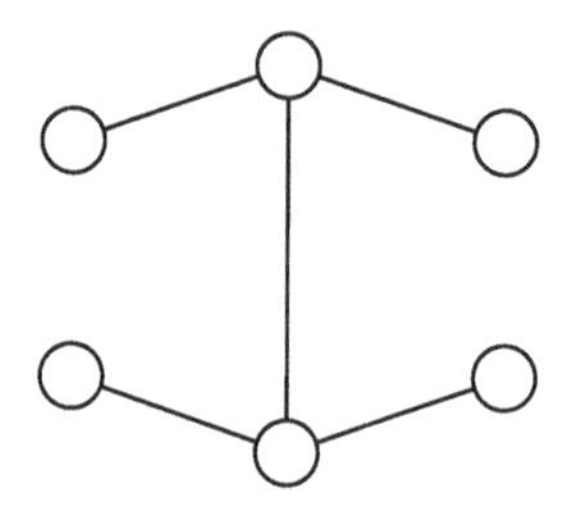

FIGURE 6.2: Example: Reduction of an instance of CLIQUE (on the left, graph G, $k = 4$) to an instance of VERTEX-COVER (on the right, graph $\overline{G}$, size of the cover $|V| - k = 2$).

if $\mathcal{I}_2$ has a solution, then the vertices that are not part of the covering set form a clique in the original graph G. This concludes the proof. □

6.4.4 Scheduling problems

Scheduling is the activity that consists of mapping an application onto a target platform and of assigning execution times to its constitutive parts. The application can often be represented as a task graph, where nodes denote computational tasks and edges model precedence constraints between tasks. For each task, an assignment (choose the processor that will execute the task) and a schedule (decide when to start the execution) are determined. The goal is to obtain an efficient execution of the application, which translates into optimizing some objective function. The traditional objective function in the scheduling literature is the minimization of the total execution time, or *makespan*; however, we will see examples with other objectives, such as those of the case study devoted to online scheduling (Chapter 15).

Traditional scheduling assumes that the target platform is a set of p identical processors, and that no communication cost is paid. In that context, a task graph is a directed acyclic vertex-weighted graph $G = (V, E, w)$, where the set V of vertices represents the tasks, the set E of edges represents precedence constraints between tasks ($e = (u, v) \in E$ if and only if $u \prec v$, where $\prec$ is the precedence relation), and the weight function $w : V \longrightarrow \mathbb{N}^*$ gives the weight (or duration) of each task. Task weights are assumed to be positive integers. A schedule σ of a task graph is a function that assigns a start time to each task: $\sigma : V \longrightarrow \mathbb{N}^*$ such that $\sigma(u) + w(u) \leqslant \sigma(v)$ whenever $e = (u, v) \in E$. In other words, a schedule preserves the *precedence constraints* induced by the precedence relation $\prec$ and embodied by the edges of the precedence graph. If $u \prec v$, then the execution of u begins at time $\sigma(u)$ and requires $w(u)$ units of time, and the execution of v at time $\sigma(v)$ must start after the end of the execution of u. Obviously, if there were a cycle in the task graph, no schedule could exist, hence, the restriction to acyclic graphs and, thus, the focus on Directed Acyclic Graphs (DAGs).

There are other constraints that must be met by schedules, namely, *resource constraints.* When there is an infinite number of processors (in fact, when there are as many processors as tasks), the problem is *with unlimited processors*, and denoted $P\infty|prec|Cmax$ in the literature [44]. We use the shorter notation SCHED(∞) in this book; each task can be assigned to its own processor. When there is a fixed number $p < n$ of available processors, the problem is *with limited processors*, and the general problem is denoted SCHED(p). SCHED(2) represents the scheduling problem with only two processors. Note that SCHED(∞) is equivalent to SCHED(q) for any value $q \geqslant n$, where n is the number of tasks. In the case with limited processors, a problem is defined by the task graph and the number of processors p. An allocation function $\mathsf{alloc} : V \longrightarrow \mathcal{P}$ is then required, where $\mathcal{P} = \{1, \ldots, p\}$ denotes the set of available processors. This function assigns a target processor to each task. The resource constraints simply specify that no processor can be allocated more than one task at the same time:

$$\mathsf{alloc}(T) = \mathsf{alloc}(T') \Rightarrow \begin{cases} \sigma(T) + w(T) \leqslant \sigma(T') \\ \text{or } \sigma(T') + w(T') \leqslant \sigma(T). \end{cases}$$

This condition expresses the fact that if two tasks T and T' are allocated to the same processor, then their executions cannot overlap in time.

The *makespan* MS(σ, p) of a schedule σ that uses p processors is its total execution time: $\text{MS}(\sigma, p) = \max_{v \in V}\{\sigma(v) + w(v)\}$ (assuming that the first task(s) is (are) scheduled at time 0). The makespan is the total execution time, or finish time, of the schedule. Let $\text{MS}_{opt}(p)$ be the value of the makespan of an optimal schedule with p processors: $\text{MS}_{opt}(p) = \min_\sigma \text{MS}(\sigma, p)$. Because schedules respect precedence constraints, we have $\text{MS}_{opt}(p) \geqslant w(\Phi)$ for all paths Φ in G (weights extend to paths in G as usual). We also have $\text{Seq} \leqslant p \times \text{MS}_{opt}(p)$, where $\text{Seq} = \sum_{v \in V} w(v) = \text{MS}_{opt}(1)$ is the sum of all task weights.

While SCHED(∞) has polynomial complexity (simply traverse the graph and start each task as soon as possible using a fresh processor), problems with a fixed amount of resources are known to be difficult. Letting DEC be the decision problem associated with SCHED, and INDEP the restriction of DEC to independent tasks (no precedence constraints), i.e., $E = \emptyset$, well-known complexity results are summarized below:

- INDEP(2) is NP-complete but can be solved by a pseudopolynomial algorithm. Moreover, $\forall \varepsilon > 0$, INDEP(2) admits a $(1+\varepsilon)$-approximation whose complexity is polynomial in $\frac{1}{\varepsilon}$ (see Section 8.1.5, p. 187).
- INDEP is NP-complete in the strong sense (see Exercise 7.10, p. 152) but can be approximated up to some constant factor (see Exercise 9.5, p. 215). Moreover, $\forall \varepsilon > 0$, there is a $(1 + \varepsilon)$-approximation algorithm for this problem [50].
- DEC(2) (and hence DEC) is NP-complete in the strong sense (see Exercise 7.11, p. 152).

All these results are gathered here for the sake of comprehensiveness. The impatient reader who wonders what is the meaning of NP-complete *in the strong sense* may refer to Section 6.6, p. 145, and to understand what is *an approximation algorithm*, she/he may have a quick look at Section 8.1, p. 179 right now.

Scheduling and certificates

Scheduling problems provide a nice illustration of the attention that must be paid to certificates. Consider the DEC decision problem, namely, scheduling a task graph with p processors and a given deadline D. For the schedule to be valid, both precedence and resource constraints must be enforced. The question is to decide whether there exists a schedule whose makespan does not exceed the deadline.

A naive verification of the schedule is to describe which tasks are executed onto which processors at each time step. Unfortunately, this description may lead to a certificate of exponential size; the time steps range from 1 to D, and the size of the scheduling problem is $O(n + p + \log W + \log D)$, where $W = \sum_{v \in V} w(v)$.

A polynomial size verification of the schedule can be easily obtained using *events*, which are time steps where a new task begins or ends. There is a polynomial number of such events ($2n$), and for each of them we perform a polynomial number of checks. From the definition of the schedule, we first construct the ordered list of events in polynomial time. The basic idea is to maintain the set of tasks that have been completed and the set of processors that are currently idle. If the event corresponds to starting a new task, we check that all its predecessors have been completed, and that the target processor belongs to the set of idle processors (and then we remove it from this set). If the event corresponds to completing a task, we mark the task accordingly, and we re-insert the target processor into the set of idle processors. Note that if several events take place at the same time step, we should start with those that correspond to task completions. We perform these checks one event after the other until we reach the last one, which corresponds to the completion of the last task, and which much take place not later than D.

In summary, we see that the weights of the tasks (given by the function w) prevent us from using a naive verification of the validity of a schedule at each step of its execution. This is because the makespan is not polynomial in the problem size.

6.4.5 Other famous NP-complete problems

We have initiated discussions with the 2-PARTITION problem (Definition 6.1) that is one of the most widely used problems to perform reductions, since it turns out to be NP-complete while being quite simple in its formulation. The NP-completeness of 2-PARTITION will be shown in Exercise 7.20, p. 155, but from now on, we assume that this problem is indeed NP-complete.

The COLOR problem (see Definition 6.3) given in Section 6.2.2 is also NP-complete and the proof is the purpose of Exercise 7.7, p. 151. Other problems will discuss variants of this graph coloring problem.

Another useful problem is HC (Hamiltonian Cycle, see Definition 6.4). We have already shown that HC is in NP (see Section 6.2.2). For the completeness, we refer the interested reader to involved reduction in [27]. There is a nice reduction from 3-SAT in the first edition of the book, and the current edition performs a reduction from VERTEX-COVER.

Starting from HC, it is easy to prove that TSP (see Definition 6.5) also is NP-complete. It is clear that TSP is in NP; a certificate is an ordered list of vertices. The reduction comes from HC. Let $\mathcal{I}_1$ be an instance of HC: This is a graph $G = (V, E)$. We build the following instance $\mathcal{I}_2$ of TSP. The graph $G' = (V, E')$ has the same set of vertices as G, but it is a complete graph. We set $k = 0$, i.e., we want to find a cycle of weight 0. Finally, for $e \in E'$ we define the cost function w such that $w(e) = 0$ if $e \in E$, and $w(e) = 1$ otherwise. This reduction is obviously of polynomial time, and the equivalence of solutions is straightforward. Note that this last NP-completeness result comes from the fact that TSP is a weighted version of HC.

For a reference list of problems known to be NP-complete, we refer the reader to the book by Garey and Johnson [38].

6.5 Importance of problem definition

In this section, we point out subtleties in problem definitions. A parameter can be either fixed for the problem or part of the problem instance. Consider the problem CLIQUE introduced in Section 6.4.2. Given a graph $G = (V, E)$, we introduce the notion of β-clique of size k, where β is a rational such that $0 < \beta \leqslant 1$ [83]; a β-clique is a subgraph of G of size k (k vertices), with edge density at least β. The edge density is the ratio of the number of edges in the subgraph over the number of edges in a clique of size k, i.e., $\binom{k}{2}$. We can now define a variant of the CLIQUE problem:

DEFINITION 6.10 (BCLIQUE). Let $G = (V, E)$ be a graph, β be a rational number such that $0 < \beta \leqslant 1$, and k be an integer such that $1 \leqslant k \leqslant |V|$. Does there exist a β-clique of size k in G?

In the BCLIQUE problem, β is part of the instance. Therefore, we can do a trivial reduction from CLIQUE, letting $\beta = 1$, to prove that it is NP-complete. However, we may define the problem in a different way, where β is given. For a constant β such that $0 < \beta \leqslant 1$, we define:

DEFINITION 6.11 (BCLIQUE(β)). Let $G = (V, E)$ be a graph and k be an integer such that $1 \leqslant k \leqslant |V|$. Does there exist a β-clique of size k in G?

We have CLIQUE = BCLIQUE(1). However, the NP-completeness of CLIQUE does not imply the NP-completeness of BCLIQUE(β) for any value of β. We prove this NP-completeness for any fixed value $0 < \beta < 1$ in the following theorem:

THEOREM 6.4. *BCLIQUE(β) is NP-complete for any rational number $\beta = \frac{p}{q}$, where p and q are positive integer constants and $p < q$.*

Proof. It is clear that BCLIQUE(β) is in NP, and the reduction comes logically from the classical CLIQUE problem. The idea is to construct an auxiliary graph $G' = (V', E')$ and to prove that G has a clique of size k if and only if $G \cup G'$ has a β-clique of size $|V'| + k$.

We build the set of vertices V' of size $|V'| = 4(|V|^2 + k^2)q - k$, containing vertices v'_1 to $v'_{|V'|}$. For $1 \leqslant i \leqslant |V'|$ and $j \in [i+1, i+|V|] \mod |V'|$, we add an edge between v_i and v_j. Therefore, each node has $2|V|$ edges, and we have added a total of $|V||V'|$ edges. Next, we add random edges in order to have a total of $K = \frac{p}{q}\binom{|V'|+k}{2} - \binom{k}{2}$ edges between the $|V'|$ vertices. Because $|V'|+k$ is a multiple of $2q$, K is an integer. Moreover, $|V'|$ is large enough so that we can prove that $\binom{|V'|}{2} \geqslant K \geqslant |V||V'|$ (see [83]), i.e., there were initially fewer than K edges, and we can have a total of K edges without exceeding the maximum number of edges in $|V'|$.

There remains to prove that G has a clique of size k if and only if $G \cup G'$ has a $\frac{p}{q}$-clique of size $|V'| + k$. Suppose first that there is a clique C of size k in G. We consider the subgraph Q of $G \cup G'$ containing vertices $C \cup V'$. We have $|Q| = |V'| + k$ and the number of edges is $K + \binom{k}{2} = \frac{p}{q}\binom{|V'|+k}{2}$; therefore, Q is a $\frac{p}{q}$-clique by definition.

Suppose now that there is a $\frac{p}{q}$-clique Q of size $|V'| + k$ in $G \cup G'$. We first construct a $\frac{p}{q}$-clique Q' such that $|Q'| = |Q|$ and $V' \subset Q'$. Since $|Q| > |V'|$, $|V' \setminus Q| \leqslant |V|$, and each vertex in $V' \setminus Q$ cannot be connected to more than $|V| - 1$ vertices of $V' \setminus Q$. Moreover, each vertex of $V' \setminus Q$ is of degree at least $2|V|$ and, therefore, it is connected to at least $|V|+1$ vertices of Q, while vertices of $Q \cap V$ are connected to at most $|V|-1$ vertices (all of them from V). Therefore, we can replace $|V' \setminus Q|$ vertices of $Q \cap V$ with the remaining vertices of V', with no reduction in the edge density. We obtain a $\frac{p}{q}$-clique Q' such that $|Q'| = |Q|$ and $V' \subset Q'$. Then, $|Q' \cap V| = k$. To see that $Q' \cap V$ is a clique of size k in V, consider the density of G'; it is K by construction. If $Q' \cap V$ does not contribute $\binom{k}{2}$ edges, then Q' cannot have density $\frac{p}{q}$. Therefore, $Q' \cap V$ is a clique of size k, hence concluding the proof. □

In scheduling problems (see Section 6.4.4, p. 140), the same distinction is often implicitly made, whether the number of processors p is part of the problem instance or not. For instance, if all tasks are unit-weighted, DEC is

NP-complete (with p in the problem instance), while DEC(2) can be solved in polynomial time and DEC(3) is an open problem [38].

6.6 Strong NP-completeness

The last technical discussion of this chapter is related to *weak* and *strong* NP-completeness. This refinement of the NPC problem class applies to problems involving numbers, such as 2-PARTITION, but also TSP, because of edge weights.

Consider a decision problem P, and let I be an instance of this problem. We have already discussed how to compute $size(I)$, the size of the instance, encoded in binary. We now define $max(I)$, which is the *maximum* size of the instance, typically corresponding to the problem instance with integers coded in unary. To give an example, consider an instance I of 2-PARTITION with n integers $a_1, \ldots, a_n$. As already discussed, we can have $size(I) = n + \sum_{1 \leqslant i \leqslant n} \log(a_i)$, or any similar (polynomially related) expression. Now we can have $max(I) = n + \sum_{1 \leqslant i \leqslant n} a_i$, or $max(I) = n + \max_{1 \leqslant i \leqslant n} a_i$, or any similar (polynomially related) expression.

Then, given a polynomial p, we define P_p, the problem P restricted to p, as the problem restricted to instances such that $max(I)$ is smaller than $p(size(I))$, i.e., the size of the instance coded in unary is bounded applying p to the binary size of the instance. A problem P (in NP) is NP-complete *in the strong sense* if and only if there exists a polynomial p such that P_p remains NP-complete. Otherwise, if the problem restricted to p can be solved in polynomial time, the problem is NP-complete *in the weak sense*; intuitively, in this case, the problem is difficult only if we do not bound the size of the input in the problem instance.

Note that for a graph problem such as the bipartite graph problem, there are no numbers, so $max(I) = size(I)$ and the problem is NP-complete in the strong sense. For problems with numbers (including weighted graph problems), one must be more careful. Coming back to 2-PARTITION, we have seen in Section 6.2.1 that it can be solved by a dynamic-programming algorithm running in time $O(n \sum_{i=1}^{n} a_i)$, or equivalently in time $O(max(I))$. Therefore, any instance I of 2-PARTITION can be solved in time polynomial in $max(I)$, which is the definition of a pseudopolynomial problem. And 2-PARTITION is not NP-complete in the strong sense (one says it is NP-complete in the weak sense).

To conclude this section, we introduce a problem with numbers that is NP-complete in the strong sense: 3-PARTITION. The name of this problem is misleading because this problem is different from partitioning n integers into three sets of same size.

DEFINITION 6.12 (3-PARTITION). Given an integer B, and $3n$ integers $a_1, \ldots, a_{3n}$, can we partition the $3n$ integers into n triplets, each of sum B? We can assume that $\sum_{i=1}^{3n} a_i = nB$ (otherwise, there is no solution), and that $B/4 < a_i < B/2$ (so that one needs exactly three elements to obtain a sum B).

Contrary to 2-PARTITION, 3-PARTITION is NP-complete in the strong sense [38].

6.7 Why does it matter?

We conclude this chapter with a discussion on polynomial problems. Why focus on polynomial problems? If the size of the data is in n, from a practical perspective it is much better to have an algorithm in $(1.0001)^n$, which is exponential, than a polynomial-time algorithm in n^{1000}. In such a case, the polynomial-time algorithm is still slower than the exponential one for $n = 10^9$, and, therefore, the exponential algorithm is faster in any practical situation. However, n^{1000} is not practical either. In general, polynomial algorithms have a small degree, typically not exceeding 4 and almost always smaller than 10.

Polynomial-time algorithms are likely to be efficient algorithms, so when confronted with a new problem, the first thing we do is to look for an algorithm that would solve it in polynomial time. If we succeed, we are finished. If we do not succeed, we have another way to go—prove that the problem is NP-complete. Then the chance of somebody else coming later and providing an optimal solution to the problem is very small because it is very unlikely that P = NP. In other words, if we can show that our problem is more difficult than one (hence all) of these famous NP-complete problems, then we show strong evidence of the intrinsic difficulty of the problem.

Of course, proving a problem NP-complete does not make it go away. One needs to keep a constructive approach, such as proposing an algorithm that provides a near-optimal solution in polynomial time (or again, proving that no such approximation algorithm exists). This is the subject of Chapter 8.

6.8 Bibliographical notes

As already mentioned, our approach to NP-completeness is original. See the book by Garey and Johnson [38] for a comprehensive treatment of NP-completeness and a famous catalog of NP-complete problems. A very intuitive proof of Cook's theorem is given in the book by Wilf [108]. A theory-oriented approach with Turing machines and complexity results is available in the book by Papadimitriou [82]. The more adventurous reader can investigate the book by Arora and Barak [4].

Chapter 7

Exercises on NP-completeness

This chapter presents a set of exercises related to NP-completeness. The aim of most questions is to prove the NP-completeness of a new problem, so that the reader will become more familiar with this reasoning.

We assume in the first sections that 2-PARTITION is NP-complete, as stated in Section 6.4.5, so that we can warm up with easy reductions in Section 7.1. We also provide thematic exercises around graph coloring (Section 7.2) and scheduling (Section 7.3). More involved reductions are proposed in Section 7.4. Finally, the NP-completeness of 2-PARTITION is proved in Section 7.5.

Some of the exercises contain hints. Additional hints can be found on page 155 at the beginning of Section 7.6, which is the section where solutions to all of the exercises are given.

7.1 Easy reductions

Exercise 7.1: Wheel (solution p. 156)

Prove that the following decision problem is NP-complete. Given a graph $G = (V, E)$ and an integer $K \geqslant 3$, does G include a wheel of size K, i.e., a set of $K + 1$ vertices w, v_1, v_2, ..., v_K such that E contains the set of edges (v_K, v_1), $\{(v_i, v_{i+1})\}_{1 \leqslant i < K}$, and $\{(v_i, w)\}_{1 \leqslant i \leqslant K}$ (w is the center of the wheel)?

Exercise 7.2: Knights of the round table (solution p. 156)

Prove that the following decision problem is NP-complete. Given n knights, and a set of pairs of knights who are enemies, is it possible to arrange the knights around a round table so that two enemies do not sit side by side?

Exercise 7.3: Variants of CLIQUE (solution p. 157)

Prove that the two following variants of CLIQUE are also NP-complete problems.

1. TWO-CLIQUES: Let $G = (V, E)$ be a graph and k be an integer such that $1 \leqslant k \leqslant |V|$. Do there exist two disjoint cliques of size k in G (i.e., two disjoint complete subgraphs of G with k vertices)?
2. CLIQUE-REG-GRAPH: Let $G = (V, E)$ be a graph whose vertices are all of the same degree and k be an integer. Does there exist a clique of size k in G?

Exercise 7.4: Path with vertex pairs (solution p. 158)

Prove that the following decision problem is NP-complete. Let $G = (V, E)$ be a directed graph (therefore, in this graph, the arc (u, v) is different from the arc (v, u)). Let s and t be two vertices of G and let $P = \{(a_1, b_1), \ldots, (a_n, b_n)\}$ be a list of pairs of vertices of G. Does there exist a directed path from s to t in G that includes at most one vertex from each of the pairs in the list P?

(*Hint:* Build a reduction from 3-SAT.)

Exercise 7.5: VERTEX-COVER with even degrees (solution p. 158)

Prove that the following decision problem is NP-complete. Let $G = (V, E)$ be a graph whose vertices all have an even degree, and let k be a positive integer. Is there a subset of the vertices of G that covers all the edges of G and whose size is at most k?

Exercise 7.6: Around 2-PARTITION (solution p. 159)

Reminder: All integers in the 2-PARTITION problem and its variants must be (strictly) positive. Furthermore, we assume that 2-PARTITION is NP-complete.

Prove that the four following decision problems are NP-complete:

1. 2PARTNEVEN: Given $2n$ integers $a_1, a_2, \ldots, a_{2n}$, is there a subset I of $\{1, \ldots, 2n\}$ such that $\sum_{i \in I} a_i = \sum_{i \notin I} a_i$?
2. 2PARTEQ: Given $2n$ integers $a_1, a_2, \ldots, a_{2n}$, is there a subset I of $\{1, \ldots, 2n\}$ such that $\sum_{i \in I} a_i = \sum_{i \notin I} a_i$, with $|I| = n$?
3. 2PARTODDEVEN: Given $2n$ integers $a_1, a_2, \ldots, a_{2n}$, is there a subset I of $\{1, \ldots, 2n\}$ such that $\sum_{i \in I} a_i = \sum_{i \notin I} a_i$ with, for any $j \in \{1, \ldots, n\}$, exactly one integer between a_{2j-1} and a_{2j} belongs to I?
4. PARTITIONIN3: Given n integers $a_1, a_2, \ldots, a_n$, are there three disjoint subsets I_1, I_2, and I_3 of $\{1, \ldots, n\}$ such that $\sum_{i \in I_1} a_i = \sum_{i \in I_2} a_i = \sum_{i \in I_3} a_i$?
5. Are the above decision problems NP-complete in the strong sense or the weak sense?

7.2 About graph coloring

We recall the definition of COLOR (see Definition 6.3): Given a graph $G = (V, E)$ and an integer k ($1 \leqslant k \leqslant |V|$), can we color G with at most k colors? A graph coloring is valid if any two vertices connected with an edge are not assigned the same color.

We define two variants of COLOR. The first is a restriction on the number of colors, and the second is a restriction on the structure of the graphs considered.

DEFINITION 7.1 (N-COLOR)**.** Given a graph $G = (V, E)$, can we color G with at most N colors?

Note that in the definition of N-COLOR, N is not part of the problem instance, as k in COLOR.

DEFINITION 7.2 (3-COLOR-PLAN)**.** Given a planar graph $G = (V, E)$ (i.e., we can draw it without having intersecting edges), can we color G with at most three colors?

Exercise 7.7: COLOR (solution p. 160)

Prove that problem COLOR is NP-complete.

Exercise 7.8: 3-COLOR (solution p. 162)

1. Propose a polynomial-time algorithm to solve 2-COLOR.
2. Prove that problem 3-COLOR is NP-complete.
   (*Hint:* Use the widget in Figure 7.1 to build a reduction from 3-SAT. We denote the three colors by 0, 1, and 2. One can show that the widget has the two following properties:
   (a) If $x = y = z = 0$, then $v = 0$.
   (b) For any other input (x, y, z), v can be colored 1 or 2.)

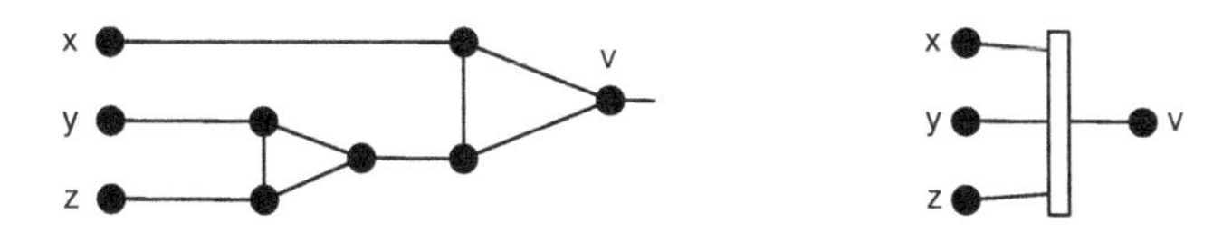

FIGURE 7.1: The widget used for proving the NP-completeness of 3-COLOR (on the left) and its representation (on the right).

Exercise 7.9: 3-COLOR-PLAN (solution p. 163)

Prove that problem 3-COLOR-PLAN is NP-complete.

(*Hint:* We propose to do a reduction from 3-COLOR using the widget presented on the right. First prove that this widget has the two following properties:

1. In any valid 3-coloring of the widget, vertices a and a' have the same color, and vertices b and b' have the same color.
2. If a and a' have the same (given) color and if b and b' have the same (given) color, then the coloring can be completed into a valid 3-coloring.)

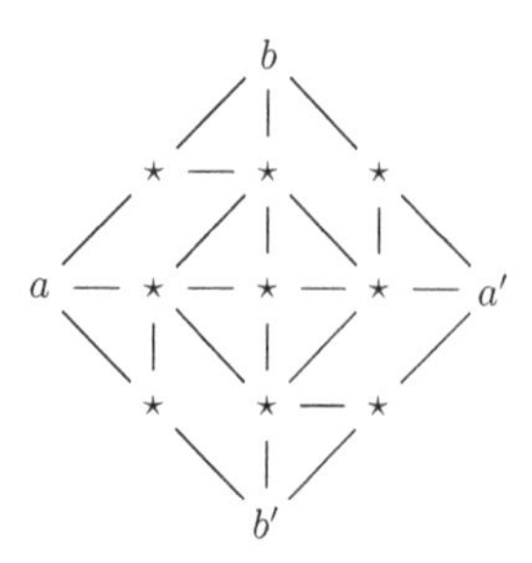

7.3 Scheduling problems

Exercise 7.10: Scheduling independent tasks with p processors (solution p. 166)

This exercise deals with the complexity of scheduling independent tasks with p processors. This is problem INDEP(p) defined in Section 6.4.4: Given a set $V = \{v_1, v_2, \ldots, v_n\}$ of n independent tasks, where each task v_i has a nonnegative integer weight $w(v_i) = a_i$ ($1 \leqslant i \leqslant n$), given p processors, and a time bound K, does there exist a valid schedule σ whose makespan does not exceed the bound K? Here there are no precedence constraints, so we search for a partition of the n tasks onto the p processors such that the maximal load of any processor (which is the sum of the weights of the tasks assigned to it) does not exceed the bound K. Prove that INDEP(p) is NP-complete in the strong sense.

Exercise 7.11: Scheduling with two processors (solution p. 166)

This exercise deals with the complexity of scheduling a graph of tasks with two processors. This is problem DEC(p) defined in Section 6.4.4: Given a task graph $G = (V, E, w)$, a number p of processors, and a bound on the execution time K, does there exist a valid schedule σ (that respects both precedence and resource constraints) whose makespan does not exceed the bound K? Prove that DEC(p) is NP-complete in the strong sense.

7.4 More involved reductions

Exercise 7.12: Transitive subchain (solution p. 167)

In a directed graph, a k-tuple of vertices $(x_1, \ldots, x_k)$ is a *transitive subchain* of length k if and only if, for any $1 \leqslant i < j \leqslant k$, $(x_i, x_j) \in E$.

Prove that the following decision problem is NP-complete: Let $G = (V, E)$ be a directed graph. Does G contain a transitive subchain of length at least equal to $\lfloor |V|/2 \rfloor$?

(*Hint:* The reduction can be made from 3-SAT. For an instance of 3-SAT with clauses $C_1, \ldots, C_k$ where, for any $i \in [1; k]$, $C_i = (x_i^1 \vee x_i^2 \vee x_i^3)$, one can build an instance of TRANSITIVE SUBCHAIN where $V = \{C_0\} \bigcup_{1 \leqslant i \leqslant k} \{C_i, x_i^1, x_i^2, x_i^3\}$. The set E of edges must be carefully defined.)

Exercise 7.13: INDEPENDENT SET (solution p. 168)

In a graph $G = (V, E)$, an *independent set* is a subset V' of the set of vertices, $V' \subset V$, such that for every two vertices u, v in V', there is no edge connecting the two, i.e., $(u, v) \notin E$. A maximum independent set is a largest independent set for a given graph, i.e., $|V'|$ is maximum. Prove that finding a maximum independent set is NP-complete.

Exercise 7.14: DOMINATING SET (solution p. 169)

Prove that the following decision problem is NP-complete: Given a graph $G = (V, E)$ and an integer $K \geqslant 3$, does G include a dominating set of size K, i.e., a subset D of V of size K such that for any vertex $u \in V \setminus D$, there exists a vertex $v \in D$ such that $(u, v) \in E$?

Exercise 7.15: Carpenter (solution p. 170)

We are given a set of n wooden sticks whose lengths, $a_1, a_2, \ldots, a_n$, are integers. These sticks are connected by hinges. The i-th stick, of length a_i, is connected at one end to the $(i-1)$-th stick and at the other end to the $(i+1)$-th stick. The problem is to fold the set of wooden sticks so that the total width does not exceed a given bound k. Figure 7.2 gives an example.

Prove that the following decision problem is NP-complete: Given n wooden stick lengths and an integer k, does there exist a folding of width at most k?

Exercise 7.16: k-center (solution p. 171)

Let $G = (V, E)$ be a complete graph whose edges are weighted by a weight function w that satisfies the triangle inequality: $w(u, v) \leqslant w(u, x) + w(x, v)$

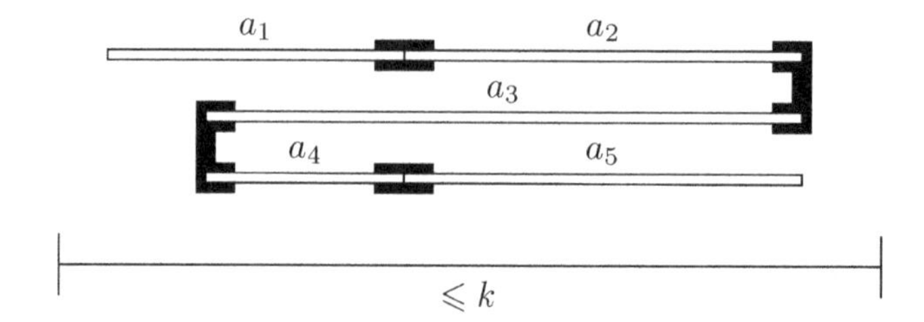

FIGURE 7.2: Example of a folding of a set of wooden sticks (the hinges are in black).

for any u, v, and x in V. Let k be a positive integer ($k \geqslant 1$). For any subset S of the vertices ($S \subset V$) and any vertex v not in S ($v \in V \setminus S$), we define $connect(v, S)$ as the minimum weight of an edge linking v to a vertex of S: $connect(v, S) = \min_{s \in S} w(v, s)$. A k-center is a subset S of V of cardinal at most k such that $center(S) = \max_{v \in V \setminus S} connect(v, S)$ is minimum.

Prove that the following decision problem is NP-complete: Given a complete weighted graph G with a weight function that satisfies the triangle inequality, and an integer $k \geqslant 1$, does there exist a k-center in G?

Exercise 7.17: Variants of 3-SAT (solution p. 172)

Prove that the two following variants of 3-SAT are NP-complete problems:

1. 3-SAT NAE (*not all equal*): The three literals of a clause cannot take the same value.
2. 3-SAT OIT (*one in three*): Exactly one literal per clause has the value TRUE.

Exercise 7.18: Variants of SAT (solution p. 174)

We call SAT-N the problem SAT restricted to formulas that do not contain more than N occurrences of each variable.

1. Prove that SAT-3 is at least as hard as SAT. Prove that for any $N \geqslant 3$, SAT-N is NP-complete.
2. Let x be a variable occurring in an instance F of SAT-2. Exhibit a formula equivalent to F and in which the variable x does not appear. Propose a polynomial-time algorithm to solve SAT-2.

7.5 2-PARTITION is NP-complete

Exercise 7.19: SUBSET-SUM (solution p. 175)

The problem SUBSET-SUM is defined as follows: Given a finite set S of positive integers and an integer t, is there a subset S' of S such that $\sum_{x \in S'} x = t$? Prove that the SUBSET-SUM problem is NP-complete, using a reduction from 3-SAT.

(*Hint:* From a set of clauses $C_0, \ldots, C_{m-1}$ on the variables $x_0, \ldots, x_{n-1}$, we build a set S of integers as follows. First, we define the values b_{ij} and b'_{ij} for $0 \leqslant i \leqslant n-1$ and $0 \leqslant j \leqslant m-1$:

$$b_{ij} = \begin{cases} 1 & \text{if } x_i \text{ appears in } C_j \\ 0 & \text{otherwise} \end{cases} \quad \text{and} \quad b'_{ij} = \begin{cases} 1 & \text{if } \overline{x_i} \text{ appears in } C_j \\ 0 & \text{otherwise.} \end{cases}$$

Then, let $v_i = 10^{m+i} + \sum_{j=0}^{m-1} b_{ij} 10^j$, $v'_i = 10^{m+i} + \sum_{j=0}^{m-1} b'_{ij} 10^j$, for $0 \leqslant i \leqslant n-1$, and $s_j = 10^j$, $s'_j = 2 \times 10^j$, for $0 \leqslant j \leqslant m-1$. S is defined as $S = \{v_i, v'_i\}_{0 \leqslant i \leqslant n-1} \cup \{s_j, s'_j\}_{0 \leqslant j \leqslant m-1}$. To fully define the instance of SUBSET-SUM, one still needs to define an integer t such that there exists a subset $S' \subseteq S$ whose sum is t if and only if there exists an instantiation for which all clauses are true.)

Exercise 7.20: NP-completeness of 2-PARTITION (solution p. 177)

Prove that the 2-PARTITION problem is NP-complete: Given n integers $a_1, \ldots, a_n$, is there a subset I of $\{1, \ldots, n\}$ such that $\sum_{i \in I} a_i = \sum_{i \notin I} a_i$?

7.6 Solutions to exercises

Additional hints

- Exercise 7.1 "Wheel": Try a reduction from HC (Hamiltonian Cycle).
- Exercise 7.2 "Knights of the round table": Try a reduction from HC (Hamiltonian Cycle).
- Exercise 7.3 "Variants of CLIQUE": Try a reduction from CLIQUE.
- Exercise 7.4 "Path with vertex pairs": Try a reduction from 3-SAT.
- Exercise 7.5 "VERTEX-COVER with even degrees": Try a reduction from VERTEX-COVER.
- Exercise 7.6 "Around 2-PARTITION": Try reductions from 2-PARTITION.
- Exercise 7.7 "COLOR": Try a reduction from 3-SAT.
- Exercise 7.8 "3-COLOR": Try a reduction from COLOR.

- Exercise 7.9 "3-COLOR-PLAN": Try a reduction from COLOR.
- Exercise 7.10 "Scheduling independent tasks with p processors": Try a reduction from 3-PARTITION.
- Exercise 7.11 "Scheduling with two processors": Try a reduction from 3-PARTITION.
- Exercise 7.12 "Transitive subchain": Try a reduction from 3-SAT.
- Exercise 7.13 "INDEPENDENT SET": Try a reduction from 3-SAT or CLIQUE.
- Exercise 7.14 "DOMINATING SET": Try a reduction from VERTEX-COVER.
- Exercise 7.15 "Carpenter": Try a reduction from 2-PARTITION.
- Exercise 7.16 "k-center": Try a reduction from DOMINATING SET.
- Exercise 7.17 "Variants of 3-SAT": Try building reductions from 3-SAT. To a clause $C_i = a_i \vee b_i \vee c_i$ associate the clauses:
  - $a_i \vee b_i \vee x_i$ and $c_i \vee \overline{x_i} \vee f$ for 3-SAT NAE;
  - $a_i \vee x_i \vee y_i$, $\overline{b_i} \vee x_i \vee x_i'$, and $\overline{c_i} \vee y_i \vee y_i'$ for 3-SAT OIT.
- Exercise 7.18 "Variants of SAT": Try reductions from SAT.
- Exercise 7.20 "NP-completeness of 2-PARTITION": Try a reduction from SUBSET-SUM.

Solution to Exercise 7.1: Wheel

The certificate is nothing but the list of vertices, and one can check that a certificate is valid in linear time. Therefore, this decision problem belongs to NP.

We build a reduction from HC (Hamiltonian Cycle, see Definition 6.4, p. 130). We thus start from an instance $\mathcal{I}_1 = (V, E)$ of HC. From $\mathcal{I}_1$, we build an instance $\mathcal{I}_2$ of wheel by adding a vertex w to $\mathcal{I}_1$ and one edge between w and each of the vertices of $\mathcal{I}_1$. Finally, we let $K = |V|$. The reduction is obviously polynomial.

If $\mathcal{I}_1$ contains a Hamiltonian cycle, then $\mathcal{I}_2$ contains a wheel of size $|V|$. Reciprocally, we assume that $\mathcal{I}_2$ contains a wheel of size $|V|$. Then, let w' be its center. If $w \neq w'$, as w' is linked to any vertex in the wheel (by definition of a wheel) and as, by construction, w is linked to any vertex of V, w and w' can exchange their roles: w' takes the place of w in the circle of size $|V|$ and w becomes the center. Therefore, $\mathcal{I}_2$ contains a wheel of center w and of size $|V|$ and, thus, $\mathcal{I}_1$ contains a Hamiltonian cycle.

Therefore, the instance $\mathcal{I}_1$ of HC has a solution if and only if instance $\mathcal{I}_2$ of wheel has one. The latter problem is thus in NPC.

Solution to Exercise 7.2: Knights of the round table

This problem trivially belongs to NP.

We craft the reduction from HC (Hamiltonian Cycle, see Definition 6.4, p. 130). Let $\mathcal{I}_1$ be an instance of HC. Then, $\mathcal{I}_1$ is a graph (V, E). From it, we build an instance $\mathcal{I}_2$ of our problem. We have a knight in $\mathcal{I}_2$ for each vertex of $\mathcal{I}_1$. Two knights are enemies if and only if there are no edges in E between the two corresponding vertices of V.

We now show that there is a Hamiltonian cycle in $\mathcal{I}_1$ if and only if there is a valid sitting for the knights of instance $\mathcal{I}_2$.

Let us first assume that there is a Hamiltonian cycle in $\mathcal{I}_1$. Then, we arrange the knights in the order of their corresponding vertices along the Hamiltonian cycle. As there are no edge in E between two vertices corresponding to two enemy knights, two vertices of $\mathcal{I}_1$ corresponding to enemy knights cannot be consecutive in the Hamiltonian cycle. Therefore, the knight arrangement is valid.

Reciprocally, we assume that there is a valid knight sitting for instance $\mathcal{I}_2$. Then, the ordering of the knights around the table defines a Hamiltonian cycle among the corresponding vertices.

Solution to Exercise 7.3: Variants of CLIQUE

The two problems belong to NP for the same reasons for which CLIQUE belongs to NP. For both problems, we craft a reduction from CLIQUE (Definition 6.8, p. 138).

1. We start from an instance $\mathcal{I}_1$ of CLIQUE. $\mathcal{I}_1$ is thus a graph G. We build an instance $\mathcal{I}_2$ of TWO-CLIQUES, which is made up of exactly two distinct copies of G. The size of $\mathcal{I}_2$ is equal to two times the size of $\mathcal{I}_1$, and the reduction is thus polynomial.

   Then, if there is a clique of size k in $\mathcal{I}_1$, there is one in each of the two copies of G in $\mathcal{I}_2$. Reciprocally, if $\mathcal{I}_2$ contains two disjoint cliques of size k, because $\mathcal{I}_2$ only contains two copies of G (we have not added any edges between these two copies), then each clique is included in one of the copies, which means that G, and thus $\mathcal{I}_1$, contains (at least) one clique of size k. Therefore, the instance $\mathcal{I}_1$ of CLIQUE contains a clique of size k if and only if instance $\mathcal{I}_2$ of TWO-CLIQUES contains two cliques of size k.

2. We start from an instance $\mathcal{I}_1$ of CLIQUE, that is, a graph $G = (V, E)$.

   Let $\delta(G)$ be the maximum degree of a vertex of G. In other words, $\delta(G) = \max_{v \in V} \text{degree}(v)$. We build an instance $\mathcal{I}_2$ of CLIQUE-REG-GRAPH as follows: We create $\delta(G)$ copies of G; for each vertex $v \in V$, we create $\delta(G) - \text{degree}(v)$ new vertices, each of them being linked to each of the $\delta(G)$ copies of v. This way we obtain a graph whose vertices all have a degree $\delta(G)$.

   It is then easy to see that the new vertices and new edges do not create new nontrivial cliques and do not enlarge the size of existing cliques. Then, there is a clique of size k in $\mathcal{I}_1$ if and only if there is a clique of size k in $\mathcal{I}_2$.

Solution to Exercise 7.4: Path with vertex pairs

A certificate for this problem is a path. It is then easy to check in polynomial time whether this path answers the question. Therefore, the problem is in NP.

We start from an instance $\mathcal{I}_1$ of 3-SAT. $\mathcal{I}_1 = C_1 \wedge \cdots \wedge C_m$ where $C_k = (l_{1,k} \vee l_{2,k} \vee l_{3,k})$. Let $x_1, \ldots, x_n$ be the variables appearing in $\mathcal{I}_1$. From $\mathcal{I}_1$ we build an instance $\mathcal{I}_2 = (V, E, s, t, P)$ of "path with vertex pairs." First, we define the set of vertices: $V = \{l_{i,j} | (i,j) \in [1;3] \times [1;m]\} \cup \{s,t\}$. Then, we define the set of edges: $E = \{(s, l_{1,i})\}_{i \in [1;3]} \cup \{(l_{k,i}, l_{k+1,i'})\}_{k \in [1;m-1], i \in [1;3], i' \in [1;3]} \cup \{l_{m,i}, t\}_{i \in [1;3]}$. Finally, let $P = \{(x_i, \overline{x_i})\}_{i \in [1;n]}$. The size of $\mathcal{I}_2$ is polynomial in the size of $\mathcal{I}_1$, and the reduction is thus polynomial.

Let us assume that there exists a solution to $\mathcal{I}_1$, that is, an instantiation of the variables of $\mathcal{I}_1$ such that all of its clauses are true. Then, we build a path in $\mathcal{I}_2$ from s to t, which goes through exactly one true literal per clause. There is at least one such literal per clause as we start from a solution of $\mathcal{I}_1$. Furthermore, this path does not contain the two vertices x_j and $\overline{x_j}$ of a pair of P because both literals cannot be simultaneously true.

Reciprocally, let us assume that $\mathcal{I}_2$ has a solution, i.e., that there exists a path $\mathcal{P}$ from s to t that does not include both vertices of a forbidden pair. Then, we define an instantiation of $\mathcal{I}_1$ by assigning the value TRUE to any literal whose corresponding vertex is included in $\mathcal{P}$. This partial instantiation is correct. Indeed, the path $\mathcal{P}$ cannot contain both vertices x_i and $\overline{x_i}$ for a given value of $i \in [1;n]$ because P includes the pair $(x_i, \overline{x_i})$. We complete this partial instantiation arbitrarily (i.e., we give whatever value to any still undefined variable). As the $(i+1)$-th vertex of $\mathcal{P}$ corresponds to a literal of clause C_i, for $i \in [1;m]$, this instantiation sets all the clauses of $\mathcal{I}_1$ to true.

Therefore, instance $\mathcal{I}_1$ of 3-SAT has a solution if and only if instance $\mathcal{I}_2$ of "path with vertex pairs" has one. The latter problem is thus NP-complete.

Solution to Exercise 7.5: VERTEX-COVER with even degrees

This problem is in NP because the problem VERTEX-COVER is in NP.

We build a reduction from VERTEX-COVER. Let $\mathcal{I}_1 = (G, k)$ be an instance of VERTEX-COVER. From it we build an instance $\mathcal{I}_2$ of VERTEX-COVER with even degrees. First, we remark that there is an even number of vertices whose degree is odd, because $\sum_{v \in V} \text{degree}(v) = 2|E|$. We then build the instance $\mathcal{I}_2 = (G', k+2)$ by adding to G three new vertices, x, y, and z one edge between x and any odd-degree vertex of G and the edges (x,y), (x,z), and (y,z). This construction is illustrated in Figure 7.3. The size of $\mathcal{I}_2$ is obviously polynomial in the size of $\mathcal{I}_1$, and the reduction is thus polynomial.

Let us assume that there is a solution for $\mathcal{I}_1$, i.e., a cover C of the edges of G containing at most k vertices. Then, $C \cup \{x, y\}$ is a cover of G' containing

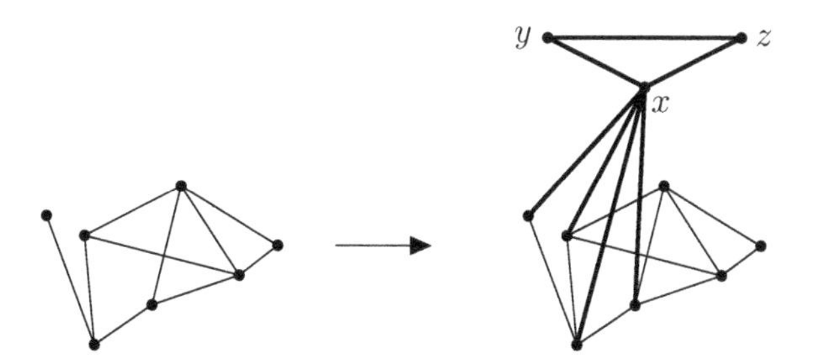

FIGURE 7.3: Reduction from an instance of VERTEX-COVER (left) to an instance of VERTEX-COVER with even degrees.

at most $k+2$ vertices and is thus a solution for $\mathcal{I}_2$.

Reciprocally, we assume that there exists a solution for $\mathcal{I}_2$, i.e., that there exists a cover C' of G' containing at most $k+2$ vertices. Then, as the vertices y and z are linked only to the vertex x and to each other, at least two of the three vertices x, y, and z must belong to C'. Therefore, $C' \setminus \{x, y, z\}$ is a cover of G containing at most k vertices.

Thus, there exists a solution for any instance $\mathcal{I}_1$ of VERTEX-COVER if and only if there exists a solution to the corresponding instance $\mathcal{I}_2$ of VERTEX-COVER with even degrees. The latter problem is thus NP-complete.

Solution to Exercise 7.6: Around 2-PARTITION

We first remark that the four decision problems are in NP, the reasoning being the same as in 2-PARTITION. These problems being variants of 2-PARTITION, all the reductions will be made from 2-PARTITION or from variants of it.

1. Let $\mathcal{I}_1$ be an instance of 2-PARTITION including n integers $a_1, \ldots, a_n$. If n is even, we let $\mathcal{I}_2 = \mathcal{I}_1$. Otherwise, we let $\mathcal{I}_2 = \mathcal{I}_1 \cup \{2S, 2S, 4S\}$, where $\sum_{i=1}^{n} a_i = S$. The equivalence is then straightforward.

2. Let $\mathcal{I}_1$ be an instance of 2-PARTITION including n integers $a_1, \ldots, a_n$. From $\mathcal{I}_1$, we create the instance $\mathcal{I}_2$ including the elements $a_1+1, a_2+1, \ldots, a_n+1$ and n elements equal to 1.

   If $\mathcal{I}_1$ admits a solution I, then we obtain a partition of $\mathcal{I}_2$ of weight $S/2+n$ and of size n by taking the elements $\{a_i+1\}_{i \in I}$ and $n-|I|$ elements of weight 1, where $\sum_{i=1}^{n} a_i = S$.

   Reciprocally, any solution I' of $\mathcal{I}_2$ is such that $\sum_{i \in I'}(a_i+1)+n-|I'| = S/2+n$, as $|I'|$ must be equal to n. Then, $\sum_{i \in I'} a_i = S/2$ and I' defines a solution of $\mathcal{I}_1$.

3. Let $\mathcal{I}_1$ be an instance of 2PARTEQ including $2n$ integers $a_1, \ldots, a_{2n}$. We then build an instance $\mathcal{I}_2$ of 2PARTODDEVEN: $\mathcal{I}_2 = (a_1, B, a_2, B, \ldots,$

a_{2n}, B) with $B > \sum_{i=1}^{2n} a_i$. The equivalence is straightforward once one realizes that, in any solution of $\mathcal{I}_2$, there must be exactly n elements equal to B in each subset.

4. Let $\mathcal{I}_1$ be an instance of 2-PARTITION including n integers $a_1, \dots, a_n$. Let $S = \sum_{i=1}^{n} a_i$. We can assume that, for any i in $[1;n]$, $a_i \leqslant S/2$; otherwise, the instance has trivially no solution. We build the instance $\mathcal{I}_2$ of PARTITIONIN3: $\mathcal{I}_2 = \mathcal{I}_1 \cup \{a_{n+1} = S/2\}$. In any solution of $\mathcal{I}_2$ one of the three subsets must exclusively contains a_{n+1}. The two other subsets define the solution to $\mathcal{I}_1$.

5. None of these problems are NP-complete in the strong sense because the reduction comes from 2-PARTITION.

Solution to Exercise 7.7: COLOR

We first prove that COLOR is in NP. An instance $\mathcal{I}$ of COLOR is a couple (G, k) of size $|V| + |E| + \mathit{log}\ k$ with $k \leqslant |V|$. For the certificate, we can take a list of the colors of the vertices. One can check in linear time $O(|V| + |E|)$ whether for each edge $(x, y) \in E$ the condition $\mathit{color}(x) \neq \mathit{color}(y)$ holds and whether there are at most k colors used.

We now build a reduction from 3-SAT.

Let $\mathcal{I}_1$ be an instance of 3-SAT, including p clauses, $C_1, \dots, C_p$, and n variables, $x_1, \dots, x_n$. Without loss of generality, we can ignore "easy" instances. Therefore, we assume that $n \geqslant 4$. From this instance, we build an instance $\mathcal{I}_2$ of COLOR, i.e., a graph $G = (V, E)$. G includes the $3n + p$ vertices: $x_1, \dots, x_n$, $\overline{x}_1, \dots, \overline{x}_n$, $y_1, \dots, y_n$, $c_1, \dots, c_p$. G exactly includes the edges:

1. $(x_i, \overline{x}_i)$ for $i \in [1;n]$;
2. (y_i, y_j), (y_i, x_j), and $(y_i, \overline{x}_j)$ for $(i, j) \in [1;n] \times [1;n]$ and $i \neq j$;
3. (x_i, c_k) if and only if $x_i \notin C_k$;
4. $(\overline{x}_i, c_k)$ if and only if $\overline{x}_i \notin C_k$.

The size of instance $\mathcal{I}_2$ is polynomial in the size of $\mathcal{I}_1$ as $|V| = O(n + p)$. Figure 7.4 presents an example of such a reduction.

We now prove that $\mathcal{I}_1$ has a solution if and only if instance $\mathcal{I}_2$ can be colored with $n + 1$ colors.

$\Rightarrow$ We assume that $\mathcal{I}_1$ has a solution. There exists an assignment of the variables such that each clause C_i, $1 \leqslant i \leqslant p$, is true. We define the following coloring:

$$\begin{cases} \mathit{color}(x_i) = i \text{ if } x_i \text{ is true} \\ \mathit{color}(x_i) = n + 1 \text{ if } x_i \text{ is false} \\ \mathit{color}(\overline{x}_i) = i \text{ if } \overline{x}_i \text{ is true} \\ \mathit{color}(\overline{x}_i) = n + 1 \text{ if } \overline{x}_i \text{ is false} \\ \mathit{color}(y_i) = i \end{cases}$$

and, finally, for any clause C_k, $1 \leqslant k \leqslant p$, we pick arbitrarily one of the literals that set C_k to true, and we assign its color to c_k.

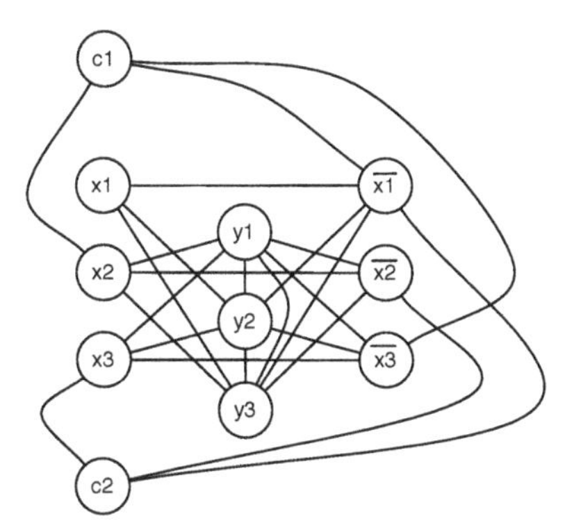

FIGURE 7.4: Instance of COLOR built from the 3-SAT instance $(x_1 \vee \overline{x_2} \vee x_3) \wedge (x_1 \vee x_2 \vee \overline{x_3})$.

We must show that *color* defines a valid coloring. The only problem could arise from the edges (x_i, c_k) or the edges $(\overline{x_i}, c_k)$. Let C_k be a clause, and let i be the color of C_k. Then, clause C_k has color i because it contains either the literal x_i, which is true, or the literal $\overline{x}_i$, which is true, and that true literal is also colored with color i. Without loss of generality, assume the true literal is x_i. Then, there is no edge between x_i and c_k. There is, however, an edge between $\overline{x_i}$ and c_k, but then the literal $\overline{x_i}$ is false and is colored with color $n+1$. Hence, the coloring satisfies this edge. For any $j \in [1;n]$, if $j \neq i$, neither x_j nor $\overline{x_j}$ has color i, and thus these vertices cannot lead to a problem with the vertex c_k. Therefore, the coloring is valid.

$\Leftarrow$ Let us now assume that there exists a coloring of $\mathcal{I}_2$ using $n+1$ colors. The vertices y_i, $1 \leqslant i \leqslant n$, form a clique. Therefore, we can permute the colors so that, for any $i \in [1;n]$, $color(y_i) = i$. There is an edge between the vertices x_i and $\overline{x}_i$, and between them and any of the vertex y_j, for $j \in [1;n]$ and $j \neq i$. Therefore, the vertices x_i and $\overline{x}_i$ are colored with different colors, and the only colors available to them are i and $n+1$. Hence, $(color(x_i), color(\overline{x}_i)$ is equal to either $(i, n+1)$ or to $(n+1, i)$. Then, we let:

$$x_i = \begin{cases} 1 \text{ if } color(x_i) = i \\ 0 \text{ if } color(x_i) = n+1 \end{cases} \quad \text{and} \quad \overline{x}_i = \begin{cases} 1 \text{ if } color(\overline{x}_i) = i \\ 0 \text{ if } color(\overline{x}_i) = n+1. \end{cases}$$

Let us consider a clause C_k, for $k \in [1,p]$. As $n \geqslant 4$, there exists an index $i \in [1;n]$ such that $x_i \notin C_k$ and $\overline{x}_i \notin C_k$. Then, there is an edge between c_k and both x_i and $\overline{x_i}$. As we have seen that at least one among x_i and $\overline{x_i}$ has $n+1$ for color, the color of c_k is not $n+1$. Then, there is a variable x_j such that either the literal x_j or the literal $\overline{x_j}$ has the same color as c_k. Then, there is no edge between that literal and c_k (because the coloring is valid). Therefore, that literal is true (its color is the color of c_k and thus not $n+1$), and it is included in the clause C_k. Therefore, this clause is true, and there is a solution to $\mathcal{I}_1$.

Solution to Exercise 7.8: 3-COLOR

1. To check whether a graph can be colored with only two colors, one just has to procede greedily. Pick arbitrarily a vertex and its color, and propagate the coloring until all vertices (in the connected components) are colored or one has reached a vertex impossible to color.

   This illustrates that the fact that COLOR is NP-complete does not imply that N-COLOR is NP-complete for a given value of N. This is because N-COLOR is a restriction of COLOR, i.e., an *easier* problem. The NP-completeness of N-COLOR can come from the parameter N.

2. 3-COLOR is in NP because COLOR is in NP.

   We first prove that the widget in Figure 7.1 satisfies property (a). Figure 7.5 presents the only two possible 3-coloring of the widget when all three entries x, y, and z have color 0. In both cases, the output v also has color 0. To prove property (b), one just has to consider one by one all possible sets of inputs and to exhibit a valid 3-coloring for which node v is colored either 1 or 2.

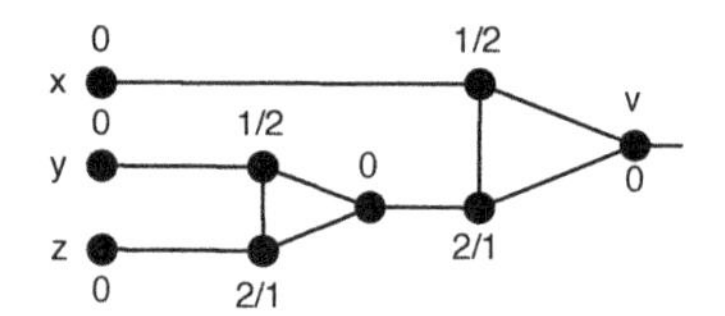

FIGURE 7.5: If x, y, and z have color 0, v also has color 0.

Using this widget, we build an instance of 3-COLOR from an instance $\mathcal{I}_1 = C_1 \wedge C_2 \wedge \cdots \wedge C_p$ of 3-SAT containing n variables $x_1, \ldots, x_n$. This construction is illustrated in Figure 7.6. First, we have two extremum vertices D and Z and the edge (D, Z). For each variable $i \in [1; n]$, we build two vertices corresponding to x_i and $\overline{x_i}$, and we add the edges $(x_i, \overline{x_i})$, (D, x_i), and $(D, \overline{x_i})$. We add to the graph p copies of the widget, one per clause. We add an edge between the vertex Z and each node v of a copy of the widget. Finally, we merge each of the three entries of the widget corresponding to a clause with the three vertices corresponding to the literals of that clause. The graph $\mathcal{I}_2$ built this way contains $2 + 2n + 6p$ vertices, and its size is thus polynomial in the size of $\mathcal{I}_1$.

We now show that $\mathcal{I}_1$ has a solution if and only if $\mathcal{I}_2$ has a 3-coloring.

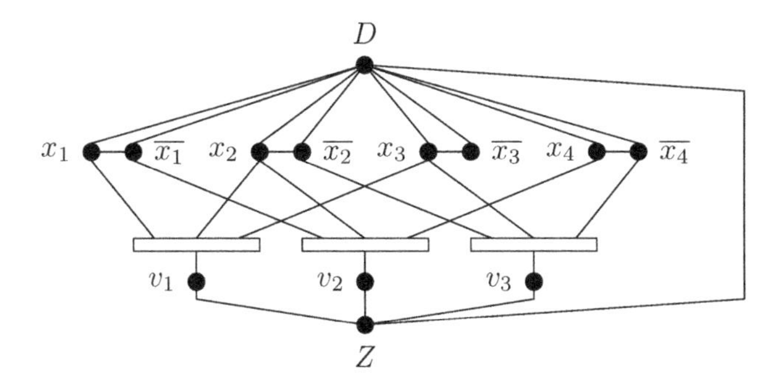

FIGURE 7.6: Instance of 3-COLOR built from the instance $(x_1 \vee x_2 \vee x_3) \wedge (\overline{x_1} \vee x_2 \vee x_4) \wedge (\overline{x_2} \vee x_3 \vee \overline{x_4})$ of 3-SAT.

Let us first assume that $\mathcal{I}_1$ has a solution. In other words, there is an assignment of truth values to variables such that all clauses are true. We define colors as follows:

$$color(x_i) = \begin{cases} 1 \text{ if } x_i \text{ is true} \\ 0 \text{ otherwise} \end{cases} \quad \text{and} \quad color(\overline{x}_i) = \begin{cases} 1 \text{ if } \overline{x}_i \text{ is true} \\ 0 \text{ otherwise.} \end{cases}$$

Because in each clause there is at least one literal that is true, then there is at least one entry of each widget that has the color 1. The second property of the widget enables us to extend the coloring of all the widgets such that none of the vertices v_j, $1 \leqslant j \leqslant p$ has the color 0. We then let $color(D) = 2$ and $color(Z) = 0$ to obtain a valid 3-coloring of the whole instance.

We now assume that there exists a 3-coloring of $\mathcal{I}_2$. Without loss of generality, we assume that $color(D) = 2$, $color(Z) = 0$. Then, for any $i \in [1; n]$, because of the edge $(x_i, \overline{x_i})$, either $(color(x_i), color(\overline{x_i})) = (0, 1)$ or $(color(x_i), color(\overline{x_i})) = (1, 0)$. As $color(Z) = 0$, the color of vertex v_j, $1 \leqslant j \leqslant p$ must be either 1 or 2. Because of the first property of the widget, this implies that at least one of the literals of each clause is not of color 0. Therefore, the assignment:

$$x_i = \begin{cases} 1 \text{ if } color(x_i) = 1 \\ 0 \text{ if } color(x_i) = 0 \end{cases}$$

is such that all clauses are true.

Solution to Exercise 7.9: 3-COLOR-PLAN

We start by proving simultaneously the two properties on the widget. To do that, we exhibit all of the valid 3-colorings of the widget.

We represent the three colors by 0, 1, and 2. We have two cases to consider: $a = b$ and $a \neq b$. We first consider the case $a = b$. Without loss of generality, we assume $a = b = 0$. Up to a permutation between colors 1 and 2, there is only a single solution to the coloring of the widget. It is presented in Figure 7.7. In Figures 7.7 through 7.10, the number between parentheses written beside a color number is the step at which the color was defined during the coloring. In Figure 7.7, at the first step we have to pick a color arbitrarily for the vertex that is linked to both a and b. We arbitrarily picked the color 1. From that point, we are able to derive the whole coloring of the widget; the coloring is fully defined and valid. We then check that we end up with $a = a' = 0$ and $b = b' = 0$.

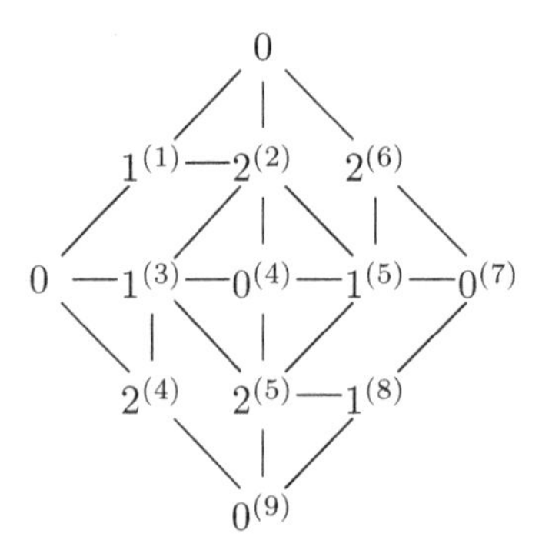

FIGURE 7.7: Coloring the widget when $a = b = 0$.

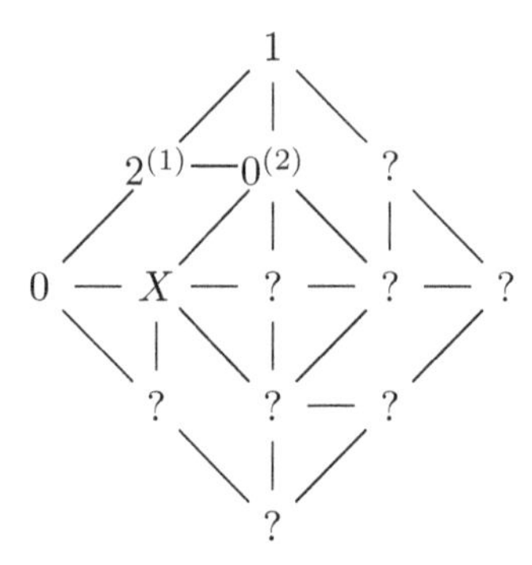

FIGURE 7.8: Starting coloring the widget when $a \neq b$.

We then consider the case $a \neq b$. Without loss of generality, we assume $a = 0$ and $b = 1$. We start coloring the widget, as shown by Figure 7.8. We have not enough constraint to decide whether the color of vertex X is 1 or 2, or whether it could be either color. We first try the case $X = 1$ and end up with the coloring displayed in Figure 7.9. In this coloring, we have $a = a' = 0$ and $b = b' = 1$. We then try the case $X = 2$ and end up with the coloring displayed in Figure 7.10. This coloring cannot be completed into a valid 3-coloring because each of the three neighbors of vertex Y has a different color.

Therefore, there are only two valid 3-coloring of the widget (up to a permutation of colors), and, in all cases, a and a' have the same color, and so have b and b'.

We now prove the NP-completeness of 3-COLOR-PLAN. First, note that 3-COLOR-PLAN is in NP because 3-COLOR is in NP. The reduction comes from 3-COLOR. Therefore, we consider an instance $\mathcal{I}_1$ of 3-COLOR. $\mathcal{I}_1$ is thus a graph $\mathcal{I}_1 = (V, E)$. We build from $\mathcal{I}_1$ an instance $\mathcal{I}_2$ of 3-COLOR-PLAN iteratively as illustrated by Figure 7.11.

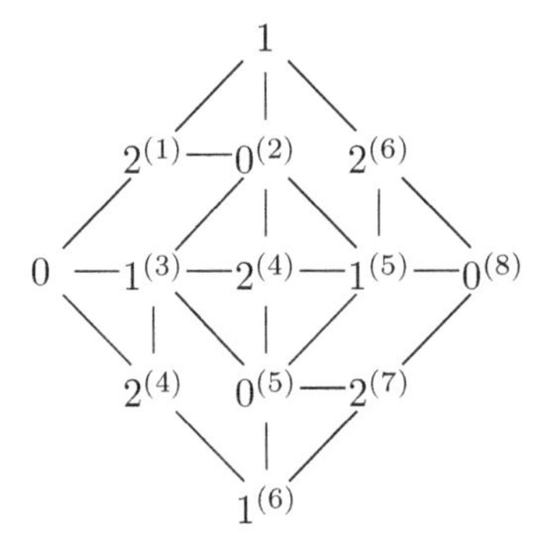

FIGURE 7.9: Coloring the widget when $a = b = 0$.

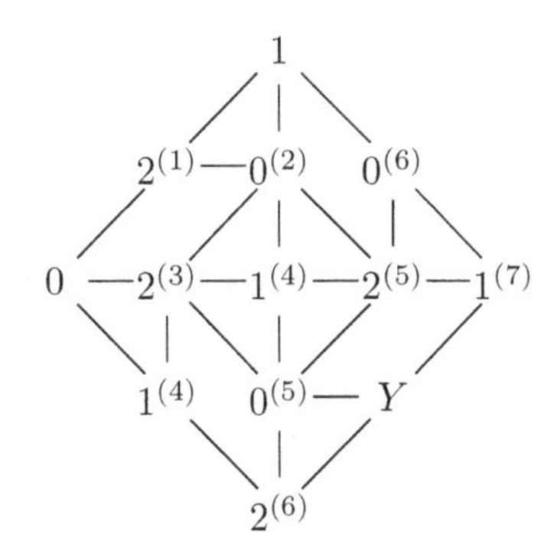

FIGURE 7.10: Starting coloring the widget when $a \neq b$.

For each edge (u, v) of E, replace each point at which this edge crosses another edge with a copy of the widget. Merge the adjacent corners of any two consecutive widgets, and merge u with the adjacent corner of the first widget. Each such transformation removes at least one intersection from the original graph, and a graph contains fewer than $|E|^2$ intersections. Therefore, the transformation eventually terminates, and the graph obtained has a size polynomial in the size of $\mathcal{I}_1$ (it contains fewer than $13|E|^2$ additional vertices and fewer than $26|E|^2$ additional edges).

Then, let us assume that $\mathcal{I}_2$ admits a 3-coloring. We use this coloring for the vertices of $\mathcal{I}_1$. Let us consider the edge (u, v) of Figure 7.11. Because of property (a), the corner of the widget to which v is connected is of the same color as u. Therefore, u and v have different colors, and the coloring is valid.

Reciprocally, let us assume that $\mathcal{I}_1$ admits a 3-coloring. We use this coloring for the vertices of $\mathcal{I}_2$ corresponding to vertices of $\mathcal{I}_1$. Because of property (b), this can be extended to a full valid 3-coloring of $\mathcal{I}_2$.

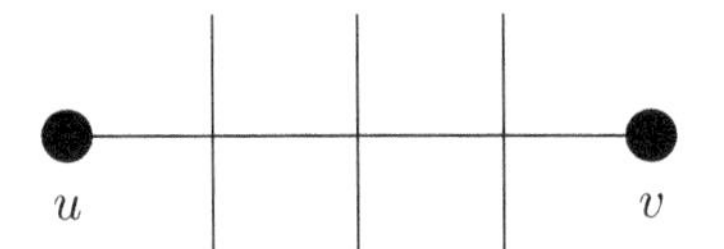

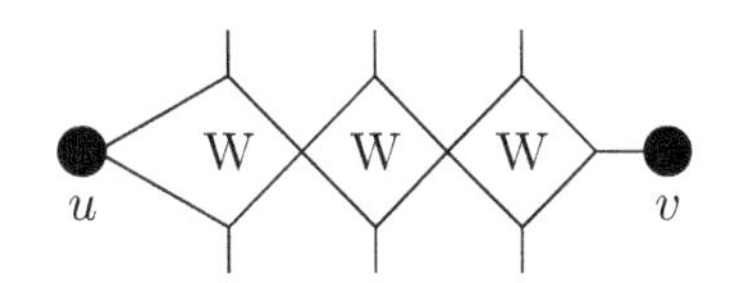

FIGURE 7.11: Elementary step in the transformation of a nonplanar graph into a planar graph. An intersection in the left-hand-side graph is removed by the insertion of a widget (right-hand-side graph).

Solution to Exercise 7.10: Scheduling independent tasks with p processors

INDEP(p) clearly belongs to NP: Given the list of tasks assigned to each processor as certificate, just check that we have indeed a partition of the original set of tasks and that no processor load exceeds the bound K.

The reduction for the strong NP-completeness of INDEP(p) is straightforward. It comes from 3-PARTITION. Consider an arbitrary instance $\mathcal{I}_1$ of 3-PARTITION, with $3n$ integers $\{a_1, a_2, \ldots, a_{3n}\}$ and bound B. The instance $\mathcal{I}_2$ of INDEP(p) is built with $3n$ independent tasks of weight a_i, $p = n$ processors, and $K = B$. Clearly, $\mathcal{I}_1$ has a solution if and only if there exists a schedule that meets the bound K, hence, if and only if $\mathcal{I}_2$ has a solution.

Solution to Exercise 7.11: Scheduling with two processors

Clearly, the problem belongs to NP. The certificate is the list of scheduling decisions, i.e., for each task, the identity of the processor that it is assigned to and the time step at which the execution begins. It is then easy to check in polynomial time that each precedence constraint is satisfied, and that no two tasks are executed simultaneously by the same processor. For instance, we can sort the tasks by their starting times and check all the conditions by scanning the sorted array.

The reduction for the strong NP-completeness comes from 3-PARTITION. Consider again an arbitrary instance $\mathcal{I}_1$ of 3-PARTITION, with $3n$ integers $\{a_1, a_2, \ldots, a_{3n}\}$ and bound B. Remember that we can assume that $\sum_{i=1}^{3n} a_i = nB$ (otherwise there is no solution), and that $B/4 < a_i < B/2$ (so that one needs exactly three elements to obtain a sum B).

The instance $\mathcal{I}_2$ of DEC(2) is built as follows: $3n$ independent tasks $\{T_1, \ldots, T_{3n}\}$, where the weight of T_i is a_i; and $3n$ other tasks, $\{X_1, Y_1, Z_1, \ldots, X_n, Y_n, Z_n\}$, all of weight B, linked by the following precedence constraints:

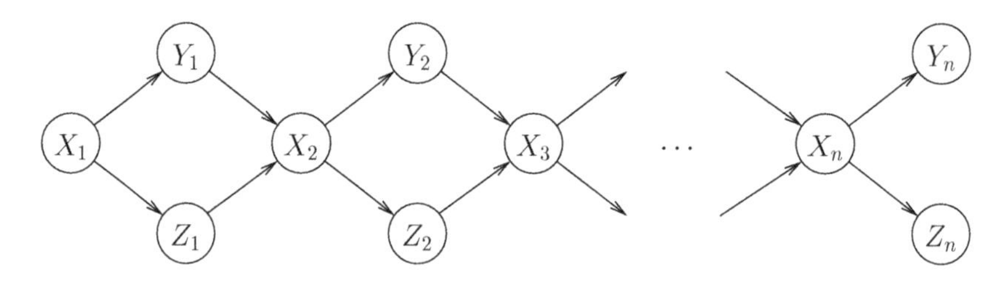

Finally, we let $K = 2nB$.

Assume first that $\mathcal{I}_1$ has a solution $I_1 \cup I_2 \cup \cdots \cup I_n$, where each I_i is composed of three numbers whose sum is B. The solution to $\mathcal{I}_2$ is the following schedule σ:

- The first processor P_1 executes all $2n$ tasks X_i and Y_i. These tasks are totally ordered along a precedence path of length $K = 2nB$.

- The second processor P_2 executes Z_i while P_1 executes Y_i. While P_1 executes a task X_i, it has a slot of size B to execute the three tasks T_j that belong to I_i.

Altogether, all precedence and resource constraints are satisfied, and σ is a valid schedule of makespan K.

Now, assume that $\mathcal{I}_2$ has a solution schedule σ. We see that the schedule σ is quite constrained. Because $X_1 \to Y_1 \to X_2 \to \cdots \to X_n \to Y_n$ is a precedence path of length K, these tasks must be processed as soon as possible, without any idle time in between. But $X_1 \to Z_1 \to X_2 \to \cdots \to X_n \to Z_n$ is another precedence path of the same length, so these tasks must be processed as soon as possible as well. This enforces that $\sigma(X_i) = (2i-2)B$ and that $\sigma(Y_i) = \sigma(Z_i) = (2i-1)B$. Up to some exchanges, we can assume that P_1 executes all the X_i and Y_i, and that P_2 executes all the Z_i. We see that the $3n$ tasks T_i are executed by P_2 during n intervals of length B, hence constituting a solution to $\mathcal{I}_1$.

Solution to Exercise 7.12: Transitive subchain

First, remark that TRANSITIVE SUBCHAIN belongs to NP. Indeed, a certificate is a k-tuple with $k \in \left[\left\lfloor \frac{|V|}{2} \right\rfloor ; |V|\right]$. To check the validity of a certificate including k vertices, one must just check the presence of $\frac{k(k-1)}{2}$ edges.

Let $G = (V, E)$ be a directed graph with $V = \{C_0\} \bigcup_{1 \leqslant i \leqslant k} \{C_i, x_i^1, x_i^2, x_i^3\}$. The three literals making clause C_i are independent (they do not share any variable). E is exactly the following set of edges:

- For each i in $[1;k]$, each j in $[1;3]$, and each l in $[i,k]$, $(x_i^j, C_l) \in E$.
- For each i in $[0;k-1]$, each j in $[1;3]$, and each l in $[i+1,k]$, $(C_i, x_l^j) \in E$.
- For each (i,j) such that $0 \leqslant i < j \leqslant k$, $(C_i, C_j) \in E$.
- For each (i,j) such that $1 \leqslant i < j \leqslant k$ and each (h,h') such that $1 \leqslant h, h' \leqslant 3$, $(x_i^h, x_j^{h'}) \in E$ if and only if $x_i^h \neq \overline{x}_j^{h'}$.

Graph G has a size polynomial in the size of the instance of 3-SAT as it contains $4k+1$ vertices (and thus fewer than $(4k+1)^2$ edges).

Figure 7.12 displays the graph built from the 3-SAT instance: $(x_1^1 \vee x_1^2 \vee x_1^3) \wedge (x_2^1 \vee x_2^2 \vee x_2^3) \wedge (x_3^1 \vee x_3^2 \vee x_3^3)$ where $x_1^1 = \overline{x}_2^2$, $x_1^3 = \overline{x}_3^3$, and $x_2^3 = \overline{x}_3^1$.

We now show that an instance $\mathcal{I}$ of 3-SAT has a solution if and only if the associated directed graph includes a transitive subchain of length at least $\left\lfloor \frac{|V|}{2} \right\rfloor$.

Let us assume that $\mathcal{I}$ has a solution. Therefore, each clause C_i, $1 \leqslant i \leqslant k$ has at least one literal whose value is true. We pick arbitrarily such a literal l_i for clause C_i for i in $[1;k]$. The set of true literals cannot contain a literal and its negation: For all $(i,j) \in [1;k]^2$, $l_i \neq \overline{l_j}$. Then, G contains a chain whose vertices are exactly those corresponding to clauses $\{C_i\}_{0 \leqslant i \leqslant k}$ or to the true literals $\{l_i\}_{1 \leqslant i \leqslant k}$. The subgraph induced by this chain is a transitive subchain. Furthermore, this chain contains $2k+1$ vertices. Hence, its length is $2k = \left\lfloor \frac{|V|}{2} \right\rfloor$ since $|V| = 4k+1$.

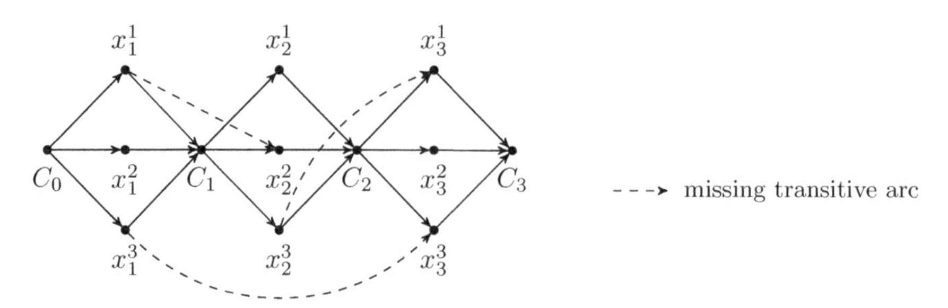

FIGURE 7.12: Graph built for the 3-SAT instance: $(x_1^1 \vee x_1^2 \vee x_1^3) \wedge (x_2^1 \vee x_2^2 \vee x_2^3) \wedge (x_3^1 \vee x_3^2 \vee x_3^3)$ where $x_1^1 = \overline{x}_2^2$, $x_1^3 = \overline{x}_3^3$, and $x_2^3 = \overline{x}_3^1$. Transitive arcs are not shown. The dashed arcs are *not* part of the graph.

Reciprocally, let us assume that G includes a transitive subchain of size $\left\lfloor \frac{|V|}{2} \right\rfloor$. Then, this subchain must include each vertex C_i, $0 \leqslant i \leqslant k$, and one literal l_j for each clause C_j for $1 \leqslant j \leqslant k$. Because of the transitivity, there cannot exist two indices i and j, $1 \leqslant i, j \leqslant k$ with $l_i = \overline{l_j}$. Therefore, the subchain defines an instantiation of the variables of $\mathcal{I}$ for which $\mathcal{I}$ is true.

Solution to Exercise 7.13: INDEPENDENT SET

An optimization problem is NP-complete if the associated decision problem is NP-complete. Here the decision problem would be, given an integer k, whether there exists an independent set of size k (this latter problem is the one called INDEPENDENT SET in the literature).

The certificate for this decision problem is the independent set, whose size is obviously polynomial in the size of the problem ($k \leqslant |V|$). To check that a certificate is valid, one has to consider all pairs of vertices in the candidate independent set and see whether the corresponding edge belongs to E. Therefore, a certificate is checked in a time quadratic in its size. Independent set is thus in NP.

To prove the completeness, we build a reduction from 3-SAT. Let $\mathcal{I}_1$ be an instance of 3-SAT: $\mathcal{I}_1 = C_1 \wedge \cdots \wedge C_k$, such that $C_i = a_{1,i} \vee a_{2,i} \vee a_{3,i}$ (where $a_{j,i} = x_j$ or $\overline{x}_j$, x_j being a variable). For each literal of each clause, we create a vertex labeled by the literal (therefore, we can have several vertices with the same label). For each clause C_i, we add an edge between any pair of its three corresponding vertices. We call the graph corresponding to a clause a widget. A widget consists of three vertices and three edges. Finally, we add an edge between a pair of opposite literals, that is, between any pair of vertices $(x_i, \overline{x_i})$. This construction is illustrated in Figure 7.13. This way, we build a graph G. This gives us an instance $\mathcal{I}_2$ of independent set once we have defined the size of the independent set we target. We chose k, the number of clauses in $\mathcal{I}_1$. The size of $\mathcal{I}_2$ is polynomial in the size of $\mathcal{I}_1$, and the reduction is thus polynomial.

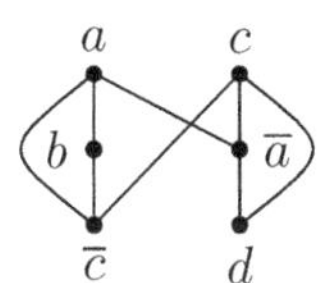

FIGURE 7.13: Reduction for the instance of 3-SAT: $(a \vee b \vee \overline{c}) \wedge (c \vee \overline{a} \vee d)$.

Let us first assume that there is a valid instantiation of $\mathcal{I}_1$. Therefore, each clause of $\mathcal{I}_1$ has at least one true literal. We define a subset S of the vertices of G as follows. For each clause C_i, for $i \in [1; k]$, we arbitrarily pick one true literal, and we put its associated vertex in S. There are exactly k vertices in S. As we take only one vertex per widget, the independence of the set cannot be violated by an edge included in a widget. The edges between widgets connect the literals which are negations of each other. Obviously, two such literals cannot simultaneously be true. Therefore, S is an independent set.

Reciprocally, let us assume that $\mathcal{I}_2$ has an independent set S of size k. As S is independent, it contains at most one vertex per widget, as each widget is a complete graph. As there are k widgets, S contains exactly one vertex per widget. Then we set to true each literal that corresponds to a vertex of S. This instantiation is well defined, as S cannot contain simultaneously the literals x_i and $\overline{x_i}$ for some $i \in [1; n]$ because their associated vertices are linked by an edge and cannot be both included in an independent set. Furthermore, this instantiation sets each clause of $\mathcal{I}_1$ to true because it contains one literal per clause of $\mathcal{I}_1$.

Instance $\mathcal{I}_1$ of 3-SAT has a solution if and only if the instance $\mathcal{I}_2$ of the decision problem associated with finding the maximum independent set has one. Therefore, this decision problem is an NP-complete problem, and INDEPENDENT SET is an NP-complete problem.

Other solution: The reduction also can be done from CLIQUE. Starting from a graph G, one takes the complementary graph, i.e., the graph having the same vertices, but in which there is an edge between any pair (u, v) of vertices if and only if there was not one in the original graph. Then, there is a clique of size k in a graph G if and only if there is an independent set of size k in its complementary graph.

Solution to Exercise 7.14: DOMINATING SET

A certificate is the dominating set D. To check whether a set D is indeed a dominating set, one must take each vertex v of $V \setminus D$ one by one and check whether there is a vertex u of D such that the graph includes the edge (u, v).

The cost of this verification is quadratic in the size of V and thus polynomial in the size of the problem. Therefore, this problem belongs to NP.

We craft a reduction from VERTEX-COVER. We start from an instance $\mathcal{I}_1 = (V, E, K)$ of VERTEX-COVER. From it we build an instance $\mathcal{I}_2$ of dominating set by adding to $\mathcal{I}_1$, for each (u, v) of E, a vertex uv and the two edges (u, uv) and (v, uv). Let $I \subset V$ be the set of isolated vertices of G, i.e., the set of the vertices that are not included in any edge. To complete the definition of $\mathcal{I}_2$, we let $K' = K + |I|$.

We now prove that instance $\mathcal{I}_1$ of VERTEX-COVER has a solution if and only if instance $\mathcal{I}_2$ of DOMINATING SET has one. This will prove that DOMINATING SET is an NP-complete problem.

We start by assuming that there is a solution for $\mathcal{I}_1$, i.e., a subset D of V of size at most K such that, for any edge (x, y) of E, $x \in D$ and/or $y \in D$. Then, $D \cup I$ is a dominating set for $\mathcal{I}_2$. First, we remark that $|D \cup I| \leqslant K + |I|$. Then, let u be any vertex of $V \setminus (D \cup I)$. By definition of I there exists at least one vertex $v \in V$ such that the edge (u, v) belongs to E. Then, as u does not belong to D and as D is a vertex cover of $\mathcal{I}_1 = (V, E)$, then, necessarily, v belongs to D.

Reciprocally, let D be a dominating set of $\mathcal{I}_2$ of size at most $K + |I|$. Any isolated vertex must be included in any dominating set (as such a vertex cannot be linked by an edge to a vertex included in the dominating set). Therefore, $I \subset D$, and $D \setminus I$ is of size at most K. Without loss of generality, we can assume that, whatever the edge $(u, v) \in E$, D does not contain the vertex uv. If this is not the case, we transform D into a dominating set of size at most $K + |I|$ that satisfies this property. Indeed, let us assume that D contains uv. If D also contains either u or v, we can just discard uv from D to obtain a new dominating set of size at most $K + |I|$ and not containing the vertex uv (remember that $\mathcal{I}_2$ includes both the edges (uv, u) and (uv, v); therefore, uv is linked to a vertex of the new dominating set). Otherwise, D contains neither u nor v. Then, we arbitrarily replace the vertex uv in D with the vertex u. This does not change the size of the set. The new set is still a dominating set as vertices v and uv are each linked by an edge to the vertex u of the new dominating set.

Without loss of generality, therefore, we now assume that D is a subset of V. To conclude, it remains to be shown that $D \setminus I$ is a vertex-cover of $\mathcal{I}_1$. To establish this result, let us consider any edge $(u, v) \in E$. Then, $\mathcal{I}_2$ includes the vertex uv. By hypothesis, uv does not belong to D. Therefore, uv must be linked to a vertex of D. As uv is included only in the two edges (uv, u) and (uv, v), then $D \setminus I$ must include u and/or v and $D \setminus I$ covers the edge (u, v) and is thus a cover.

Solution to Exercise 7.15: Carpenter

A certificate can be given as the set of hinges that are folded. For the instance drawn in Figure 7.2, the certificate would be $(2, 3)$, $(3, 4)$ to indicate that the hinge between the second and the third wooden sticks is folded, as the one between the third and the fourth stick. Then, to check that a certificate is valid, one has to compute the coordinate of the rightmost and leftmost ends of wooden sticks and compute the difference. This can be done in linear time; therefore, this problem belongs to NP.

To prove that this problem is NP-complete, we build a reduction from 2-PARTITION. Let $\mathcal{I}_1 = \{a_1, \dots, a_n\}$ be an instance of 2-PARTITION. We then build an instance $\mathcal{I}_2$ of the carpenter problem including $n+4$ sticks. $\mathcal{I}_2 = (S, S/2, a_1, \dots, a_n, S/2, S)$ where $S = \sum_{i=1}^{n} a_i$. We let $k = S$. The size of $\mathcal{I}_2$ is obviously polynomial in the size of $\mathcal{I}_1$.

Let us assume that $\mathcal{I}_1$ as a solution I: $\sum_{i \in I} a_i = \sum_{i \notin I} a_i$. Then, we fold instance $\mathcal{I}_2$ as follows. The first stick is from left to right, from abscisse 0 to S (without loss of generality). We then fold the hinge, and the second stick is from right to left, thus ending at $S/2$. Then, for any $i \in [1, n]$, if $i \in I$ the stick is laid from left to right, and right to left otherwise. The $(n+3)$-th stick is laid from left to right and the last one right to left. We focus on the sticks corresponding to integers in $\mathcal{I}_1$. As $\sum_{i \in I} a_i = \sum_{i \notin I} a_i$, the sum of the lengths of those laid from left to right is equal to the sum of the lengths of those laid from right to left. Therefore, the one before the last stick, of length $S/2$, is laid from $S/2$ to S, and the last one from S to 0. Furthermore, as $\sum_{i \in I} a_i = \sum_{i \notin I} a_i$, the sum of the sizes of any subset of I and of any subset of $[1; n] \setminus I$ is smaller than or equal to $S/2$. Therefore, no stick has an end whose abscisse is smaller than 0 or greater than S. We have built a valid solution for $\mathcal{I}_2$.

Let us now assume that $\mathcal{I}_2$ has a solution. Let I be the set of the indices of integers in $\mathcal{I}_1$ corresponding to sticks that are laid from left to right. Then, the difference of the abscisses of the second end of the second stick and the first end of the $(n+3)$-th stick is equal to $\sum_{i \in I} a_i - \sum_{i \notin I} a_i$. Without loss of generality, we can assume that the first stick is laid from left to right, from 0 to S. Then, the second stick must be laid from right to left, from S to $S/2$. The last stick, being of length S, must be laid between 0 and S, whatever its direction. If it is laid left to right, from 0 to S, the $(n+3)$-th stick is laid from $S/2$ to 0. If the last stick is laid from right to left, from S to 0, then the $(n+3)$-th stick must be laid from $S/2$ to S. In both cases, the first end of the $(n+3)$-th stick is in $S/2$. Therefore, the second stick ends in $S/2$ where the $(n+3)$-th stick starts. Therefore, $\sum_{i \in I} a_i = \sum_{i \notin I} a_i$, and we have a solution to $\mathcal{I}_1$.

Solution to Exercise 7.16: k-center

To define the decision problem associated with the minimization of a k center, we need a value c. Then, the decision problem is: Is there a k center S such that $center(S) \leqslant c$? The certificate for this problem is a set of size $k \leqslant |V|$ and is thus polynomial in the size of the problem. Checking that a candidate k-center satisfies the condition mandates at worst that all edges in the (complete) graph are scanned. This has a cost polynomial in the size of the graph. Therefore, this decision problem belongs to NP.

We craft a reduction from DOMINATING SET (see Exercise 7.14, p. 153).

Let $\mathcal{I}_1 = (G = (V, E), k)$ be an instance of DOMINATING SET. We build an instance $\mathcal{I}_2$ of k-center as follows. We build a complete graph $G' = (V, E')$. We define a function w that associates a weight with any edge $e \in E'$: $w(e) = 1$ if $e \in E$, and $w(e) = 2$ otherwise. As all the distances in G' are equal to either 1 or 2, the value of any k-center is then either 1 or 2. Therefore, we let $c = 1$. Instance $\mathcal{I}_2$ is then fully defined, and its size is polynomial in the size of $\mathcal{I}_1$. The reduction is thus polynomial. This construction is illustrated in Figure 7.14.

Let us assume we have a solution for instance $\mathcal{I}_1$ of DOMINATING SET, that is, a dominating set D of size k for $\mathcal{I}_1$. Then, we take D as the k-center. Indeed, by definition of a dominating set, for any vertex u in $V \setminus D$, there is a vertex $v \in D$ such that $(u, v) \in E$. Then, the edge (u, v) has a weight 1 in G'.

Reciprocally, let us assume that $\mathcal{I}_2$ admits a k-center S satisfying the property $center(S) \leqslant 1$. Then, let us consider any vertex u in $V \backslash S$. By definition of the k center, there exists a vertex $v \in S$ such that the edge (u, v) of G' has a weight of 1, which means that the edge (u, v) belongs to S. Therefore, in G, any vertex in $V \setminus S$ is connected to at least one vertex of S. Thus, S is a dominating set of size k of $\mathcal{I}_1$.

Instance $\mathcal{I}_1$ of DOMINATING SET has a solution if and only if instance $\mathcal{I}_2$ of the decision problem associated with k-center has one. Therefore, this decision problem is NP-complete, and k-center is an NP-complete problem.

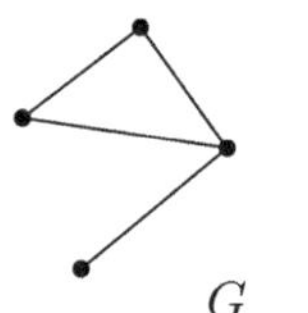

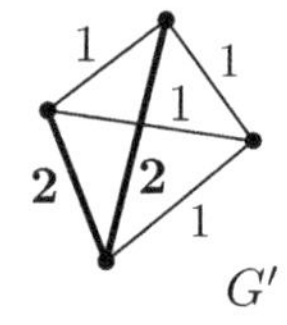

FIGURE 7.14: Example of a reduction from DOMINATING SET to k-center.

Solution to Exercise 7.17: Variants of 3-SAT

The two problems 3-SAT NAE and 3-SAT OIT belong to NP because 3-SAT does. For both problems, we craft a reduction from 3-SAT.

1. 3-SAT NAE

   We start from an instance $\mathcal{I}_1$ of 3-SAT. For any clause $C_i = a_i \vee b_i \vee c_i$, we include in $\mathcal{I}_2$ the two clauses $a_i \vee b_i \vee x_i$ and $c_i \vee \overline{x_i} \vee f$ where f is common to all the clauses created this way. $\mathcal{I}_2$ has twice as many clauses as $\mathcal{I}_1$, and its number of variables is increased by one (f) plus the number of clauses. Therefore, the size of $\mathcal{I}_2$ is polynomial in the size of $\mathcal{I}_1$.

We now show that the instance $\mathcal{I}_1$ of 3-SAT admits an instantiation if and only if the instance $\mathcal{I}_2$ of 3-SAT NAE does.

Let us assume that there exists an instantiation of the variables of instance $\mathcal{I}_1$ of 3-SAT such that all of its clauses are true. Then, for $\mathcal{I}_2$, we take the same values as for the variables of $\mathcal{I}_1$. We complete this assignment by having $x_i = \overline{a_i \vee b_i}$ and by setting f to FALSE. Indeed, if either a_i or b_i is true, then clause $a_i \vee b_i \vee x_i$ is true with x_i being false, and clause $c_i \vee \overline{x_i} \vee f$ is true because of $\overline{x_i}$ while f is false. If a_i and b_i are both false, then c_i is true because all clauses of $\mathcal{I}_1$ are true, including $a_i \vee b_i \vee c_i$. Then, clause $a_i \vee b_i \vee x_i$ is true because of x_i (its only true literal), and clause $c_i \vee \overline{x_i} \vee f$ is true because of c_i, while f is false. Therefore, in all cases, we have built a valid assignment of $\mathcal{I}_2$, i.e., one in which not all literals in a clause are true.

Reciprocally, let us assume that there is a valid instantiation of instance $\mathcal{I}_2$ of 3-SAT NAE. We have to consider two cases:

(a) f is false. We take for each variable of $\mathcal{I}_1$ its value in the instantiation of $\mathcal{I}_2$. Indeed, for the clause C_i either c_i is true, and then clause C_i is true, or c_i is false and, then, because f is also false, $\overline{x_i}$ must be true. Hence, x_i is false, and because $a_i \vee b_i \vee x_i$ is true, then either a_i or b_i is and thus also C_i.

(b) f is true. We take for each variable of $\mathcal{I}_1$ the *negation* of its value in the instantiation of $\mathcal{I}_2$. Indeed because, by definition, not all variables can be equal in a clause, the negation of a valid instantiation is also a valid instantiation. And the previous case shows that taking for the variables of $\mathcal{I}_1$ their values from an instantiation of $\mathcal{I}_2$ where f is false defines a valid solution.

2. 3-SAT OIT

We start from an instance $\mathcal{I}_1$ of 3-SAT. For any clause $C_i = a_i \vee b_i \vee c_i$, we include in $\mathcal{I}_2$ the three clauses $a_i \vee x_i \vee y_i$, $\overline{b_i} \vee x_i \vee x'_i$, and $\overline{c_i} \vee y_i \vee y'_i$. $\mathcal{I}_2$ contains three times as many clauses as $\mathcal{I}_1$. Furthermore, its number of variables is equal to that of $\mathcal{I}_1$ plus four times the number of clauses. Therefore, the size of $\mathcal{I}_2$ is polynomial in the size of $\mathcal{I}_1$.

We now show that the instance $\mathcal{I}_1$ of 3-SAT admits an instantiation if and only if the instance $\mathcal{I}_2$ of 3-SAT OIT does.

We first assume that there exists an instantiation of the instance $\mathcal{I}_1$ of 3-SAT. We take for the variables of $\mathcal{I}_2$ that are variables of $\mathcal{I}_1$ their value in the instantiation of $\mathcal{I}_1$. We then consider any clause $C_i = a_i \vee b_i \vee c_i$ of $\mathcal{I}_1$. Then, if a_i is true, we let $x_i = y_i =$ FALSE, $x'_i = b_i$, and $y'_i = c_i$. Otherwise, a_i is false. Then, b_i and/or c_i is true because we start from a valid instantiation of $\mathcal{I}_1$, i.e., one for which each clause is true. Without loss of generality (as our construction is symmetric for b_i and c_i), let us assume that b_i is true. Then, we let $x'_i = y_i = \text{FALSE}$, $x_i = \text{TRUE}$, and $y'_i = c_i$. One can then check that in all cases, we have exactly one true literal per clause of $\mathcal{I}_2$. We, thus, have defined a valid instantiation of $\mathcal{I}_2$.

Reciprocally, let us assume that there exists an instantiation of instance $\mathcal{I}_2$ of 3-SAT OIT. Then, to obtain a valid instantiation of $\mathcal{I}_1$, we have only to take for the values of its variables the values of the corresponding variables in $\mathcal{I}_2$. Indeed, if, under this instantiation, there were a clause $C_i = a_i \vee b_i \vee c_i$ that was false, then a_i, b_i, and c_i would be false. Then, because of the clauses $\overline{b_i} \vee x_i \vee x'_i$ and $\overline{c_i} \vee y_i \vee y'_i$, we would have $x_i = y_i = \text{FALSE}$, as exactly one literal per clause is true under the instantiation of $\mathcal{I}_2$. Then, the clause $a_i \vee x_i \vee y_i$ of $\mathcal{I}_2$ would be false, which would contradict the hypothesis that we are dealing with a valid instantiation of $\mathcal{I}_2$.

Solution to Exercise 7.18: Variants of SAT

1. Whatever the value N, the decision problem SAT-N belongs to the class NP because SAT does.

   We start from an instance $\mathcal{I}_1$ of SAT. To each variable x_i appearing k times in $\mathcal{I}_1$, with $k > 3$, we apply the following transformation. First, we replace the j-th occurrence of the variable x_i with the variable $y_{i,j}$. Then, we add to the instance the k clauses:

$$y_{i,1} \vee \overline{y_{i,2}}, \quad y_{i,2} \vee \overline{y_{i,3}}, \quad \ldots, \quad y_{i,k-1} \vee \overline{y_{i,k}}, \quad y_{i,k} \vee \overline{y_{i,1}}.$$

   Each of the new variables $y_{i,j}$ appears exactly three times: once replacing an occurrence of x_i, and the two other times in the new clauses (as $y_{i,j}$ and $\overline{y_{i,j}}$). Furthermore, all these variables have the same value because of the new clauses. Indeed, if $y_{i,j}$ is true, then $y_{i,j-1}$ must also be true (or $y_{i,k}$ if $j = 1$). Reciprocally, if $y_{i,j}$ is false, then $y_{i,j+1}$ must also be false (or $y_{i,1}$ if $j = k$). As the set of new clauses corresponding to variable x_i forms a cycle of constraints, all the variables $y_{i,j}$ have the same value in a valid instantiation of the new formula. Hence, the two formulas are equivalent.

   Finally, if $\mathcal{I}_1$ contains n variables and m clauses, the new instance contains at most nm variables and $(m + nm)$ clauses. Therefore, any instance of SAT can be transformed in polynomial time in an equivalent instance of instance SAT-3. Therefore, SAT-3 is an NP-complete problem.

   As, for any value $N \geqslant 3$, SAT-3 is a particular case of SAT-N, then SAT-N is an NP-complete problem.

2. Let x be a variable in a formula F of SAT-2. Depending on the use of x, we transform F so that the new formula does not contain the variable x and such that there is a valid instantiation of the new formula if and only if there was one for the original formula.

   - If x (or $\overline{x}$) does appear only once, we just discard the clause in which this variable appears. Indeed, we are free to instantiate x (or $\overline{x}$) to TRUE, therefore setting the clause to TRUE.
   - If x (or $\overline{x}$) appears twice in a same clause, or in two distinct clauses, we discard the clause or clauses including x for the same reason.
   - If x and $\overline{x}$ appear in the same clause, this clause is true whatever the instantiation, and we just discard it.
   - If x and $\overline{x}$ appear in two distinct clauses, $x \vee C_1$ and $\overline{x} \vee C_2$, then the formula has a valid instantiation if and only if at least one of the two clauses C_1 and C_2 can be instantiated to TRUE while all the other clauses can be instantiated to TRUE. (If only one of these two clauses C_1 and C_2 is true, we pick the value of x so that both $x \vee C_1$ and $\overline{x} \vee C_2$ are true.)
     - If $C_1 = C_2 = \emptyset$, the original formula does not have any valid instantiation.
     - Otherwise, we remove the two clauses and replace them with the clause $C_1 \vee C_2$.

   This way, we define an algorithm in n steps, one per variable. At each step, we decide the value of a variable for a cost of m, where m is the initial number of clauses. We, therefore, can solve SAT-2 in time $O(nm)$.

Solution to Exercise 7.19: SUBSET-SUM

The certificate for the SUBSET-SUM problem is the set S', and checking the value of its sum can obviously be done in a time polynomial in the size of the problem. Hence, SUBSET-SUM belongs to NP.

It is easy to show that SUBSET-SUM is NP-complete by using a reduction from 2-PARTITION. Indeed, to do so, one just has to take for t half of the sum of the values of the elements in the instance of 2-PARTITION. However, SUBSET-SUM is actually used to prove the NP-completeness of 2-PARTITION in the next exercise, so we have to assume that we do not know yet whether 2-PARTITION is NP-complete.

We thus start from an instance $\mathcal{I}_1$ of 3-SAT and build an instance $\mathcal{I}_2$ of SUBSET-SUM as proposed in the hint. In order to fully understand the proposed construction, we consider all the v_is, v_i's, s_js, and s_j's as integers with $n + m$ (decimal) digits.

- For any $i \in [0; n-1]$, the n most significant digits of these integers are used to identify variables. These digits are all null for v_i and v_i' except for the digit standing for variable i, namely, the $(1+m+i)$ digit that is equal to 1. This identifies v_i and v'_i as integers representing the variable i (they respectively represent the literals x_i and $\overline{x_i}$).

- For any $i \in [0; m-1]$, the m least significant digits of these integers are used to identify clauses. These digits are all zero for v_i except for the digits corresponding to clauses in which the literal x_i appears. In other words, if the literal x_i appears in clause C_j, then the $1+j$-th digit of v_i is equal to 1. The same is true for v'_i and the literal $\overline{x_i}$.

As we want to be able to build a valid instantiation of $\mathcal{I}_1$ from the solution of $\mathcal{I}_2$, we want in the solution of SUBSET-SUM exactly one of the two integers corresponding to a variable x_i, i.e., exactly one integer between v_i and v'_i, is taken. To ensure this, we want the n most significant digits of t to be equal to 1.

Suppose we have an instantiation of $\mathcal{I}_1$. Let us consider the integer T obtained by taking, for any $i \in [1; n]$, the integer v_i if x_i is true in this instantiation and by taking the integer v'_i otherwise. Formally, $T = \sum_{k=0}^{n-1}[x_k v_k + (1-x_k)v'_k]$ (by assuming that TRUE is encoded by the value 1 and FALSE by the value 0). The n most significant digits of T are equal to 1. Let us focus on one of the m least significant digits of T, say, the digit of rank j (the least significant digit is said to be of rank 0). This digit corresponds to clause C_j. Its value is equal to the number of literals in C_j that are true in the instantiation considered. So, this value is between 1 and 3 in a solution to the instance of 3-SAT considered. We want to avoid the cases where one of these digits is zero. We cannot, however, impose that all of these digits are equal to one, because such a property is not necessarily satisfied in any valid instantiation of any instance of 3-SAT. This is where the integers s_j and s'_j, for $j \in [0; m]$, come into play; we use them to complement the value obtained for the digit of rank j of T. We take them so as to reach a target value, but it is a target that cannot be reached if the digit is originally zero. Taking s_j and/or s'_j or none of them enables one to add a value of 0, 1, 2, and 3. Hence, 4 is the lowest value that is not reachable if the digit of T is zero. Furthermore, if the value of the digit of T is:

- 3, taking s_j enables to reach 4 (=3+1);
- 2, taking s'_j enables to reach 4 (=2+2);
- 1, taking both s_j and s'_j enables to reach 4 (=1+1+2).

Therefore, we take for t the integer whose n most significant digits are equal to 1 and the m least significant ones are equal to 4:

$$t = \underbrace{11\cdots11}_{n}\ \underbrace{44\cdots44}_{m}.$$

So, from the instance $\mathcal{I}_1$ of 3-SAT, we build the instance $\mathcal{I}_2$ of SUBSET-SUM where $S = \{v_i | 0 \leqslant i \leqslant n-1\} \cup \{v'_i | 0 \leqslant i \leqslant n-1\} \cup \{s_j | 0 \leqslant j \leqslant m-1\} \cup \{s_j | 0 \leqslant j \leqslant m-1\}$, and $t = \underbrace{11\cdots11}_{n}\ \underbrace{44\cdots44}_{m}$.

From what precedes, if there is a solution of the instance $\mathcal{I}_1$ of 3-SAT, then we build a solution of the instance $\mathcal{I}_2$ of SUBSET-SUM by putting into the set S', for any $i \in [0; n-1]$, the integer v_i if the variable x_i is true, and the integer v'_i otherwise. We complete S' by taking, for any $j \in [0; m-1]$, the integer s_j if the clause C_j has exactly three true literals; the integer s'_j if it has two true literals; and both integers s_j and s'_j otherwise.

Reciprocally, assume that there exists a subset S' of S such that $\sum_{x\in S'} x = t$. Then, for any $i \in [0; n-1]$, S' contains exactly one of the two integers v_i and v'_i as the n most significant digits of t are equal to 1. Then, we set x_i to TRUE if v_i belongs to S' and to FALSE otherwise. We need to prove that under this instantiation any clause C_j, for $j \in [0; m-1]$ is set to TRUE. The digit of rank j of t is equal to 4. As we have seen, this implies that S' contains at least one integer whose digit of rank j is equal to 1. Then, the corresponding literal is set to TRUE; it is included by definition in the clause C_j and thus sets C_j to TRUE.

Therefore, the instance $\mathcal{I}_1$ of 3-SAT has a solution if and only if instance $\mathcal{I}_2$ of SUBSET-SUM has one, and SUBSET-SUM is an NP-complete problem.

Solution to Exercise 7.20: NP-completeness of 2-PARTITION

We already know that 2-PARTITION is in NP. To prove the completeness, we craft a reduction from SUBSET-SUM.

Let $\mathcal{I}_1$ be an instance of SUBSET-SUM. Given a finite set $S = \{a_1, \ldots, a_n\}$ of positive integers, and an integer t, is there a subset S' of S such that $\sum_{x\in S'} x = t$? We build an instance of 2-PARTITION as follows.

- If $t \leqslant \sum_{i=1}^{n} a_i < 2t$, let $a_{n+1} = 2t - \sum_{i=1}^{n} a_i$. We build an instance $\mathcal{I}_2$ of 2-PARTITION with $a_1, \ldots, a_n, a_{n+1}$. Therefore, $\sum_{i=1}^{n+1} a_i = 2t$. If $\mathcal{I}_1$ has a solution S', we have $\sum_{x\in S'} x = t$, and S' is also a solution for instance $\mathcal{I}_2$, since $\frac{1}{2}\sum_{i=1}^{n+1} a_i = t$. Let us now assume that $\mathcal{I}_2$ has a solution I. Then, either $n+1 \in I$, or $n+1 \in \{1, \ldots, n+1\} \setminus I$. The set S' that does not contain a_{n+1} is a solution to $\mathcal{I}_1$.

- If $2t \leqslant \sum_{i=1}^{n} a_i$, let $a_{n+1} = \sum_{i=1}^{n} a_i - 2t$, and we build $\mathcal{I}_2$ as before. We have $\sum_{i=1}^{n+1} a_i = 2\sum_{i=1}^{n} a_i - 2t$. As before, solutions are equivalent to $\sum_{i\in I} a_i = \sum_{i\in\{1,\ldots,n+1\}\setminus I} a_i = \sum_{i=1}^{n} a_i - t$. Let J be the subset that does not contain a_{n+1}. Then, we have the equivalence with $S' = \{a_i\}_{i\in\{1,\ldots,n+1\}\setminus J}$, since $\sum_{x\in S'} x = \sum_{i=1}^{n} a_i - \sum_{i\in J} a_i = t$.

Therefore, 2-PARTITION is NP-complete.

7.7 Bibliographical notes

Many exercises in this chapter are quite classical and can be found in several textbooks. Exercises 7.7, 7.8, and 7.9 (graph coloring) and Exercise 7.15 (carpenter) come from the book by Kozen [65].

Chapter 8

Beyond NP-completeness

At the conclusion of Chapter 6, we stated that proving a problem is NP-complete does not make it go away. The subject of this chapter is to go beyond NP-completeness and to describe the various approaches that can be taken when confronted with an NP-complete problem.

The first approach (see Section 8.1) is the most elegant. When deriving *approximation algorithms*, we search for an approximate solution, but we also guarantee that it is of good quality. Of course, the approximated solution must be found in polynomial time.

The second approach (see Section 8.2) is less ambitious. Given an NP-complete problem, we show how to characterize particular instances that have polynomial complexity.

The third approach (see Section 8.3) often provides useful lower bounds. The idea is to cast the optimization problem under study in terms of a *linear program*. While solving a linear program with integer variables is NP-complete, solving a linear program with rational variables has polynomial complexity (we are restricted to rational variables because of the impossibility of efficiently encoding real numbers). The difficulty is then to reconstruct a solution of the integer linear program from an optimal solution of that program with rational variables. This is not always possible, but this method at least provides a lower bound on any optimal integer solution.

We briefly introduce, in Section 8.4, *randomized algorithms* as a fourth approach that solves "most" instances of an NP-complete problem in polynomial time.

Finally, we provide in Section 8.5 a detailed discussion of *branch-and-bound* and *backtracking* strategies, where one explores the space of all potential solutions in a clever way. While the worst-case exploration may require exponential time, on average, the optimal solution is found in "reasonable" time.

8.1 Approximation results

In this section, we first define polynomial-time approximation algorithms and (fully) polynomial-time approximation schemes (PTAS and FPTAS). Then, we give some examples of approximation and inapproximability results.

8.1.1 Approximation algorithms

In Chapter 6, we have defined the NP-completeness of problems and exhibited several NP-complete decision problems. As discussed in Section 6.3.4, the target problem is often an optimization problem that has been restricted to a decision problem so that we can prove its NP-completeness.

If the optimal solution of an optimization problem cannot be found in polynomial time, one may want to find an approximate solution in polynomial time.

DEFINITION 8.1. A λ-approximation algorithm is an algorithm whose execution time is polynomial in the instance size and that returns an approximate solution guaranteed to be, in the worst case, at a factor λ away from the optimal solution.

For instance, for each instance $\mathcal{I}$ of a minimization problem, the solution of the approximation algorithm for instance $\mathcal{I}$ must be smaller than or equal to λ times the optimal solution for instance $\mathcal{I}$.

The closer λ to 1, the better the approximation algorithm. We categorize some particular approximation algorithms for which λ is close to 1 as polynomial-time approximation schemes.

DEFINITION 8.2. A *Polynomial-Time Approximation Scheme* (PTAS) is such that for any constant $\lambda = 1+\varepsilon > 1$, there exists a λ-approximation algorithm, i.e., an algorithm that is polynomial in the instance size and guaranteed at a factor λ.

Note that the algorithm may not be polynomial in $1/\varepsilon$ and thus have a high complexity when ε gets close to zero. A Fully PTAS is such that the algorithm is polynomial both in the instance size and in $1/\varepsilon$.

DEFINITION 8.3. A *Fully Polynomial-Time Approximation Scheme*, or FPTAS, is such that for any constant $\lambda = 1 + \varepsilon > 1$, there exists a λ-approximation algorithm that is polynomial in the instance size *and* in $1/\varepsilon$.

The difference between PTAS and FPTAS is simply that the $\forall\varepsilon$ quantifier changes sides. For a PTAS, ε is a fixed constant, so that $2^{\frac{1}{\varepsilon}}$ is a constant as well. On the contrary, the complexity of an FPTAS scheme must be polynomial in $\frac{1}{\varepsilon}$. Of course, having an FPTAS is a stronger property than having a PTAS (i.e., FPTAS $\Rightarrow$ PTAS).

Finally, we define asymptotic PTAS and FPTAS, which add a constant to the approximation scheme. We define formally only the APTAS for a minimization problem, and the definition can easily be extended for maximization problems and AFPTAS.

DEFINITION 8.4. An *Asymptotic Polynomial-Time Approximation Scheme*, or APTAS, is such that for any constant $\lambda = 1 + \varepsilon > 1$, there exists an algorithm, polynomial in the instance size, such that $C_{APTAS} \leqslant \lambda C_{opt} + \beta$ (for a minimization problem), where C_{APTAS} is the cost of the solution of the algorithm, C_{opt} is the cost of an optimal solution, and β is a constant that may depend on ε but should be independent of the problem size.

In the following, we discuss several approximation algorithms, and we show how to prove that an algorithm is an approximation algorithm (possibly an (A)PTAS or (A)FPTAS) or how to prove that a problem cannot be approximated in polynomial time up to any fixed constant λ.

8.1.2 Vertex cover

We consider here the classical vertex cover problem, which was shown to be NP-complete in Section 6.4.3. We discuss a weighted version of this problem in Section 8.3.

We first recall the definition of the vertex cover problem in its optimization problem formulation. Given a graph $G = (V, E)$, we want to find a set of vertices of minimum size that is covering all edges (i.e., any edge in E includes at least one of the vertices of the set).

We consider the following greedy algorithm to solve the problem, called **greedy-vc**. Initialize $S = \emptyset$. Then, while some edges are not covered (i.e., neither of their end vertices are in set S), pick one edge $e = (u, v)$, add both vertices u and v to set S, and mark all edges including u or v as covered. It is clear that **greedy-vc** returns a valid vertex cover, and that it is polynomial in the size of the instance. We now prove that it is a 2-approximation algorithm for the vertex cover optimization problem.

THEOREM 8.1. **Greedy-vc** *is a 2-approximation algorithm for vertex cover.*

Proof. Let A be the set of edges selected by the greedy algorithm. Two edges of A cannot have a common vertex, and, therefore, the size of the cover of this algorithm is $C_{\textbf{greedy-vc}} = 2|A|$. However, all edges selected greedily are independent, and each of them must be covered in any solution; hence, an optimal solution has at least $|A|$ vertices: $C_{opt} \geqslant |A|$. We deduce that $C_{\textbf{greedy-vc}} \leqslant 2 \times C_{opt}$, which concludes the proof. □

Note that this approximation factor of 2 is achieved, for instance, if G consists of two vertices joined by an edge. There is a polynomial-time algorithm that is a $2 - \frac{\log(\log(|V|))}{2\log(|V|)}$ approximation [79], but, for instance, we do not know any polynomial-time algorithms that would be a 1.99 approximation (the problem is still open).

8.1.3 Traveling salesman problem (TSP)

Let $G = (V, E)$ be a complete graph and $w : E \to \mathbb{N}$ be a cost function. The TSP problem consists of finding a cycle $\mathcal{C}$ going through each vertex once and only once, with $\sum_{e \in \mathcal{C}} w(e) \leqslant k$. The decision problem, in which k is a fixed integer, is NP-complete, as mentioned in Section 6.4.5. For the optimization problem, the goal is to minimize k.

First, we prove that TSP cannot be approximated unless P = NP. Then, we propose an approximation algorithm in the particular case where the cost function follows the triangle inequality.

Inapproximability of TSP

THEOREM 8.2. *For any constant $\lambda \geqslant 1$, there does not exist any λ-approximation algorithm for TSP unless P = NP.*

To prove such a result, the methodology is often as follows. The idea consists of assuming that there is a λ-approximation algorithm for the target problem (by definition, this is a polynomial-time algorithm). Then, one uses this approximation algorithm to solve in polynomial time a problem that is known to be NP-complete. For TSP, we show how any instance of problem Hamiltonian Cycle (HC, see Definition 6.4) can be solved in polynomial time using any approximation algorithm for TSP.

Proof. Let us assume that there is a λ-approximation algorithm for TSP. We consider an instance $\mathcal{I}_{hc}$ of HC, which is a graph $G = (V, E)$, with $n = |V|$. Then, we build an instance $\mathcal{I}_{tsp}$ of TSP as follows. In the complete graph, we build a cost function such that $w(e) = 1$ if $e \in E$, and $w(e) = \lambda n + 1$ otherwise. The size of $\mathcal{I}_{tsp}$ is obviously polynomial in the size of $\mathcal{I}_{hc}$.

We use the λ-approximation algorithm to solve $\mathcal{I}_{tsp}$. Let C_{algo} be its solution. This solution is such that $C_{algo} \leqslant \lambda C_{opt}$, where C_{opt} is the optimal solution.

We consider the two following cases:

- If $C_{algo} \geqslant \lambda n + 1$, then $C_{opt} > n$. This means that instance $\mathcal{I}_{hc}$ has no solution. Indeed, a Hamiltonian Cycle for $\mathcal{I}_{hc}$ would be a solution of cost n for $\mathcal{I}_{tsp}$.

- Otherwise, $C_{algo} < \lambda n + 1$, and therefore the solution of $\mathcal{I}_{tsp}$ is not using any edge not in E (otherwise, the cost would be at least $\lambda n + 1$). This solution is therefore a Hamiltonian Cycle for $\mathcal{I}_{hc}$, which means that instance $\mathcal{I}_{hp}$ has a solution.

Therefore, the result of the algorithm for $\mathcal{I}_{tsp}$ allows us to conclude whether there is a Hamiltonian Cycle in $\mathcal{I}_{hc}$, which concludes the proof. □

Note that we assumed that λ is constant, but we can even have $\lambda = 1 + 2^{-n}$, since the algorithm would still be polynomial in the instance size (λ can be encoded in logarithmic size, hence in $O(n)$). However, Theorem 8.2 does not forbid the existence of a $2^{2^{-n}}$-approximation algorithm.

Approximation algorithm with triangle inequality

We now assume that the cost function w satisfies the triangle inequality, i.e., for all vertices $v_1, v_2, v_3 \in V$, $w(v_1, v_3) \leqslant w(v_1, v_2) + w(v_2, v_3)$.

The approximation algorithm **spanning-tsp** works as follows. First, we build a minimum spanning tree T of the graph G, which can be done in polynomial time with a greedy algorithm (remove edges by nonincreasing costs while keeping a connected graph, see Section 3.4). Then, we perform a tree traversal of T (once a node u is visited, one completely visits the subtree rooted at one of the children of u before starting to visit any subtree rooted at another child). Each edge of T is visited exactly twice. We extract a solution for TSP, i.e., a Hamiltonian Cycle, by recording the order in which vertices are visited for the first time. From this ordered list of vertices, we build a cycle by taking the edges that link consecutive vertices (recall that the graph is complete).

We now prove that this algorithm is a 2-approximation.

THEOREM 8.3. Spanning-tsp *is a 2-approximation algorithm for the traveling salesman problem with the triangle inequality.*

Proof. The optimal cost C_{opt} is at least equal to the sum of the costs of the edges in the minimum spanning tree T, denoted by $w(T)$. Indeed, an optimal solution is a cycle. If we remove an edge from an optimal solution, we obtain a spanning tree, and T is a spanning tree of minimum weight. Therefore, $C_{opt} \geqslant w(T)$.

Now, we consider the cost of the solution returned by the algorithm. We denote this solution by S and its cost by $C_{\textbf{spanning-tsp}}$. Let O be the order in which the vertices are visited in the traversal of T. Vertices that are not leaves of T appear several times in O. S is obtained from O by keeping only the first occurrence of each vertex. Because of the triangular inequality, deleting a vertex from O does not increase the cost of the associated path. (Suppose we delete the vertex y in the sequence (x, y, z) of O; this is equivalent to replacing the two edges (x, y) and (y, z) with the single edge (x, z).) Hence, $C_{\textbf{spanning-tsp}}$ is less than or equal to the cost of the path associated with O. Furthermore, the path associated with O contains each edge exactly twice, and its cost is exactly $2 \times w(T)$. Therefore, $C_{\textbf{spanning-tsp}} \leqslant 2C_{opt}$, which proves the approximation result. □

8.1.4 Bin packing

In this section, we introduce a new classical problem that is the bin packing problem.

DEFINITION 8.5 (BP – Bin Packing). Given n rational numbers (also called objects) $a_1, \ldots, a_n$, with $0 < a_i \leqslant 1$, for $1 \leqslant i \leqslant n$, can we partition them in k bins $B_1, \ldots, B_k$ of capacity 1, i.e., for each $1 \leqslant j \leqslant k$, $\sum_{i \in B_j} a_i \leqslant 1$?

First, we prove the NP-completeness of this problem, then we exhibit several approximation results.

NP-completeness of BP

THEOREM 8.4. *BP is NP-complete.*

Proof. It is straightforward to see that BP is in NP: A certificate is the list, for each bin, of the indices of the numbers it contains.

The reduction comes from 2-PARTITION. We consider an instance $\mathcal{I}_1$ of 2-PARTITION, with n integers $b_1, \ldots, b_n$. We build the following instance $\mathcal{I}_2$ of BP: For $1 \leqslant i \leqslant n$, $a_i = \frac{2b_i}{S}$, with $S = \sum_{i=1}^{n} b_i$, and we set $k = 2$.

It is then straightforward to see that the size of the new instance is polynomial and to check the equivalence of solutions. □

Inapproximability of BP

THEOREM 8.5. *For all $\varepsilon > 0$, there does not exist any $(\frac{3}{2} - \varepsilon)$-approximation algorithm for BP unless P = NP.*

Proof. Let us assume that there is a $(\frac{3}{2} - \varepsilon)$-approximation algorithm for BP. We then exhibit a polynomial algorithm to solve 2-PARTITION.

Given an instance of 2-PARTITION, we execute the algorithm for BP with the a_i as defined earlier. If there exists a 2-PARTITION of the b_i, the algorithm returns at most $2 \times (\frac{3}{2} - \varepsilon) = 3 - 2\varepsilon$ bins, so it returns two bins. Otherwise, the algorithm returns a solution with at least three bins. Thanks to the polynomial approximation algorithm, we can solve 2-PARTITION in polynomial time, which implies that P = NP. This concludes the proof. □

Approximation algorithms for BP

We start with a simple greedy algorithm in which we select objects in a random order, and, at each step, we place the object either in the last used bin where it fits (**next-fit** algorithm) or in the first used bin where it fits (**first-fit** algorithm); otherwise (i.e., the object is not fitting in any used bin), we create a new bin and place the object in this new bin. We prove below that **next-fit** (and, hence, **first-fit**) is a 2-approximation algorithm for the BP problem.

THEOREM 8.6. Next-fit *is a 2-approximation algorithm for BP.*

Proof. Let $A = \sum_{i=1}^{n} a_i$. We have a lower bound on the cost of the optimal solution (the number of bins used by the optimal solution): $C_{opt} \geqslant \lceil A \rceil$.

Now we bound the cost of **next-fit** as follows. If we consider two consecutive bins, the sum of the objects that they contain is strictly greater than 1;

otherwise, we would not have created a new bin. Therefore, if $C_{\textbf{next-fit}} = K$, and B_k is the k-th bin of the solution returned by **next-fit**, for $1 \leqslant k \leqslant K$, then by summing the contents of two consecutive bins, we get

$$\sum_{k=1}^{K-1} \left(\sum_{i \in B_k} a_i + \sum_{i \in B_{k+1}} a_i \right) > K - 1 \,.$$

Moreover, by definition of A, we have $\sum_{k=1}^{K-1} \left(\sum_{i \in B_k} a_i + \sum_{i \in B_{k+1}} a_i \right) \leqslant 2A$, and, therefore, $K - 1 < 2A \leqslant 2\lceil A \rceil$. Finally, $C_{\textbf{next-fit}} = K \leqslant 2\lceil A \rceil \leqslant 2C_{opt}$, which concludes the proof. □

Note that the approximation ratio is tight for the **next-fit** algorithm. Consider an instance of BP with $4n$ objects such that $a_{2i-1} = \frac{1}{2}$ and $a_{2i} = \frac{1}{2n}$, for $1 \leqslant i \leqslant 2n$. Then, if **next-fit** chooses the objects in the sequential order, its solution uses $2n$ bins (one object a_{2i-1} and one object a_{2i} in each bin), while the optimal solution uses only $n + 1$ bins (for $1 \leqslant i \leqslant 2n$, the $2n$ objects a_{2i} in one bin and two objects a_{2i-1} in each of the other n bins).

The previous algorithms can be qualified as *online* algorithms because no sorting is done on the objects, and we can pack them in the bins when they arrive, on the fly. If we have the knowledge of all objects before executing the algorithm, we can refine the algorithm by sorting the objects beforehand. Such algorithms are called *offline* algorithms. The **first-fit-dec** algorithm sorts the objects by nonincreasing size (**dec** stands for decreasing), and then it applies the **first-fit** rule: The object is placed in the first used bin in which it fits; otherwise, a new bin is created.

THEOREM 8.7. *$C_{\textbf{first-fit-dec}} \leqslant \frac{3}{2}C_{opt} + 1$, where $C_{\textbf{first-fit-dec}}$ is the cost returned by the* **first-fit-dec** *algorithm, and C_{opt} is the optimal cost.*

Note that this is not an approximation algorithm as defined above because of the "+1" in the expression, which corresponds to one extra bin that the **first-fit-dec** algorithm may use. This is rather an *asymptotic* approximation algorithm, which is similar to an A(F)PTAS scheme. Indeed, the constant 1 is independent of the problem size, and the algorithm is asymptotically a $\frac{3}{2}$-approximation.

Proof. We split the a_i in four categories:

$$A = \left\{ a_i > \frac{2}{3} \right\} \quad B = \left\{ \frac{2}{3} \geq a_i > \frac{1}{2} \right\} \quad C = \left\{ \frac{1}{2} \geq a_i > \frac{1}{3} \right\} \quad D = \left\{ \frac{1}{3} \geq a_i \right\}$$

Case 1: There is at least one bin containing only objects of category D in the solution of **first-fit-dec**. In this case, at most one bin (the last one) has a sum of objects of less than $\frac{2}{3}$, and it contains only objects of category D. Indeed, if the objects of D of the last bin have not fit in the previous bins, it means that each bin (except the last one) has a sum of objects of at least $\frac{2}{3}$.

Therefore, if we ignore the last bin, $C_{opt} \geqslant \sum_{i=1}^{n} a_i \geqslant \frac{2}{3}(C_{\textbf{first-fit-dec}} - 1)$, which concludes the proof for this case.

Case 2: There is no bin with only objects of category D. In this case, we can ignore the objects of category D because they are added into the bins at the end of the algorithm, and they do not lead to the creation of new bins. We now prove that the solution of **first-fit-dec** for the objects of A, B, and C is optimal. Indeed, in any solution, objects of A are alone in a bin, and there are at most two objects of B and C in a bin, with at most one object of B. The **first-fit-dec** algorithm is placing first each object A and B in a separate bin, then it does the best matching of objects C, because they are placed in the bins by decreasing order. In this case, **first-fit-dec** is optimal. □

Note that the reasoning does not hold if the categories are made differently, with, for instance, $\frac{1}{4}$ instead of $\frac{1}{3}$. Indeed, we can then fit three objects of category C in a single bin, and the reasoning does not hold anymore. However, we point out that it is also possible to prove that $C_{\textbf{first-fit-dec}} \leqslant \frac{11}{9} C_{opt} + 1$, and we refer to [112] for further details. The idea of the proof is similar, but more categories of objects are considered, and the algorithm turns out to be much more complex.

Without allowing an extra bin, we can finally prove that **first-fit-dec** is a $\frac{3}{2}$-approximation algorithm.

THEOREM 8.8. **First-fit-dec** *is a* $\frac{3}{2}$*-approximation algorithm for the bin packing problem.*

Proof. Let $k = C_{\textbf{first-fit-dec}}$ be the cost returned by the **first-fit-dec** algorithm, and let $j = \lceil \frac{2}{3}k \rceil$. Bins are numbered from 1 to k, and we consider two cases.

Case 1: If bin j contains an object a_i such that $a_i > \frac{1}{2}$, then if $j' < j$, there is an object $a_{i'}$ in bin j' such that $a_{i'} \geqslant a_i > \frac{1}{2}$. This is true for $1 \leqslant j' < j$, and, therefore, there are at least j objects of size greater than $\frac{1}{2}$ that should be placed in distinct bins. This implies that the optimal cost C_{opt} is greater than j.

Case 2: None of the bins $j' \geqslant j$ contains any object of size strictly greater than $\frac{1}{2}$; there are at least two objects per bin, except for bin k that may contain only one object, hence $2(k-j)+1$ objects in bins $j, j+1, \ldots, k$. None of these objects fits into bins $1, 2, \ldots, j-1$, by definition of **first-fit-dec**. We show below that $2(k-j)+1 \geqslant j-1$, and by combining $j-1$ of these objects with each of the first $j-1$ bins, we obtain that the sum of the a_is is strictly greater than $j-1$, i.e., C_{opt} is greater than j. In order to prove the inequality $2(k-j)+1 \geqslant j-1$, we show that $j = \lceil \frac{2}{3}k \rceil \leqslant \frac{2}{3}(k+1)$. Let $y = j - \frac{2}{3}k$. Note that j and k are integers, and $0 \leqslant y < 1$. Moreover, $k = \frac{3}{2}j - \frac{3}{2}y$. If j is even, then $\frac{3}{2}j$ is an integer; therefore, $\frac{3}{2}y$ is an integer strictly smaller than $\frac{3}{2}$, i.e., $\frac{3}{2}y \leqslant 1$ and $y \leqslant \frac{2}{3}$. Otherwise, $\frac{3}{2}y + \frac{1}{2}$ is an integer, and because $y < 1$, we have $\frac{3}{2}y + \frac{1}{2} < 2$, i.e., $\frac{3}{2}y + \frac{1}{2} \leqslant 1$ and $y \leqslant \frac{1}{3}$. Altogether, $\lceil \frac{2}{3}k \rceil = j = \frac{2}{3}k + y \leqslant \frac{2}{3}k + \frac{2}{3}$.

In both cases, we have

$$C_{opt} \geqslant j = \left\lceil \frac{2}{3}k \right\rceil \geqslant \frac{2}{3}C_{\textbf{first-fit-dec}} \, ,$$

which concludes the proof. □

8.1.5 2-PARTITION

We discuss approximation algorithms for the 2-PARTITION problem. The optimization problem associated with 2-PARTITION is the following: Given n integers $a_1, \ldots, a_n$, find a subset I of $\{1, \ldots, n\}$ such that $\max\left(\sum_{i \in I} a_i, \sum_{i \notin I} a_i\right)$ is minimum. Note that the minimum is always at least $\max(P_{max}, P_{sum}/2)$, where $P_{max} = \max_{1 \leqslant i \leqslant n} a_i$ and $P_{sum} = \sum_{i=1}^{n} a_i$.

This problem is similar to a scheduling problem with two identical processors. There are n independent tasks $T_1, \ldots, T_n$, and task T_i $(1 \leqslant i \leqslant n)$ can be executed on one of the two processors in time a_i. The goal is to minimize the total execution time. The processors are denoted by P_1 and P_2.

We start by analyzing two greedy algorithms for this problem. Then, we show how to derive a PTAS for 2-PARTITION and even an FPTAS.

Greedy algorithms

The two natural greedy algorithms are the following. We choose tasks in a random order (online algorithm, **greedy-online**) or sorted by nonincreasing execution time (offline algorithm, **greedy-offline**), and we assign the chosen task to the processor that has the lowest current load.

The idea of sorting in the offline algorithm is that a task with a large execution time, if considered at the end of the algorithm, may unbalance the entire execution. However, the offline version requires that all execution times are known beforehand. The online algorithm can be applied in a problem where tasks arrive dynamically (for instance, scheduling user jobs on a biprocessor server).

THEOREM 8.9. **Greedy-online** *is a* $\frac{3}{2}$*-approximation algorithm, and* **greedy-offline** *is a* $\frac{7}{6}$*-approximation algorithm for the 2-PARTITION problem. Moreover, these approximation ratios are tight.*

Proof. First, we consider the **greedy-online** algorithm. Let us assume that processor P_1 finishes the execution at time $M_1 \geqslant M_2$ (where M_2 is the time at which P_2 finishes its execution), and that T_j is the last task executed on P_1. We have $M_1 + M_2 = P_{sum}$. Moreover, since the greedy algorithm chose processor P_1 to execute task T_j, it means that $M_1 - a_j \leqslant M_2$; otherwise, T_j would have been scheduled on P_2. Finally, the cost of **greedy-online** is such that:

$$C_{\textbf{online}} = M_1 = \frac{1}{2}(M_1 + (M_1 - a_j) + a_j) \leqslant \frac{1}{2}(M_1 + M_2 + a_j) = \frac{1}{2}(P_{sum} + a_j),$$

and since $C_{opt} \geqslant P_{sum}/2$ and $C_{opt} \geqslant a_i$ for $1 \leqslant i \leqslant n$, we have $C_{\mathbf{online}} \leqslant C_{opt} + \frac{1}{2}C_{opt} = \frac{3}{2}C_{opt}$, which concludes the proof.

For the offline version of the greedy algorithm, we start as before, but we refine the inequality $a_j \leqslant C_{opt}$. If $a_j \leqslant \frac{1}{3}C_{opt}$, we obtain the approximation ratio of the theorem, i.e., $C_{\mathbf{offline}} \leqslant \frac{7}{6}C_{opt}$. We focus now on the case where $a_j > \frac{1}{3}C_{opt}$. Then, $j \leqslant 4$. Indeed, if a_j were the fifth task, because the tasks are sorted by nonincreasing execution times, there would be at least five tasks of time at least $\frac{1}{3}C_{opt}$, and any schedule would need to schedule at least three of these tasks on the same processor, leading to an execution time strictly greater than C_{opt}, and hence a contradiction. Then, we note that, in this case, the cost $C_{\mathbf{offline}}$ when we restrict to the scheduling of the first four tasks is identical to the cost when scheduling all tasks. Finally, it is easy to check (exhaustively) that **greedy-offline** is optimal when scheduling at most four tasks. We conclude that $C_{\mathbf{offline}} = C_{opt}$ in this case, which ends the proof.

Finally, we prove that the ratios are tight. For **greedy-online**, we consider an instance with two tasks of time 1 and one task of time 2. The greedy algorithm schedules the tasks in time 3 (each task of time 1 on a distinct processor, then the task of time 2 after one of those), while the optimal algorithm takes a time 2 (with the two first tasks on the same processor). For **greedy-offline**, we consider an instance with two tasks of time 3 and three tasks of time 2. The greedy algorithm schedules each task of time 3 on a distinct processor, leading to a total execution time of 7, while the optimal solution consists of grouping those two tasks on the same processor, with a total time of 6. □

PTAS: A $(1+\varepsilon)$-approximation algorithm

THEOREM 8.10. *$\forall\varepsilon > 0$, there is a $(1+\varepsilon)$-approximation algorithm for the 2-PARTITION problem. In order words, 2-PARTITION has a PTAS.*

Proof. We consider an instance $\mathcal{I}$ of 2-PARTITION, $a_1, \ldots, a_n$ (recall that the a_is can be interpreted as the execution time of tasks), and $\varepsilon > 0$.

We classify the tasks into two categories. Let $L = \max(P_{max}, P_{sum}/2)$. The *big* tasks are in the set $T_{big} = \{i \mid a_i > \varepsilon L\}$, while the *small* tasks are in the set $T_{small} = \{i \mid a_i \leqslant \varepsilon L\}$. We consider an instance $\mathcal{I}^*$ of the problem with the tasks of T_{big}, and $\lfloor \frac{S}{\varepsilon L} \rfloor$ tasks of identical size εL, where $S = \sum_{i \in T_{small}} a_i$.

The proof goes as follows. We show that the optimal schedule for instance $\mathcal{I}^*$ has a cost C^*_{opt} close to the cost C_{opt} of the optimal schedule for instance $\mathcal{I}$, i.e., $C^*_{opt} \leqslant (1+\varepsilon)C_{opt}$. Moreover, it is possible to compute the optimal schedule for instance $\mathcal{I}^*$ in a polynomial time. Building upon this schedule, we finally construct a solution to the original instance $\mathcal{I}$, with a guaranteed cost.

First, we prove that $C^*_{opt} \leqslant (1+\varepsilon)C_{opt}$. Let *opt* be an optimal schedule for instance $\mathcal{I}$, of cost C_{opt}. Then, let S_1 (resp. S_2) be the sum of the small tasks in this optimal schedule on processor P_1 (resp. P_2). We build a new schedule *sched** in which the big tasks of the optimal schedule *opt* remain on the same processors, but small tasks are replaced with $\left\lceil \frac{S_i}{\varepsilon L} \right\rceil$ tasks of size εL on processor P_i, for $i = 1, 2$. Because

$$\left\lceil \frac{S_1}{\varepsilon L} \right\rceil + \left\lceil \frac{S_2}{\varepsilon L} \right\rceil \geqslant \left\lfloor \frac{S_1 + S_2}{\varepsilon L} \right\rfloor = \left\lfloor \frac{S}{\varepsilon L} \right\rfloor ,$$

we have scheduled at least as many tasks of size εL as the total number of small tasks in instance $\mathcal{I}^*$. Moreover, the execution time on processor P_i, for $i = 1, 2$, has been increased of at most

$$\left\lceil \frac{S_i}{\varepsilon L} \right\rceil \times \varepsilon L \; - \; S_i \leqslant \varepsilon L ,$$

which means that the cost of this schedule is such that $C^*_{sched} \leqslant C_{opt} + \varepsilon L$. Moreover, this schedule is a schedule for instance $\mathcal{I}^*$ and, therefore, $C^*_{sched} \geqslant C^*_{opt}$. Finally, $C^*_{opt} \leqslant C_{opt} + \varepsilon L \leqslant C_{opt} + \varepsilon \times C_{opt}$, which concludes the proof that $C^*_{opt} \leqslant (1 + \varepsilon) C_{opt}$.

Next, we discuss how to find an optimal schedule for instance $\mathcal{I}^*$. First, we provide a bound on the number of tasks in $\mathcal{I}^*$. Because we replaced small tasks of $\mathcal{I}$ with tasks of size εL, we have not increased the total execution time, which is at most $P_{sum} \leqslant 2L$. Each task of $\mathcal{I}^*$ has an execution time of at least εL (small tasks), so there are at most $\frac{2L}{\varepsilon L} = \frac{2}{\varepsilon}$ tasks. Note that this is a constant number because ε is a constant. Moreover, we note that the size of $\mathcal{I}^*$ is polynomial in the size of instance $\mathcal{I}$ (because the size of $\mathcal{I}^*$ is a constant). We can optimally schedule $\mathcal{I}^*$ by trying all $2^{\frac{2}{\varepsilon}}$ possible schedules and keeping the best one. Of course, this algorithm is not polynomial in $1/\varepsilon$, but it is polynomial in the size of the instance $\mathcal{I}$ because it is a constant.

Now we have an optimal schedule *opt** for instance $\mathcal{I}^*$, of cost C^*_{opt}, and we aim to build a schedule *sched* for instance $\mathcal{I}$. For $i = 1, 2$, we let $L^*_i = B^*_i + S^*_i$ be the total execution time of processor P_i in the schedule *opt**, where B^*_i (resp. S^*_i) is the time spent on big (resp. small) tasks. Then, we build the schedule *sched* in which the big tasks are kept on the same processor as in *opt**, and we greedily assign small tasks to processors. First, we assign small tasks to processor P_1 until their processing time does not exceed $S^*_1 + 2\varepsilon L$. Then, we schedule the remaining small tasks to processor P_2. Let us prove now that once all small tasks have been scheduled, the execution time has not increased by more than $2\varepsilon L$.

Because small tasks have a size of at most εL, the greedy algorithm assigns at least a total of $S^*_1 + \varepsilon L$ small tasks on processor P_1. Then, there are at most a total of $S - (S^*_1 + \varepsilon L)$ small jobs to assign to processor P_2. However, by construction of $\mathcal{I}^*$, we have $S^*_1 + S^*_2 = \varepsilon L \left\lfloor \frac{S}{\varepsilon L} \right\rfloor > S - \varepsilon L$ and, therefore, $S - (S^*_1 + \varepsilon L) \leqslant S^*_2$, and the execution time of P_2 in the new schedule *sched* is not greater than in the schedule *opt**.

The schedule *sched* is a schedule for instance $\mathcal{I}$, which is built in polynomial time. The cost of this schedule is at most $C_{sched} \leqslant C^*_{opt} + 2\varepsilon L$. We use the previous result that $C^*_{opt} \leqslant (1+\varepsilon)C_{opt}$ and the fact that $L \leqslant C_{opt}$ to conclude that $C_{sched} \leqslant C^*_{opt} + 2\varepsilon L \leqslant (1+3\varepsilon)C_{opt}$. This is true for all ε, so we can apply this algorithm with $\varepsilon/3$ to obtain the desired ratio. □

Note that a simpler proof can be done by using an optimal schedule for the big tasks, of cost C_{big}, and the **greedy-online** algorithm introduced above. Once the small tasks have been scheduled greedily on the two processors, there are two cases. If the total time has not changed, i.e., it is C_{big}, it is optimal. Otherwise, the processor that ends the execution is executing a small task a_j. This means that before the greedy choice of scheduling task a_j onto this processor, the finishing time of the processor was less than $P_{sum}/2$; otherwise, task a_j would have been assigned to the other processor because of the greedy choice. Finally, the cost of the schedule returned by this algorithm is at most $P_{sum}/2 + a_j \leqslant L + \varepsilon L \leqslant (1+\varepsilon)C_{opt}$.

A PTAS provides an approximate solution that is as close to the optimal as one wants. The only downside is that the algorithm running time increases with the quality of the approximate solution. Some readers may thus be puzzled by the idea of having a PTAS or an FPTAS for an NP-complete problem whose objective function takes values in a discrete set, such as 2-PARTITION. Indeed, a PTAS, for such a problem, enables one to obtain an optimal solution whenever one is ready to pay the cost. Let us consider any given instance $\mathcal{I}$ of 2-PARTITION. Let S be the sum of the elements of $\mathcal{I}$. If $\varepsilon < \frac{1}{S}$, then any $1+\varepsilon$ approximation produces an optimal solution. Indeed, $(1+\varepsilon)C_{opt} < (1+\frac{1}{S})C_{opt} \leqslant C_{opt} + 1$ because $C_{opt} \leqslant S$ and because the objective function can take only integral values. This may be surprising at first sight, but it does not contradict anything we have written so far. One should not forget that the running time of an FPTAS is polynomial in the size of $\frac{1}{\varepsilon}$, that is, in our example, in the size of S. The running time of a PTAS can even be exponential in the size of $\frac{1}{\varepsilon}$. Finding the optimal solution for 2-PARTITION in time exponential in the size of S is quite simple. One generates all the subsets of $\mathcal{I}$ and computes the sum of the elements of each subset. If $\mathcal{I}$ includes n elements, there are $2^n = O(2^S)$ subsets of $\mathcal{I}$. The sum of the elements of each of them is computed in time $O(n)$ and thus $O(S)$. Therefore, readers should not be surprised that, for a given value of ε, an algorithm whose running time is polynomial in the size of the instance can find an optimal solution to 2-PARTITION.

FPTAS for 2-PARTITION

We have provided a PTAS for 2-PARTITION, but the algorithm finds an optimal schedule for instance $\mathcal{I}^*$ (i.e., an optimal schedule of the big tasks), and this is not polynomial in $1/\varepsilon$. Below, we provide an FPTAS, i.e., a $(1+\varepsilon)$-approximation algorithm that is polynomial in the size of $\mathcal{I}$ and in $1/\varepsilon$.

THEOREM 8.11. *$\forall \varepsilon > 0$, there is a $(1+\varepsilon)$-approximation algorithm for the 2-PARTITION problem that is polynomial in $1/\varepsilon$. In order words, 2-PARTITION has an FPTAS.*

Proof. The idea of the proof is to encode the schedules as *vector sets*, in which the first (resp. second) element of a vector represents the running time of the first (resp. second) processor. Formally, for $1 \leqslant k \leqslant n$, VS_k is the set of vectors representing schedules of tasks $a_1, \ldots, a_k$: $VS_1 = \{[a_1, 0], [0, a_1]\}$, and we build VS_k from VS_{k-1} as follows. For all $[x, y] \in VS_{k-1}$, we add $[x + a_k, y]$ and $[x, y + a_k]$ to VS_k. The optimal schedule is represented by a vector $[x, y] \in VS_n$, and it is such that $\max(x, y)$ is minimized.

The approximation algorithm enumerates all possible schedules, but some of them are discarded on the fly so that we keep a polynomial algorithm.

Let $\Delta = 1 + \frac{\varepsilon}{2n}$. We partition the square $P_{sum} \times P_{sum}$ following the power of Δ, from 0 to Δ^M. We have $M = \lceil \log_\Delta(P_{sum}) \rceil = \left\lceil \frac{\ln(P_{sum})}{\ln(\Delta)} \right\rceil \leqslant \left\lceil \left(1 + \frac{2n}{\varepsilon}\right) \ln(P_{sum}) \right\rceil$. Indeed, note that if $z \geqslant 1$, then $\ln(z) \geqslant 1 - \frac{1}{z}$.

The idea of the algorithm consists of building the vector sets but adding a new vector to a set only if there are no other vectors in the same square of the partitioned $P_{sum} \times P_{sum}$ square. Because M is polynomial in $1/\varepsilon$ and in $\ln(P_{sum})$, and the size of instance $\mathcal{I}$ is greater than $\ln(P_{sum})$, the algorithm is polynomial both in the size of $\mathcal{I}$ and in $1/\varepsilon$. We need to prove that this algorithm is a $(1+\varepsilon)$-approximation to conclude the proof.

First, let us formally describe the algorithm. Initially, $VS_1^\# = VS_1$. Then, for $2 \leqslant k \leqslant n$, we build $VS_k^\#$ from $VS_{k-1}^\#$ as follows. For all $[x, y] \in VS_{k-1}^\#$, we add $[x + a_k, y]$ (resp. $[x, y + a_k]$) to $VS_k^\#$ if and only if there is no vector from $VS_k^\#$ in the same square. Note that two vectors $[x_1, y_1]$ and $[x_2, y_2]$ are in the same square if and only if $\frac{x_1}{\Delta} \leqslant x_2 \leqslant \Delta x_1$ and $\frac{y_1}{\Delta} \leqslant y_2 \leqslant \Delta y_1$.

We keep at most one vector per square at each step, which gives an overall complexity in $n \times M^2$, which is polynomial both in the size of instance $\mathcal{I}$ and in $1/\varepsilon$.

Next, we prove that for all $1 \leqslant k \leqslant n$ and $[x, y] \in VS_k$ there exists $[x^\#, y^\#] \in VS_k^\#$ such that $x^\# \leqslant \Delta^k x$ and $y^\# \leqslant \Delta^k y$. The proof is done recursively. The result is trivial for $k = 1$. If we assume that the result is true for $k - 1$, then let us consider $[x, y] \in VS_k$. Either $x = u + a_k$ and $y = v$ (case 1), or $x = u$ and $y = v + a_k$ (case 2), with $[u, v] \in VS_{k-1}$. By recursion hypothesis, there exists $[u^\#, v^\#] \in VS_{k-1}^\#$ with $u^\# \leqslant \Delta^{k-1} u$ and $v^\# \leqslant \Delta^{k-1} v$. For case 1, note that $[u^\# + a_k, v^\#]$ may not be in $VS_k^\#$, but we know that there is at least one vector in the same square in $VS_k^\#$; there exists $[x^\#, y^\#] \in VS_k^\#$ such that $x^\# \leqslant \Delta\left(u^\# + a_k\right)$ and $y^\# \leqslant \Delta v^\#$. Finally, we have $x^\# \leqslant \Delta^k u + \Delta a_k \leqslant \Delta^k (u + a_k) = \Delta^k x$ and $y^\# \leqslant \Delta v^\# \leqslant \Delta^k y$, and case 2 is symmetrical. This proves the result.

For $k = n$, we can deduce that $\max(x^\#, y^\#) \leqslant \Delta^n \max(x, y)$. There remains to be proven that $\Delta^n \leqslant (1+\varepsilon)$, where $\Delta^n = \left(1 + \frac{\varepsilon}{2n}\right)^n$. We rearrange the last inequality and study the function $f(z) = \left(1 + \frac{z}{n}\right)^n - 1 - 2z$, for $0 \leqslant z \leqslant 1$, $f'(z) = \frac{1}{n} n \left(1 + \frac{z}{n}\right)^{n-1} - 2$. We deduce that f is a convex function, and that its minimum is reached in $\lambda_0 = n\left(\sqrt[n-1]{2} - 1\right)$. Moreover, $f(0) = -1$ and $f(1) = \left(1 + \frac{1}{n}\right)^n - 3 \leqslant 0$. Because f is convex, and $f(z) \leqslant 0$ for $z = 0$ and $z = 1$, we can deduce that $f(z) \leqslant 0$ for $0 \leqslant z \leqslant 1$. This concludes the proof. □

8.2 Polynomial problem instances

When confronted with an NP-complete problem, one algorithmic solution consists of finding good approximation algorithms. While some problems may have good approximation schemes, such as PTAS or FPTAS (see Section 8.1), some problems cannot be approximated. However, with a slight change of the problem parameters (constant value for a parameter, different rule of the game, etc.), it may be possible to find a good approximation algorithm or even to be able to solve the problem in pseudopolynomial or polynomial time.

The analysis of a problem is comprehensive when we are able to identify at which point the problem becomes NP-complete and then at which point the problem cannot be approximated any more. We refine the problem complexity as follows:

- The class P consists of all optimization problems that can be solved in polynomial time.

- The class FPTAS consists of all optimization problems that have an FPTAS, and it contains P.

- The class PTAS consists of all optimization problems that have a PTAS, and it contains FPTAS.

- The class APX consists of all optimization problems that have a polynomial-time approximation algorithm with a constant ratio, and it contains PTAS.

- Finally, the class NP contains APX: Some problems may be in NP but not in APX.

We also consider the class of problems that can be solved in pseudopolynomial time, PPT. This class includes P but none of the other previous classes. Some problems that can be solved in pseudopolynomial time may not have an FPTAS or may not even be in APX. The problems of (i) finding a pseudopolynomial-time algorithm to solve the problem exactly and (ii) finding good polynomial-time approximation algorithms are not correlated.

In the following, we illustrate how the problem can move from one category to another when parameters are modified. In particular, we check whether the problem can be solved in polynomial time or in pseudopolynomial time, and if there is no polynomial-time algorithm to solve the problem, we investigate polynomial-time approximation algorithms.

8.2.1 Partitioning problems

First, we provide both the optimization and decision versions of the partitioning problem that we consider, and then we investigate variants of the problem.

Optimization problem (PART-OPT). Let $a_1, \ldots, a_n$ be n positive integers. The goal is to partition these integers into p subsets $A_1, \ldots, A_p$, in order to minimize the maximum (over all subsets) of the sum of the integers in a subset:

$$\min \left(\max_{1 \leqslant j \leqslant p} \sum_{i \in A_j} a_i \right) .$$

Decision problem (PART-DEC). The associated decision problem is the following: Let $a_1, \ldots, a_n$ be n positive integers. Given a bound K, is it possible to partition these integers into p subsets $A_1, \ldots, A_p$, such that the sum of the integers in each subset does not exceed K? In other words,

$$\text{for all } 1 \leqslant j \leqslant p, \quad \sum_{i \in A_j} a_i \leqslant K .$$

We can easily prove, from a reduction from 3-PARTITION, that PART-DEC is NP-complete in the strong sense. No pseudopolynomial algorithm is known to solve PART-DEC. However, PART-OPT is a classical scheduling problem. The goal is to schedule n independent tasks onto p processors, where a_i is the execution time of task T_i, for $1 \leqslant i \leqslant n$, and the goal is to minimize the total execution time. There is a PTAS to approximate this problem [49].

One way to simplify the problem is to restrict it to the case $p = 2$. The problem is then equivalent to 2-PARTITION, and it can be solved in pseudopolynomial time using a dynamic-programming algorithm (see Section 6.2.1). Moreover, this problem is in the class FPTAS, as was shown in Section 8.1.5.

In order to identify polynomial instances of this problem, we consider the following variants:

1. We consider the case in which all integers are equal, i.e., $a_1 = a_2 = \cdots = a_n = a$. In this case, we can find the solution to the optimization problem, which is simply $\left\lceil \frac{n}{p} \right\rceil \times a$. Therefore, we also can solve the decision problem in polynomial time, even in constant time.

2. We change the rule of the game. The subsets must contain only continuous elements, for instance, $[a_i, a_{i+1}, \ldots, a_{i'}]$. The subsets are then

intervals, and the problem can be solved in polynomial time. It is the classical chains-on-chains partitioning problem (see Chapter 11), which can be solved, for instance, with a dynamic-programming algorithm in time $O(n^2 \times p)$.

If we consider the problem as a scheduling problem where we must schedule n tasks onto p processors, we can conclude that the problem becomes difficult (NP-complete) as soon as the tasks are different (the case of identical tasks is case 1) and as soon as we are allowed any mapping (no fixed ordering to enforce, such as in case 2). Moreover, while the problem is in PPT and has an FPTAS with $p = 2$ processors, it is no longer in PPT for an arbitrary number of processors and has only a PTAS.

For a deeper analysis of partitioning problems, the interested reader can refer to the chains-on-chains partitioning case study (Chapter 11).

8.2.2 Assessing problem complexity

In this section, we mention two classical approaches when facing NP-complete problems and aiming at identifying polynomial instances. We illustrate these approaches with two different problems.

The first problem is a routing problem, which is discussed extensively in Chapter 13. Given a directed graph $G = (V, E)$ and a set of terminal pairs $\mathcal{R} = \{R_i = (s_i, t_i)\}$, the goal is to connect as many pairs as possible using edge-disjoint simple paths. In a solution $\mathcal{A}$, each $R_i \in \mathcal{A}$ must be assigned a simple path π_i from s_i to t_i in G so that no two paths π_i and π_j, where $R_i \in \mathcal{A}$, $R_j \in \mathcal{A}$ and $i \neq j$, have an edge in common.

The goal is to maximize $|\mathcal{A}|$, the cardinality of $\mathcal{A}$, i.e., the number of connected terminal pairs. It turns out that this routing problem is NP-complete, and Chapter 13 presents approximation algorithms. But how can we find polynomial instances? A first idea is to bound the number of terminal pairs with a constant, but this does not work, as it turns out that the problem remains NP-complete with only two terminal pairs [35]. Another idea is to restrict the problem to some special classes of graphs. We show in Chapter 13 that the problem is polynomial for linear chains and stars, regardless of the number of terminal pairs.

The second problem is a geometric problem, which is investigated in Chapter 14. How can we partition the unit square into p rectangles of given area $s_1, s_2, \ldots, s_p$ (such that $\sum_{i=1}^{p} s_i = 1$) so as to minimize the sum of the p half perimeters of the rectangles? In Chapter 14, we explain the relevance of this problem to parallel computing, and we show that it is NP-complete. What can we do here? The problem becomes polynomial if we restrict to same-size rectangles [64], but this is very restrictive. Another approach is to change the rules of the game and ask for some specific partitioning of the unit square. Indeed, we show in Chapter 14 that the problem becomes polynomial when restricting to column-based partitioning, i.e., imposing that the rectangles are arranged along several columns within the unit square. Going further in that direction, we show that the optimal column-based partitioning is indeed a good approximation of the general solution. We hope that this short discussion will urge the reader to read the full case study of Chapter 14.

8.3 Linear programming

Sometimes the solution of an NP-complete problem can be expressed as the solution of an integer linear program. Once we have written an optimization problem as an integer linear program, we can do three things:

1. Solve the integer linear program to obtain optimal solutions for (very) small instances.
2. Relax the integer linear program into a (rational) linear program and solve it to obtain a bound on the optimal solution for the original problem.
3. Relax the integer linear program into a (rational) linear program, solve the latter program to obtain a rational solution, and build an integral solution from the rational one.

We first introduce the necessary notions and definitions (Section 8.3.1). Then we describe several *rounding* approaches to transform a solution of a relaxed linear program into a solution of the original integer linear program (Section 8.3.2).

8.3.1 Formal definition

Linear programming is a mathematical method in which an optimization problem is expressed as the minimization (or maximization) of a linear function whose arguments are constrained by a set of affine equations and inequalities.

DEFINITION 8.6 (Linear program). A linear program is an optimization problem of the form:

$$\text{MINIMIZE} \quad c^T \cdot x \quad \text{SUBJECT TO}$$
$$Ax \leqslant b \quad \text{and} \quad x \geqslant 0$$

where x is an (unknown) vector of variables of size n, A is a (known) matrix of coefficients of size $m \times n$, and b and c are the two (known) vectors of coefficients of respective size m and n (and where c^T is the transpose of vector c).

An *integer linear program* is a linear program whose variables can take only integral values. A *mixed linear program* is a linear program in which some variables must take integral values and some can take rational values.

In the above formal definition, linear programs are given under a canonical form. Therefore, the formal definition of linear programs may look more restrictive than the informal definition we gave right before the formal one. In fact, both definitions are equivalent:

- A maximization problem with the objective function $c^T \cdot x$ is equivalent to a minimization problem with the objective function $-c^T \cdot x$.
- An equality $d^T \cdot x = e$ is equivalent to the set of two inequalities:

$$\begin{cases} d^T \cdot x \leqslant & e \\ -d^T \cdot x \leqslant & -e. \end{cases}$$

- A variable that can take both positive and negative values can be equivalently replaced by the difference of two nonnegative variables.

An example: Weighted vertex cover

In Section 8.1.2, we have seen the classical version of the vertex cover problem. Given a graph $G = (V, E)$, we want to return a set U of vertices ($U \subset V$) of minimum size that is covering all edges, i.e., such that for each edge $e = (i, j) \in E$, $i \in U$ and/or $j \in U$.

Here, we consider the weighted version of this problem. We assign a weight w_i to each vertex $i \in V$. The problem is then to minimize $\sum_{i \in U} w_i$, where U is once again a vertex cover. This problem amounts to the classical one if $w_i = 1$ for all $i \in V$ and is also NP-complete.

We express this minimization problem as an integer linear program. We introduce a set of Boolean variables, one for each vertex, stating whether the corresponding vertex belongs to the cover. Let x_i be the variable associated with vertex $i \in V$. We will have $x_i = 1$ if i belongs to the cover ($i \in U$) and $x_i = 0$ otherwise.

$$\text{MINIMIZE} \sum_{i \in V} x_i w_i \quad \text{SUBJECT TO}$$

$$\begin{cases} \forall (i,j) \in E & -x_i - x_j \leqslant -1 \\ \forall i \in V & x_i \leqslant 1 \\ \forall i \in V & x_i \geqslant 0 \end{cases} \tag{8.1}$$

We now show that solving the Integer Linear Program (8.1), with $x_i \in \{0, 1\}$, is absolutely equivalent to solving the minimum weighted vertex cover problem for the graph G.

One can easily check that, if U is an optimal solution to the weighted vertex cover problem, then, by letting $x_i = 1$ for any vertex i in U and $x_j = 0$ for any vertex j not in U, one builds a solution to the above linear program for which the objective function takes the value of the cost of the cover U.

Reciprocally, consider an optimal solution to the Integer Linear Program (8.1), with $x_i \in \{0, 1\}$. From this solution, we build a subset U of V as follows. For any vertex i of V, i belongs to U if and only if $x_i = 1$. For any edge

$e = (i,j) \in E$ we have $-x_i - x_j \leqslant -1$, which is equivalent to $x_i + x_j \geqslant 1$. In other words, either x_i or x_j or both variables are equal to 1 (remember that here the x_is are integer variables). Therefore, at least one of the two vertices i and j is a member of U, and U is thus a cover. The objective function is obviously the cost of the cover U. Therefore, U is a cover of minimum weight.

Complexity

In the general case, the decision problem associated with the problem of solving integer linear programs is an NP-complete problem [58, 38]. However, (rational) linear programs can be solved in polynomial time [93]. Hence, the motivation, when confronted with an NP-complete problem, is to express it as an integer or mixed linear program and then to solve this program as if it were a rational linear program. This method is called *relaxation*. However, the solution obtained this way may be meaningless. For instance, in the case of the linear program for the weighted vertex cover problem (Linear Program (8.1)), one of the variables x_i can have a value different from 0 and 1, which does not make any sense because a vertex cannot be *partially* included in the solution. The problem then becomes how to build an integral solution from a rational one. We now focus on this problem, which is called *rounding*.

8.3.2 Relaxation and rounding

Rounding to the nearest integer

The simplest rounding method is the rounding of any rational variable to the nearest integer. (Obviously, this method is not fully defined because one will still have to decide how to handle variables whose values are of the form $z + 0.5$ where z is an integer.) We illustrate this method with the weighted vertex cover problem.

Algorithm **lp-wvc** is defined as follows. First, solve the Linear Program (8.1) over the rationals rather than on the integers, and let $\{x_i^*\}_{i \in V}$ be the found optimal solution. Then, any vertex i of V belongs to the cover U if and only if $x_i^* \geqslant \frac{1}{2}$. In other words, we build from the x_i^*s the Boolean variables x_is, by: $x_i = 1 \Leftrightarrow x_i^* \geqslant \frac{1}{2}$. Not only is Algorithm **lp-wvc** correct, it is even an approximation algorithm, as we now prove.

THEOREM 8.12. **lp-wvc** *is a* 2*-approximation algorithm for weighted vertex cover.*

Proof. First, we check that **lp-wvc** returns a cover. Let $(i,j) \in E$ be an edge. Then, because the x_i^*s are a rational solution to the linear program, we have $x_i^* + x_j^* \geqslant 1$, and at least one of them is greater than or equal to $1/2$. Therefore, in the solution of our problem, we have either $x_i = 1$ or $x_j = 1$ (we also can have $x_i = x_j = 1$). Therefore, the edge (i,j) is covered, $x_i + x_j \geqslant 1$.

To prove that the algorithm is a 2-approximation, we compare the cost of the algorithm $C_{\textbf{lp-wvc}} = \sum_{i \in V} x_i w_i$ with the cost of an optimal solution C_{opt}.

The result comes from two observations: (i) For all i, we have $x_i \leqslant 2x_i^*$ (whether i has been chosen to be part of the cover or not), and (ii) the optimal solution of the linear program over the integers has necessarily a higher cost than the rational solution (the integer solution is a solution to the rational problem). Because $\sum_{i \in V} x_i^* w_i$ is an optimal solution to the rational problem, $C_{opt} \geqslant \sum_{i \in V} x_i^* w_i$. Finally, we have

$$C_{\mathbf{lp\text{-}wvc}} = \sum_{i \in V} x_i w_i \leqslant \sum_{i \in V} (2x_i^*) w_i \leqslant 2C_{opt},$$

which concludes the proof. □

Threshold rounding

We do not have any a priori guarantee that the rounding to the nearest integer will produce a valid integer solution. We illustrate this potential problem with the set cover problem.

DEFINITION 8.7 (SET-COVER). Let V be a set. Let $\mathcal{S}$ be a collection of k subsets of V: $\mathcal{S} = \{S_1, \ldots, S_k\}$ where, for $1 \leqslant i \leqslant k$, $S_i \subset V$. Let K be an integer, with $K < k$. Is there a subcollection of at most K elements of $\mathcal{S}$ that covers all elements of V?

SET-COVER is an NP-complete problem [58, 38]. It easily can be coded as an integer linear program. Let $\delta_{i,j}$ be a Boolean constant indicating whether the element $v \in V$ belongs to the subset $s \in \mathcal{S}$. As previously, variable x_s indicates whether the set $s \in \mathcal{S}$ belongs to the solution. The following integer linear program then searches for a minimum set cover. The first inequality just states that, whatever the element v of V, at least one of the subsets containing v must be picked in the solution.

$$\text{MINIMIZE} \sum_{s \in \mathcal{S}} x_s \quad \text{SUBJECT TO}$$

$$\begin{cases} \forall v \in V & -\sum_{s \in S} \delta_{v,s} x_s \leqslant -1 \\ \forall s \in \mathcal{S} & x_s \leqslant 1 \\ \forall s \in \mathcal{S} & -x_s \leqslant 0 \end{cases} \tag{8.2}$$

Now, consider the following particular instance of minimum cover: $V = \{a, b, c, d\}$ and $\mathcal{S} = \{S_1 = \{a, b, c\}, S_2 = \{a, b, d\}, S_3 = \{a, c, d\}, S_4 = \{b, c, d\}\}$. One can easily see that any two elements of $\mathcal{S}$ define an optimal solution. We write explicitly the Linear Program (8.2) for that instance:

$$\text{MINIMIZE} \quad x_{S_1} + x_{S_2} + x_{S_3} + x_{S_4} \quad \text{SUBJECT TO}$$

$$\begin{cases} & -x_{S_1} - x_{S_2} - x_{S_3} \leqslant -1 \\ & -x_{S_1} - x_{S_2} - x_{S_4} \leqslant -1 \\ & -x_{S_1} - x_{S_3} - x_{S_4} \leqslant -1 \\ & -x_{S_2} - x_{S_3} - x_{S_4} \leqslant -1 \\ \forall s \in \{S_1, S_2, S_3, S_4\} & x_s \leqslant 1 \\ \forall s \in \{S_1, S_2, S_3, S_4\} & -x_s \leqslant 0. \end{cases} \tag{8.3}$$

By summing the first four inequalities, we obtain $x_{S_1} + x_{S_2} + x_{S_3} + x_{S_4} \geqslant \frac{4}{3}$. Hence, the optimal value of the objective function is not smaller than $\frac{4}{3}$. Then, one can check that $x^*_{S_1} = x^*_{S_2} = x^*_{S_3} = x^*_{S_4} = \frac{1}{3}$ defines an optimal solution of the relaxed (rational) version of the Linear Program (8.3). Rounding this optimal rational solution to the nearest integer would lead to $x_{S_1} = x_{S_2} = x_{S_3} = x_{S_4} = 0$, which, obviously, does not define a cover. To circumvent this problem, rather than to round each variable to the nearest integer, one can use a generalization of this technique: threshold rounding. When variables are 0-1 variables, that is, when variables can take only the values 0 or 1, one first sets a threshold and then rounds to 1 exactly those variables whose values are not smaller than the threshold. This technique leads to an approximation algorithm for the minimum set cover problem.

THEOREM 8.13. *Let $\mathcal{P} = (V, \mathcal{S})$ be an instance of the minimum set cover problem in which each element of V belongs to at most p elements of $\mathcal{S}$. Then, solving the Linear Program (8.3) over the rationals and rounding the solution with the threshold $\frac{1}{p}$ builds a cover whose size is at most p times the optimal.*

Proof. Let us consider an optimal solution x^* of the relaxed linear program. Let v be any element of V. By definition of p, v belongs to $q \leqslant p$ elements of $\mathcal{S}$: $S_{\sigma(1)}, \ldots, S_{\sigma(q)}$. The Linear Program (8.2) contains the constraint $-x^*_{S_{\sigma(1)}} - x^*_{S_{\sigma(2)}} - \cdots - x^*_{S_{\sigma(q)}} \leqslant -1$. Therefore, there exists at least one $i \in [1, q]$ such that $x_{S_{\sigma(i)}} \geqslant \frac{1}{q} \geqslant \frac{1}{p}$ and the solution contains at least one element of $\mathcal{S}$ that includes v, namely, $S_{\sigma(i)}$. Thus, the solution is a valid cover. Then, for any element s of $\mathcal{S}$, $x_s \leqslant p \times x^*_s$. Indeed, if $x^*_s \geqslant \frac{1}{p}$, then $x_s = 1$ and $x_s = 0$ otherwise. This completes the proof for the approximation ratio. □

Randomized rounding

In the previous two approaches, the value of a variable in a rational solution was considered to be a deterministic indication of what should be the value of this variable in an integer solution. In the randomized rounding approach, the fractional part of such a value is interpreted as a probability.

Let us consider a nonintegral component x_i^* of an optimal rational solution x^*, and let y_i^* be its fractional part: $x_i^* = \lfloor x_i^* \rfloor + y_i^*$, with $0 < y_i^* < 1$. Then, in randomized rounding, y_i^* is considered to be the probability that, in the integral solution, x_i will be equal to $\lceil x_i^* \rceil$ rather than to $\lfloor x_i^* \rfloor$. In practice, using any uniform random generator over the interval $[0, 1]$, one generates a number $r \in [0, 1]$. If $r \geqslant y_i^*$, then we let $x_i = \lceil x_i^* \rceil$, and $x_i = \lfloor x_i^* \rfloor$ otherwise.

Iterative rounding

In all the previously described rounding approaches, a single relaxed linear program is solved, and then one tries to build an integral solution from the rational solution. A potential problem of these approaches is that the assignment of a particular value to one of the variables may force the value of some other variables in any valid solution. For instance, let us go back to the example showing that rounding to the nearest integer could lead to nonfeasible solutions to the minimum cover problem. There, setting $x_{S_1} = 0$ and $x_{S_2} = 0$ imposes that $x_{S_3} = x_{S_4} = 1$ (because, respectively, a and b must be covered). Rounding to the nearest integer ignores this implication and leads to an infeasible solution. A way to avoid such a problem is to assign values only to a subset of the variables and then solve the relaxed version of the linear program while taking into account the assignments made so far. This way, we obtain a new rational solution where fewer variables have nonintegral values. The process is then iterated until an integral solution is built (or the transformed linear program has no solution). The smaller the number of variables assigned at each iteration, the higher the probability to end up with a valid solution but also the higher the number of iterations, the complexity, and the execution time.

8.4 Randomized algorithms

In this section, we briefly explore how randomized algorithms can help deal with NP-complete problems. We restrict ourselves to a randomized algorithm to solve the NP-complete HC problem (recall that HC stands for Hamiltonian Cycle, see Definition 6.4, p. 130). Given an undirected graph, the algorithm incrementally builds a cycle, taking random decisions on the next vertex to visit to augment the current path. The algorithm will indeed output a Hamiltonian cycle with high probability as soon as the graph contains enough edges. We will quantify this last statement in what follows.

8.4.1 The algorithm

Consider a graph $G = (V, E)$. How can we build a Hamiltonian cycle in G by taking random decisions? The first idea is to grow a path iteratively by picking any neighbor of the current path head that has not been picked so far. Start by picking a vertex, say v_1, at random, and make it the head of the path. Then, pick any neighbor of v_1, say v_2, and make it the new head of the path. Progress likewise at each step; pick any neighbor v_{k+1} of the current path head v_k, and make it the new head of the path. But what if v_{k+1} is equal to some vertex v_i, $1 \leqslant i \leqslant k-1$, that is already present in the path? Then, the algorithm can perform a *rotation*, as illustrated by Figure 8.1.

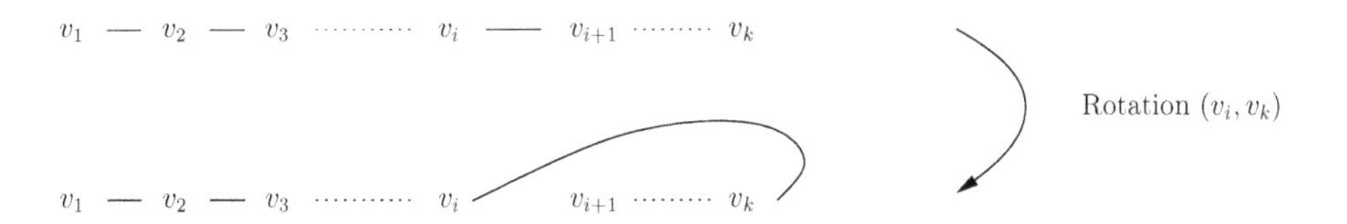

FIGURE 8.1: Rotation (v_k, v_i) of the path. The new head is v_{i+1}.

We obtain the following algorithm, where at each step we pick at random a neighbor u of the current path head v_k among the set of edges originating from v_k that have not been used so far. At the beginning, no edge has been used yet.

8.4.2 Results

What is the probability that Algorithm 8.1 will successfully build a Hamiltonian cycle for G? We would like to express this probability as a function of $n = |V|$, the number of vertices in G. Note that there exist exactly $2^{n(n-1)/2}$ different graphs with n vertices because there are $\binom{n}{2}$ possible edges that can or cannot be added to the graph.

THEOREM 8.14. *There exist constants c and d such that if we pick at random a graph G with n vertices and at least $c \log n$ edges, then with probability at least $1 - \frac{1}{n}$, Algorithm 8.1 will find a Hamiltonian cycle during its first $dn \log n$ steps.*

Proving this theorem is not difficult. This requires, however, some basic knowledge about probability theory (binomial distributions and Markov bound essentially) that is out of the scope of this chapter. We refer the reader to [78] for a proof and many more details about random graphs. We limit ourselves to some comments. First, the randomized algorithm does not give any insight on the P versus NP problem, nor does it help solve all instances of the HC problem. However, on the positive side, we have a fast algorithm that

```
Input: graph G = (V, E) with n vertices
Output: a Hamiltonian cycle in G or failure
1  foreach v ∈ V do
2  └ unused(v) := {(v, u) | (v, u) ∈ E}
3  pick a vertex at random and make it the head of the path
4  while true do
5  │ let (v_1, ..., v_k) be the current path (with head v_k)
6  │ if unused(v_k) = ∅ then return failure
7  │ else let (v_k, u) be the first element in unused(v_k)
8  │ delete edge (v_k, u) from unused(v_k) and unused(u)
9  │ if u ∉ {v_1, ..., v_{k-1}} then
10 │ │ add u to the path and let v_{k+1} = u be the new path head
11 │ else
12 │ │ let i be such that v_i = u
13 │ │ if k = n and v_i = v_1 then return {v_1, ..., v_n}
14 └ └ else rotate (v_k, v_i) and let v_{i+1} be the new path head
```

ALGORITHM 8.1: Randomized algorithm for the HC problem.

solves HC in most instances, as soon as the graph has enough edges. This is expected news, as we expect a random graph to be connected and then to have large cliques, or a Hamiltonian cycle, when its number of edges grow. But the beauty of Theorem 8.14 is to quantify this observation.

8.5 Branch-and-bound and backtracking

In this last section, we introduce branch-and-bound and backtracking techniques. The principle is to represent as a tree the search space (i.e., all candidate solutions) and then to explore this tree and remove branches that either lead to no valid solution or lead to solutions that are less good. Such algorithms return exact solutions to an NP-complete problem. For decision problems, the technique is called backtracking, while it is called branch-and-bound for optimization problems. While there is no guarantee on the execution time of such algorithms (the worst case may well be exponential because we may need to explore the entire search space), they are offering practical and often efficient solutions to deal with NP-complete problems.

We first present a small example of a backtracking algorithm with the n-queens problem. Then, we investigate branch-and-bound with the knapsack problem. Finally, we discuss some more complex graph algorithms.

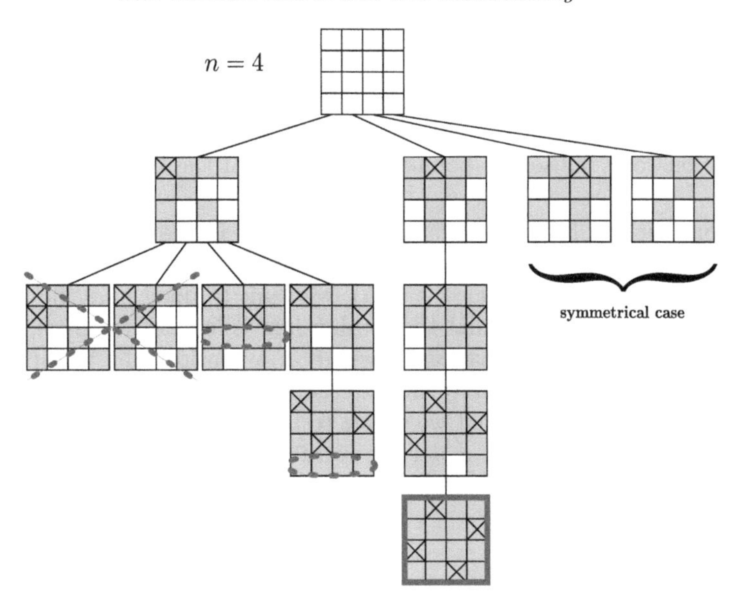

FIGURE 8.2: The n-queens backtracking tree.

8.5.1 Backtracking: The n queens

In a chess game, a queen can move as far as she wants: horizontally, vertically, or diagonally. We consider a chess board with n rows and n columns. The problem is to place n queens on this chess board so that none of them can attack any other in one move.

In any solution, there is exactly one queen per row. Therefore, the search space is of size n^n. However, because of the many constraints, many solutions can be discarded. The idea of the backtracking algorithm is to place a queen on the first row (n possible choices) and then perform a recursive call for the next row. We discard the choices that lead to no solution, and if no solution is found on a branch of the tree, we go up in the tree and try the next possibility (the next branch).

Figure 8.2 illustrates the tree for $n = 4$. Because the problem is symmetrical, we develop only the portion of the tree in which we place the first queen either on the first or on the second column. Once a queen has been placed, the squares on which it is not possible to place another queen have been colored. Therefore, if we place the first queen on the top left corner, the queen on the second row can be placed only on the third or fourth column. If we place it on the third column, there is no further choice for the third queen. If we place it on the fourth column, we can still place the third queen on the second column, but then there is no possibility for the last queen. However, a solution is found by exploring the second branch of the tree.

8.5.2 Branch-and-bound: The knapsack

A branch-and-bound algorithm works in two phases. The *branch* consists of splitting a set of solutions into subsets, while the *bound* consists of evaluating the solutions of a subset by bounding the value of the best solution in this subset.

We consider the knapsack problem, which was introduced in Section 4.2 and that we redefine briefly. Given a set of items $I_1, \dots, I_n$, where item I_i has a weight w_i and a value c_i ($1 \leqslant i \leqslant n$), we want to determine the items to include in the collection so that the total weight is less than a given limit W and the total value is as large as possible. We consider the variant of the problem where we have as many units of each item as we want. Let x_i be the number of units of item I_i that we decide to add into the knapsack. The goal is to maximize $\sum_{i=1}^{n} x_i \times c_i$, under the constraint $\sum_{i=1}^{n} x_i \times w_i \leqslant W$.

We consider the running example from [15]. There are four items, and the goal is to find $\max(4x_1 + 5x_2 + 6x_3 + 2x_4)$, under the constraint $33x_1 + 49x_2 + 60x_3 + 32x_4 \leqslant 130$.

The search space is represented as a tree. The leaves of the tree correspond to maximal solutions, i.e., solutions to which we cannot add any item because of the constraint on total weight. At the root of the tree, we have not chosen any item. The root has $\left\lceil \frac{W}{w_1} \right\rceil + 1$ children, which corresponds to picking, respectively, $0, 1, \dots, \left\lceil \frac{W}{w_1} \right\rceil$ units of I_1. Then, for each of these nodes, we add one child for each possible number of units of the next item that can be chosen. For the last item, we fill the knapsack by adding systematically as many units of this item as we can. A part of the tree corresponding to this example is depicted in Figure 8.3. Its height is equal to the number of different items, n. Each leaf corresponds to a solution, and the number of leaves is exponential in the problem size.

Note that we have ordered the items such that the c_i/w_i are nonincreasing, i.e., the first item has the best value/weight ratio.

Given a search space represented by a tree, the branch-and-bound algorithm works as follows. At the beginning, there is only one active node, the root of the tree. At each step, we choose an active node, and we process its children nodes. If a child has only one child itself, we traverse the branch until we eventually find a leaf or a node with at least two children. Then we evaluate the node as follows: (i) If the node is a leaf, it corresponds to a solution, and we can compute the exact value of this solution. We keep the best solution between case (i) and the previously best known solution; (ii) otherwise, we provide an upper bound on the solutions in the branch by filling the unused weight with the item that has not yet been considered and that has the best value/weight ratio as if it were a liquid, that is, as if we were allowed to use a noninteger number of items. All the nodes from case (ii) become active. Before moving to the next step (i.e., picking up a new active node), we remove the active nodes that will never lead to a better solution than one of the solutions

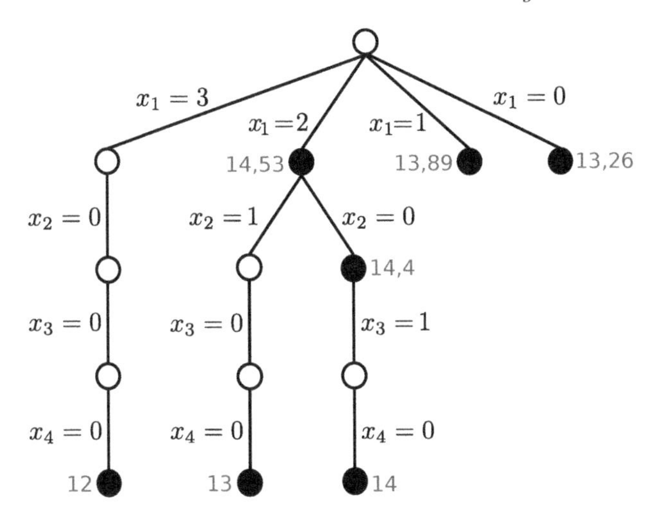

FIGURE 8.3: Branch-and-bound algorithm for the knapsack problem.

already found, i.e., if their upper bound is smaller than the value of the best solution. This corresponds to the pruning of the search space.

In the example (see Figure 8.3), we first process the child node corresponding to $x_1 = 3$. We cannot add any other item in the knapsack, so we reach a leaf of the tree. The value of the solution is $3 \times 4 = 12$. This is the best current solution. Then, we consider the second child node of the root, corresponding to $x_1 = 2$. It has two children, corresponding to $x_2 = 1$ and $x_2 = 0$ (we cannot add more than one unit of item I_2 in the knapsack). Therefore, we evaluate this node. The upper bound is computed with $x_1 = 2$, and all the remaining space $(130 - 66)$ is filled with item I_2, which is the remaining item with the best value/weight ratio. We obtain $2 \times 4 + 5/49 \times (130 - 66) = 14.53$. Because $14.53 \geqslant 12 + 1$, it may be possible to find a better solution than the current one (whose value is 12) in this tree, i.e., a solution whose value is at least 13 (solutions are integers). Therefore, this node becomes active. With $x_2 = 1$, we obtain a solution of value 13. Then, we evaluate the node for $x_2 = 0$ because there may still be a solution of value 14 in this subtree. The evaluation is done by filling the remaining space with item I_3, leading to $2 \times 4 + 6/60 \times (130 - 66) = 14.4$. This branch leads to a solution 14, with $x_3 = 1$. Because the upper bound for this subtree is 14.4, we cannot find a better solution. We evaluate the third child of the root to 13.89 and then the last child to 13.26; therefore, no better solution can be found. There are no more active nodes. The nodes of the tree colored in black are the nodes that have been evaluated.

Note that several strategies can be considered for the choice of the next active node. A depth-first search, as we have done in the example, is very practical because there are few nodes that are simultaneously active. A breadth-first search often leads to poor results. Another strategy consists of picking the active node with the best evaluation. Some hybrid strategies also can be considered. For instance, one can perform a depth-first search until a solution is found and then use a best evaluation strategy to find even better solutions. Such a strategy may allow the pruning of several branches.

Note that other strategies can be used to solve this kind of problem. The branch-and-bound algorithm is often not very efficient in the worst case. However, it often leads to efficient algorithms on average, as we detail in the next section.

8.5.3 Graph algorithms

In this section, we consider two important NP-complete graph problems that we aim to solve with backtracking algorithms. First, we investigate the problem of finding the largest independent set, and then we investigate the graph coloring problem.

8.5.3.1 Independent sets

Let $G = (V, E)$ be a graph with n vertices, numbered from 1 to n. The problem is to find the size of the largest independent set of G, i.e., a subset $S \subseteq V$ such that, for all $i, i' \in S$, $(i, i') \notin E$, and $|S|$ is maximum.

The backtracking algorithm is easy to describe and analyze for this problem. The idea is to explore all possible independent sets and to build a tree with all the solutions to the problem. The root of the tree corresponds to the empty set. The children of the root node correspond to independent sets of size 1, and we add a node only if it is an independent set. The tree is built in a depth-first traversal. First, we search for independent sets containing vertex 1, which correspond to the first child of the root, denoted $\{1\}$ (if $(1, 1) \notin E$). We then try to increase the size of this set by adding vertex 2. The children of $\{1\}$ are the independent sets of size 2 containing vertex 1. If $(1, 2) \in E$, then there is no independent set containing both 1 and 2; therefore, we do not add any node in the solution tree and proceed with vertices $3, \ldots, n$. Otherwise, we add $\{1, 2\}$ as a child node of $\{1\}$ and move to the next level of the tree, trying to add vertices $3, \ldots, n$ to this independent set and building independent sets of size 3. When no vertex can be further added, we *backtrack* up in the tree and develop all remaining branches of the solution tree. The height of the solution tree gives the maximum size of an independent set.

The solution tree has one node per independent set and, therefore, the complexity of the algorithm depends on the number of independent sets, which can be exponential: For a graph with $E = \emptyset$, this number is 2^n. However, for a clique of size n, there are only $n + 1$ independent sets. The analysis aims at determining the average complexity of the algorithm, i.e., the average number of independent sets, denoted I_n.

Let $I(G)$ be the number of independent sets of a graph $G = (V_G, E)$.

$H(G, S)$ equals 1 if S is an independent set of G, and 0 otherwise. Therefore, $I(G) = \sum_{S \subseteq V_G} H(G, S)$, and the sum contains the 2^n possible subsets of V.

The average number of independent sets I_n is then the sum over all possible graphs G with n vertices, divided by the number of such graphs, $2^{n(n-1)/2}$. We obtain

$$I_n = 2^{-n(n-1)/2} \sum_{|V_G|=n} \sum_{S \subseteq V_G} H(G, S) .$$

We can invert the two sums, and we examine $\sum_{|V_G|=n} H(G, S)$. Given a set S, this value corresponds to the number of graphs with n nodes that contain S as an independent set. If $|S| = k$, there are $k(k-1)/2$ edges that cannot exist in G, and there are $n(n-1)/2 - k(k-1)/2$ possible edges, which leads to $2^{n(n-1)/2 - k(k-1)/2}$ graphs with n vertices such that $H(G, S) = 1$. Finally, since the number of sets S with k vertices is $\binom{n}{k}$, we obtain

$$I_n = \sum_{k=0}^{n} \binom{n}{k} 2^{-k(k-1)/2} .$$

On average, the algorithm is much better than in the worst case; for instance, with $n = 40$, $I_n = 3862.9$, while $2^n > 10^{12}$. In fact, for large values of n, $I_n = O(n^{\log(n)})$ and, therefore, the average complexity of the algorithm remains subexponential.

8.5.3.2 Graph coloring

For the graph coloring problem, the backtracking algorithm leads to more efficient results on average than for the independent sets problem because it turns out that the average complexity is, in fact, constant for a fixed number of colors, even when the number of vertices tends to infinity.

Let $G = (V, E)$ be a graph with n vertices, numbered from 1 to n, and K be an integer. The K-coloring problem is to associate a color with each vertex such that two vertices connected by an edge have a different color, where K is the number of colors.

The backtracking algorithm builds all partial colorings of the graph with only a subset of vertices $\{1, \ldots, L\}$, with $1 \leqslant L \leqslant n$. The root of the tree corresponds to the coloring of the empty graph; it is represented by an empty set. It has K children nodes, corresponding to the possible colors for vertex 1. The node is labeled by the set of colors for the vertices that we consider, i.e., the children of the root are labeled $1, \ldots, K$. Similar to the backtracking algorithm for the independent sets problem, we build the tree in a depth-first traversal. We add a node 11 as a child of 1 if and only if $(1, 2) \notin E$, then we assign the lowest possible color to the third vertex, and so on. If there is no possible color for one of the vertices, or if we have successfully colored all vertices, we go up in the tree until we can try another color for one of the vertices. (Remember that the backtracking algorithm builds *all* partial colorings of the graph.)

Note that the branch of a tree may stop before a color has been assigned to each vertex, and it may happen that no valid coloring can be found. At level L of the tree, we have all partial colorings of vertices $\{1, \ldots, L\}$, and a valid coloring has been found if the tree has nodes of level n. Graph G restricted to vertices $\{1, \ldots, L\}$ is denoted $H_L(G)$ in the following.

The goal is to determine the average number of nodes $A_{n,K}$ of a backtrack tree generated when coloring a graph of size n with at most K colors. There are $2^{n(n-1)/2}$ different graphs, and we decompose the backtrack trees into levels. If G is a graph with n vertices, we denote by $P(K, H_L(G))$ the number of nodes at level L of the backtrack tree of G. It is equal to the number of correct colorings of graph $H_L(G)$ with K colors. Finally,

$$A_{n,K} = 2^{-n(n-1)/2} \sum_{|V_G|=n} \sum_{L=0}^{n} P(K, H_L(G)) \,.$$

We invert the two sums and examine $\sum_{|V_G|=n} P(K, H_L(G))$, given a level L. Note that there are exactly $2^{n(n-1)/2 - L(L-1)/2}$ graphs that share the same graph $H_L(G)$, and, therefore,

$$A_{n,K} = 2^{-n(n-1)/2} \sum_{L=0}^{n} 2^{n(n-1)/2 - L(L-1)/2} B_{L,K} = \sum_{L=0}^{n} 2^{-L(L-1)/2} B_{L,K} \,,$$

where $B_{L,K}$ is the total number of correct colorings with K colors of all graphs with L vertices. Given a coloring, we denote by s_i the number of vertices that are colored with the color i, for $1 \leqslant i \leqslant K$. Because the graphs have L vertices, we have $\sum_{i=1}^{K} s_i = L$. Moreover, an edge can connect only two vertices of different colors, and, thus, the maximum number of edges is $E_{n,K} = s_1 s_2 + s_1 s_3 + \cdots + s_1 s_K + s_2 s_3 + \cdots + s_{K-1} s_K = \sum_{1 \leqslant i < j \leqslant K} s_i s_j$. We compute this value as follows:

$$\begin{aligned} E_{n,K} &= \tfrac{1}{2} \textstyle\sum_{i \neq j} s_i s_j = \tfrac{1}{2} \left(\sum_{i,j=1}^{K} s_i s_j - \sum_{i=1}^{K} s_i^2 \right) \\ &= \tfrac{1}{2} \left(\textstyle\sum_{i=1}^{K} s_i \right)^2 - \tfrac{1}{2} \sum_{i=1}^{K} s_i^2 = \tfrac{1}{2} L^2 - \tfrac{1}{2} \sum_{i=1}^{K} s_i^2 \,. \end{aligned}$$

It is easy to check that $\sum_{i=1}^{K} s_i^2 \geqslant L^2/K$, because $L = \sum_{i=1}^{K} s_i$:

$$\begin{aligned} \textstyle\sum_{i=1}^{K} s_i^2 - L^2/K &= \textstyle\sum_{i=1}^{K} s_i^2 - 2L^2/K + L^2/K \\ &= \textstyle\sum_{i=1}^{K} \left(s_i^2 - 2L s_i/K + L^2/K^2 \right) = \sum_{i=1}^{K} \left(s_i - L/K \right)^2 \geqslant 0 \,. \end{aligned}$$

Therefore, $E_{n,K} \leqslant \frac{1}{2} L^2 - \frac{1}{2} L^2/K = L^2(1 - 1/K)/2$. The number of graphs $H_L(G)$ with the same coloring is at most $2^{L^2(1-1/K)/2}$. Because there are at most K^L different colorings (counting invalid ones), we obtain $B_{L,K} \leqslant K^L 2^{L^2(1-1/K)/2}$, and, finally,

$$A_{n,K} \leqslant \sum_{L=0}^{n} 2^{-L(L-1)/2} K^L 2^{L^2(1-1/K)/2} \leqslant \sum_{L=0}^{\infty} K^L 2^{L/2} 2^{-L^2/2K} .$$

This infinite series is converging; therefore, $A(n, K)$ is bounded for all n.

8.6 Bibliographical notes

The FPTAS for scheduling independent tasks on two processors (Section 8.1.5) is presented in [95]. Further references for approximation algorithms are the books by Ausiello et al. [5] and by Vazirani [103]. Randomized algorithms (Section 8.4) are dealt with in the books by Mitzenmacher and Upfal [78] and by Motwani and Raghavan [80]. Section 8.5.2 (branch-and-bound) is inspired from [15]. The backtracking graph algorithms (Section 8.5.3) are analyzed in [108].

Chapter 9

Exercises going beyond NP-completeness

This chapter presents a set of exercises related to Chapter 8. The main focus is on approximation results (Section 9.1), and there are also exercises dealing with linear programming, randomized algorithms, and branch-and-bound techniques (Section 9.2). Solutions are provided in Section 9.3.

9.1 Approximation results

Exercise 9.1: Single machine scheduling (solution p. 219)

We have n independent tasks, $T_1, \ldots, T_n$, to execute on a single computer. For any $1 \leqslant j \leqslant n$, the execution time of task T_j is w_j and task T_j is submitted to the system at time r_j. In other words, the processing of task T_j cannot start earlier than time r_j. We denote by C_j the completion time of task T_j, that is, the time at which its processing completes. We assume that w_j is a positive integer and r_j is a nonnegative integer.

The objective function is the sum of the completion times: $\mathcal{S} = \sum_{j=1}^{n} C_j$.

1. We assume in this question that tasks can be *preempted* (we explain below in detail what preemption is). Prove that in this framework the optimal solution is obtained by scheduling during each interval $[i; i+1]$, for any nonnegative integer i, the task whose remaining processing time is the smallest. This scheduling algorithm is called *Shortest Remaining Processing Time* first, or SRPT.

   To explain preemption, let us consider a task T_j that is executed during the time interval $[i; i+1]$ (where i is some integer) and whose processing has not completed at time $i+1$. We allow another task T_k (with $j \neq k$) to be processed during the time interval $[i+1; i+2]$. The execution of task T_j will eventually be resumed. The sum of the durations of the time intervals during which T_j is executed is equal to w_j; when the execution of T_j resumes after a preemption, it restarts from where it was interrupted.

2. We order the tasks by their completion times under an execution of algorithm SRPT using preemption. Prove that scheduling the tasks according to this order defines an approximation algorithm for the case without preemption. (We admit that the problem is NP-hard.)

Exercise 9.2: SUBSET-SUM (solution p. 221)

In Exercise 7.19 (p. 155), we proved that the decision problem SUBSET-SUM was NP-complete. Here we focus on the associated optimization problem: Given a finite set S of positive integers, and an integer t, we look for a subset S' of S such that the sum of its elements is the largest possible while being not greater than t ($\sum_{x \in S'} x \leqslant t$). We call the sum of the elements of S' the *optimal sum*.

Let $S = \{x_1, x_2, \dots, x_n\}$. In the following, we assume that all the lists are sorted in nondecreasing order. Given a list of integers L and an integer x, $L + x$ denotes the list of integers obtained by adding x to each of the elements of L. Given two (sorted) lists L and L', we denote by **Merge**(L, L') the sorted union of the elements of the two lists. Algorithm 9.1 is an attempt at finding the optimal sum.

```
1 n ← |S|
2 L_0 ← {0}
3 for i = 1 to n do
4 |  L_i ← Merge(L_{i-1}, L_{i-1} + x_i)
5 |_ Discard from L_i any element greater than t
6 return the largest element of L_n
```

ALGORITHM 9.1: Sum(S, t).

1. Does Algorithm 9.1 return the optimal sum?
2. What is the complexity of Algorithm 9.1?

In order to define an approximation algorithm for the SUBSET-SUM problem, we introduce Algorithm 9.2, which takes as input a sorted list and a threshold δ and outputs a subset of the original list such that two consecutive elements are at least at a factor $1 + \delta$ from each other. We use this algorithm to define Algorithm 9.3.

3. Evaluate the number of elements in L_i at the end of Step 6. What is the complexity of Algorithm 9.3?
4. Show that Algorithm 9.3 is a fully polynomial-time approximation scheme (FPTAS).

```
1 L' ← ⟨y_1⟩
2 last ← y_1
3 for i = 2 to m do
4     if y_i > (1 + δ)last then
5         Insert y_i at the end of L'
6         last ← y_i
7 return L'
```

ALGORITHM 9.2: Threshold(L, δ), where $L = \langle y_1, \ldots, y_m \rangle$ is sorted.

```
1 n ← |S|
2 L_0 ← {0}
3 for i = 1 to n do
4     L_i ← Merge(L_{i-1}, L_{i-1} + x_i)
5     L_i ← Threshold(L_i, ε/2n)
6     Discard from L_i each element greater than t
7 return the largest element of L_n
```

ALGORITHM 9.3: Sum-with-threshold(S, t, ε).

Exercise 9.3: SET-COVER (solution p. 223)

The decision problem SET-COVER is defined as follows: Given a set X containing n elements, m subsets of X, $S_1, \ldots, S_m$, and an integer $k \leqslant n$, is there a set of k of the subsets that covers all elements of X? In other words, is there a subset I of $\{1, \ldots, m\}$, $|I| \leqslant k$, such that for all $x \in X$, $x \in \cup_{i \in I} S_i$?

SET-COVER is an NP-complete problem [38]. The associated optimization problem is to find the smallest cover. We consider the following greedy algorithm for this optimization problem:

- Pick a subset S_j that covers the largest number of elements of X.
- Suppress from X and from the remaining S_is the elements covered by S_j, and start again.

1. Is this greedy algorithm optimal?
2. Prove that if an optimal solution contains k subsets, the greedy algorithm takes at most $O(k \ln(n))$ subsets.
3. Is this greedy algorithm a λ-approximation of SET-COVER? If so, what is the value of λ?

Exercise 9.4: VERTEX-COVER (solution p. 224)

We recall here the VERTEX-COVER problem that was introduced in Section 6.4.3 (p. 139): Let $G = (V, E)$ be a graph and k be an integer such that

$1 \leqslant k \leqslant |V|$; do there exist k vertices $v_{i_1}, \ldots, v_{i_k}$ such that each edge $e \in E$ is covered by (at least) one of the v_{i_j}, for $1 \leqslant j \leqslant k$?

We consider here the weighted version. We have a weight function on the vertices $w : V \rightarrow \mathbb{Q}+$, and the optimization problem is the search for a vertex-cover $S \subseteq V$ whose weight is minimal. To build a 2-approximation, we introduce a particular weight function where the weight is proportional to the degree (there exists a constant c such that for any vertex $v \in V$, $w(v) = c \times \text{degree}(v)$).

1. Let $\{G = (V, E), w\}$ be an instance of VERTEX-COVER such that there exists a constant c that satisfies $w(v) = c \times \text{degree}(v)$ for any vertex $v \in V$. Show that $w(V) \leqslant 2w(OPT)$ where $w(V) = \sum_{v \in V} w(v)$ and OPT is an optimal cover.

2. We know how to build a 2-approximation of any instance where the weight function is proportional to the degree. Propose a method to build a weight function w' that is proportional to the degree, not greater than w (i.e., $w'(v) \leqslant w(v)$ for all $v \in V$), and maximal.

3. We consider Algorithm 9.4. Prove that this algorithm terminates in polynomial time.

```
1  t ← 0
2  G_0 ← G
3  w_0 ← w
4  while G_t contains at least one edge do
5  |   D_t ← {u ∈ V_t : degree_t(u) = 0}
6  |   w'_t ← maximal weight function of G_t proportional to the degree
   |   and not greater than w_t
7  |   S_t ← {u ∈ V_t | w'_t(u) = w_t(u)}
8  |   G_{t+1} ← G_t \ (D_t ∪ S_t)
9  |   w_{t+1} ← w_t − w'_t
10 |_  t ← t + 1
11 return C = ⋃_{k=0}^{t−1} S_k
```

ALGORITHM 9.4: Building a vertex-cover of $G = (V, E, w)$.

4. Prove that the set C returned is a vertex cover.

5. For each vertex v in the cover C, express $w(v)$ as a function of the $w'_i(v)$s. What about the vertices not in the cover?

Note: we will let $w'_i(u) = 0$ for any vertex u that does not belong to G_i.

6. Let C^* be an optimal vertex cover for $\{G = (V, E), w\}$. At each iteration i of the "while" loop, compare $w'_i(C \cap G_i)$ and $w'_i(C^* \cap G_i)$.

7. Prove that Algorithm 9.4 is a 2-approximation for the minimum weight vertex cover.

Exercise 9.5: Scheduling independent tasks in parallel (solution p. 226)

We aim at scheduling a set $V = \{v_1, v_2, \ldots, v_n\}$ of n independent tasks, where each task v_i has a nonnegative integer weight (or size) $w(v_i) = a_i$ $(1 \leqslant i \leqslant n)$, on p identical processors (see problem INDEP(p) in Section 6.4.4). We have shown that this problem is NP-complete in the strong sense in Exercise 7.10. We extend to the p processor case both greedy algorithms introduced in Section 8.1.5 (p. 187) for the two-processor case: **greedy-online** that considers the task in an arbitrary order, and **greedy-offline** that considers the task in nonincreasing weight. Both algorithms assign the next task on the least loaded processor.

1. Determine λ such that **greedy-online** is a λ-approximation algorithm. Prove that your bound is tight.

2. Determine λ such that **greedy-offline** is a λ-approximation algorithm. Prove that your bound is tight.

Exercise 9.6: Point clustering (solution p. 228)

We consider a set $\mathcal{S}$ of n points of a metric space (distances satisfy the triangle inequality). We want to partition $\mathcal{S}$ into k groups such that the maximum diameter of a group is minimized. The diameter of a group is the maximum distance between two points in the group. In this problem, n and k are fixed given values.

1. We suppose that the optimal diameter d is known. Propose a 2-approximation algorithm.

2. Let us consider an algorithm that chooses k times as "center" the point that is the farthest from the already chosen centers. This algorithm then assigns each point to the center closer to it. Using the result of the previous question, show that this algorithm is also a 2-approximation algorithm.

Exercise 9.7: k-center (solution p. 229)

Let $G = (V, E)$ be a complete graph whose edges are weighted by a function w that satisfies the triangle inequality: $w(u, v) \leqslant w(u, w) + w(w, v)$ for any three vertices u, v, and w. For any subset S of the vertices, $S \subset V$, and each vertex v, $v \in V \setminus S$, let $connect(v, S)$ be the minimum weight of an edge linking v to a vertex of S: $connect(v, S) = \min_{s \in S} w(v, s)$. Let k be a positive integer ($k \geqslant 1$). The problem here is to find a k-center, that is, a subset S of size k of the vertices and such that $center(S) = \max_{v \in V \setminus S} connect(v, S)$ is minimal. We will build a 2-approximation of the k-center problem, i.e., a subset S of size k with $center(S) \leqslant 2OPT$, where $OPT = \min_{S \subset V, |S|=k} center(S)$.

1. We sort the edges of E by nondecreasing weights: $w(e_1) \leqslant w(e_2) \leqslant \cdots \leqslant w(e_m)$, where $m = |E|$. We let $G_i = (V, E_i)$ where $E_i = \{e_1, e_2, \ldots, e_i\}$ is the set of the first i edges. Show that solving the k-center problem is equivalent to finding the smallest index i such that G_i has a *dominating set* of size at most k, i.e., a subset D of V of size k such that for any vertex $u \in V \setminus D$ there exists a vertex $v \in D$ such that $(u, v) \in E_i$.

2. The square of a graph $G = (V, E)$, denoted $G^{(2)} = (V, E^{(2)})$, is the graph whose edges are the paths of length at most two of G: $(u, v) \in E^{(2)}$ if $(u, v) \in E$ or if there exists $w \in V$ such that $(u, w) \in E$ and $(w, v) \in E$. An *independent set* is a subset V' of the set of vertices, $V' \subset V$, such that no two vertices of V' are linked by an edge. If H is any graph and I an independent set of its square graph $H^{(2)}$, prove that $|I| \leqslant dom(H)$ where $dom(H)$ is the size of the smallest dominating set of H.

3. We study Algorithm 9.5. Prove that (i) $w(e_j) \leqslant OPT$, and (ii) this algorithm is a 2-approximation algorithm.

```
1 Build the graphs G_1^(2), G_2^(2), ..., G_m^(2)
2 for each i in [1;m] do
3     Greedily build a maximal independent set M_i of the graph G_i^(2)
      (i.e., an independent set to which no vertex can be added)
4 Let j be the smallest index in [1;m] such that |M_j| <= k
5 return M_j
```

ALGORITHM 9.5: 2-approximation k-center for graph G.

4. Prove that the approximation ratio of 2 is tight. Exhibit a graph on which the algorithm solution is indeed twice as large as the optimal one.

5. Prove that, if P $\neq$ NP, there does not exist a $(2-\varepsilon)$-approximation algorithm for the k-center for any $\varepsilon > 0$.

Exercise 9.8: Knapsack (solution p. 231)

Given a finite set X of n objects, each object $x_i \in X$ having a value $p_i \in \mathbb{N}$ and a size $a_i \in \mathbb{N}$, the problem is to find a subset Y of X whose value is maximal and whose total size $\sum_{x_i \in Y} a_i$ is not greater than a given bound B. We denote by $m^*(X)$ the optimal value. We build two approximation algorithms. The first one is a greedy algorithm (Questions 1 and 2), and the second one is a dynamic program (Questions 3 and 4).

1. We consider the greedy algorithm that adds objects to the solution by nonincreasing value of the ratio p_i/a_i. Let $m_g(X)$ denote the value of its solution. For any integer K, build an instance X_K such that $m^*(X_K)/m_g(X_K) > K$.

2. Let p_{max} be the maximal value of an object. Prove that

$$\frac{m^*(X)}{\max\{m_g(X), p_{max}\}} < 2$$

   and propose a 2-approximation algorithm.

   (*Hint.* Let j be the index of the first object not taken by the greedy algorithm. Prove that $m^*(X) \leqslant \sum_{i=1}^{j} p_i$.)

3. We now build a dynamic program solving the knapsack problem but that differs from the classic one (i.e., Algorithm 4.3, p. 86). For any k $(1 \leqslant k \leqslant n)$ and any candidate value p $(0 \leqslant p \leqslant \sum_{i=1}^{n} p_i)$, we look among all the subsets of $\{x_1, x_2, \ldots, x_k\}$ of value *equal* to p and of size at most B for one subset of minimal size. We denote by $M^*(k,p)$ the optimal solution and by $S^*(k,p)$ its size. Propose an algorithm to compute the $M^*(k,p)$s through a dynamic programming approach. What is the complexity of your algorithm?

4. For any rational $r > 1$, we consider the following approximation scheme. Let p_{max} be the maximal value of an object and let $t = \lfloor \log_2(\frac{r-1}{r}\frac{p_{max}}{n}) \rfloor$. Let X' be the instance defined by the set of n objects $\{(p'_i, a_i)\}_{i \in [1;n]}$ where $p'_i = \lfloor \frac{p_i}{2^t} \rfloor$. We use the dynamic program of Question 3 to solve the knapsack problem for instance X'. From this solution we build a solution for X; the solution for X contains the i-th object if and only if the solution for X' contains it. We then denote by $m_{AS}(X,r)$ the value of the solution for X obtained by this method.

   (a) Prove that the execution time of this method is $O(\frac{r}{r-1}n^3)$.

   (b) Show that $\dfrac{m^*(X)}{m_{AS}(X,r)} \leqslant r$.

   (*Hint:* prove that $\frac{m^*(X) - m_{AS}(X,r)}{m^*(X)} \leqslant \frac{n2^t}{p_{max}}$.)

9.2 Dealing with NP-complete problems

Exercise 9.9: Mixed integer linear program for replica placement (solution p. 234)

We consider a tree $\mathcal{T}$ whose set of nodes is $\mathcal{C} \cup \mathcal{N}$, where the clients $\mathcal{C}$ are leaves of the tree. Each client $i \in \mathcal{C}$ has $r_i \in \mathbb{N}$ requests; each node $j \in \mathcal{N}$ has a processing capacity $W_j \in \mathbb{N}$. We need to decide which node is equipped with a replica and thus becomes a server ($j \in \mathcal{R}$, where $\mathcal{R}$ is the set of replicas). Requests are subject to a distance constraint, i.e., the requests of $i \in \mathcal{C}$ can be served only by nodes j on the path from i to the root of the tree ($j \in \mathit{ancestors}(i)$), such that the number of hops between i and j, $\mathit{dist}(i,j)$, does not exceed d_i. The goal is then to minimize the cost of replicas, i.e., $\sum_{j \in \mathcal{R}} W_j$.

1. Give an integer linear program to solve the problem based on the following variables:
   - x_j is a Boolean variable equal to 1 if j is a server (for one or several clients);
   - $y_{i,j}$ is an integer variable equal to the number of requests from client i processed by node j.

2. We solve the previous linear program without enforcing that the $y_{i,j}$s are integer but allowing $y_{i,j} \in \mathbb{Q}$, for $i \in \mathcal{C}$ and $j \in \mathcal{N}$. Show how we can build an exact solution in polynomial time based on the mixed LP solution.

3. What can we say if the x_js are also rational numbers?

Exercise 9.10: A randomized algorithm for independent set (solution p. 237)

An independent set in a graph is a subset of vertices with no edge between them. Finding the largest independent set is NP-hard, as we have seen in Exercise 7.13 (p. 153). Let $G = (V, E)$ be a graph with $n = |V|$ vertices and $m = |E|$ edges. Assume that $d = \frac{2m}{n} \geqslant 1$. This exercise uses a randomized algorithm to show that there always exists an independent set with at least $\lceil \frac{n^2}{4m} \rceil$ vertices. The algorithm is the following:

Step 1: Delete each vertex of G, together with all its incident edges, independently and with probability $1 - \frac{1}{d}$.

Step 2: For each remaining edge, remove it and one of its adjacent vertices.

Show that when the algorithm terminates, it outputs an independent set whose expected size is at least $\frac{n^2}{4m}$. Deduce the result.

Exercise 9.11: Branch-and-bound applied to MAX-SAT (solution p. 237)

We consider the problem MAX-SAT, which is the optimization version of the classical SAT optimization problem (see Definition 6.6). We want to maximize the number of clauses that are satisfied.

1. Draw the complete search tree with all solutions for this problem for the formula $(x_1 \vee \overline{x_2}) \wedge (x_1 \vee x_3 \vee \overline{x_4}) \wedge (\overline{x_1} \vee x_2) \wedge (x_1 \vee \overline{x_3} \vee x_4) \wedge (x_2 \vee x_3 \vee \overline{x_4}) \wedge (x_1 \vee \overline{x_3} \vee \overline{x_4}) \wedge x_3 \wedge (x_1 \vee x_4) \wedge (\overline{x_1} \vee \overline{x_3}) \wedge x_1$. Can this formula be satisfied? What is the optimal solution to MAX-SAT?

2. When we find a solution, how can we cut some branches of the tree to reduce the search space? Propose an algorithm based on a depth-first-search strategy.

3. Can you adapt the algorithm using a breadth-first-search strategy? Compare its time complexity with the previous algorithm (in terms of number of generated vertices) in the example.

4. Apply both algorithms when the order of the variables is modified to x_4, x_3, x_1, x_2, and compare the number of visited vertices using the different search strategies.

9.3 Solutions to exercises

Solution to Exercise 9.1: Single machine scheduling

1. To prove the optimality of the *Shortest Remaining Processing Time* (SRPT) first policy, we use an exchange argument. Let $\mathcal{I}$ be an instance of the problem and let $\mathcal{A}$ be a scheduling algorithm whose output is optimal in this instance for our objective function, the minimization of the sum of completion times. We then consider the first time t at which the behaviors of $\mathcal{A}$ and of SRPT differ. Then, at time t SRPT processes a task T_i and $\mathcal{A}$ processes a task T_j, with $i \neq j$. From $\mathcal{A}$ we construct an alternative schedule $\mathcal{A}'$. This construction is illustrated in Figure 9.1. $\mathcal{A}'$ is identical to $\mathcal{A}$ before the date t and, after that date, for all tasks except T_i and T_j. After date t, during each time interval during which $\mathcal{A}$ processed either T_i or T_j, $\mathcal{A}'$ first processes T_i and, when the processing of T_i is completed, $\mathcal{A}'$ processes T_j.

   Let C_k (respectively C'_k) denote the completion time of task T_k, for any $k \in [1;n]$, under $\mathcal{A}$ (respectively $\mathcal{A}'$). Only the completion times of T_i and T_j are modified when transforming $\mathcal{A}$ into $\mathcal{A}'$. Therefore, for any $k \in [1;n] \setminus \{i,j\}$, $C_k = C'_k$. T_i is completed earlier under $\mathcal{A}'$ than under $\mathcal{A}$; therefore, $C'_i \leqslant C_i$. Furthermore, at time t, as SRPT decides to schedule T_i rather than T_j, this means that the remaining processing time of T_j is not smaller than that of T_i. Therefore, T_i is completed under $\mathcal{A}'$ not later than T_j under $\mathcal{A}$: $C'_i \leqslant C_j$. Moreover, as $\mathcal{A}$ and $\mathcal{A}'$ use the same time slots to process T_i and T_j after t, $\max\{C_i, C_j\} = \max\{C'_i, C'_j\}$. We have two cases to consider:

(a) $\max\{C_i, C_j\} = C_j$. Then, $C'_j = C_j$. Because $C'_i \leqslant C_i$, we obtain $C'_i + C'_j \leqslant C_i + C_j$.

(b) $\max\{C_i, C_j\} = C_i$. Then, $C'_j = C_i$. Because $C'_i \leqslant C_j$, we obtain $C'_i + C'_j \leqslant C_j + C_i$.

In all cases we have:

$$\sum_{k\in[1;n]} C'_k = \left(\sum_{k\in[1;n]\setminus\{i,j\}} C'_k \right) + C'_i + C'_j \leqslant \left(\sum_{k\in[1;n]\setminus\{i,j\}} C_k \right) + C_i + C_j.$$

Therefore, $\mathcal{A}'$ is optimal for instance $\mathcal{I}$ because $\mathcal{A}$ is optimal by hypothesis. If the solution produced by $\mathcal{A}'$ differs from SRPT, its first difference happens strictly later than for $\mathcal{A}$. We then iterate the transformation process we just applied to $\mathcal{A}$. We will eventually obtain a schedule identical to SRPT and whose sum of completion times is optimal, hence, proving the desired result.

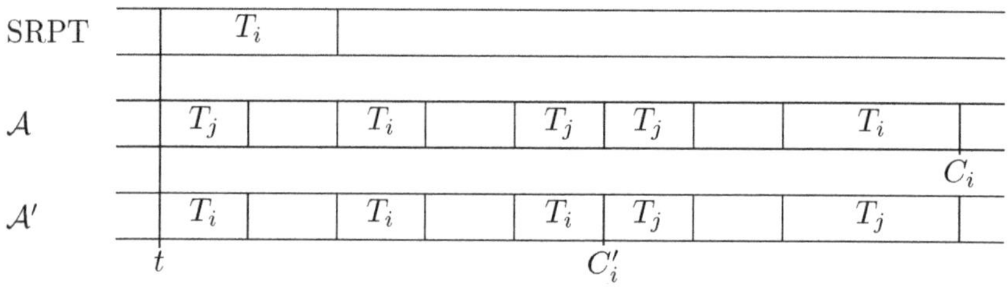

FIGURE 9.1: Exchange used to prove the optimality of SRPT.

2. We schedule the tasks in the order of their completion times under the SRPT schedule with preemption. This way we define a schedule S, under which task T_j completes at time C_j. The same task completed at time C'_j under SRPT with preemption. Under S, the tasks that complete before T_j starts are exactly the ones that complete before T_j completes under SRPT-preemption. As T_j completes at time C'_j under SRPT-preemption, all of these tasks complete by that time under SRPT-preemption. Therefore, the sum of the execution times of these tasks is not greater than C'_j (because all release dates are nonnegative). The completion time of T_j under S is equal to the release time r_j of T_j,

plus the remaining processing time of the tasks scheduled before T_j and not completed by time r_j, plus the execution time of T_j. Therefore, $C_j \leqslant r_j + C'_j + w_j \leqslant 2C'_j$ because, obviously, $r_j + w_j \leqslant C'_j$. Finally,

$$\sum_{j\in[1;n]} C_j \leqslant 2 \sum_{j\in[1;n]} C'_j = 2\mathcal{S}^{\text{SRPT-PREEMPTION}} \leqslant 2\mathcal{S}^{\text{OPT}}$$

where $\mathcal{S}^{\mathcal{A}}$ denotes the sum of completion times under algorithm $\mathcal{A}$. We establish this inequality by remarking that SRPT is optimal for minimizing the sum of completion times with preemption and that the optimal sum of completion times without preemption cannot be smaller than the optimal sum of completion times with preemption.

Solution to Exercise 9.2: SUBSET-SUM

1. In Algorithm 9.1, for any $i \in [1;n]$, L_i is the set of all partial sums of elements of $\{x_1, \ldots, x_i\}$ that are not greater than t (this is trivially proved by induction). Therefore, $\max(L_n)$ is the optimal sum, and Algorithm 9.1 always returns the optimal sum.

2. In the worst case, not a single value is discarded from L_i at Step 5, for any $i \in [1;n]$. This worst case occurs when $t \geqslant \sum_{i=1}^{n} x_i$. In the worst case, $|L_i| = 2 \times |L_{i-1}|$. This happens if there are never two distinct subsets of S that give rise to the same sum. The instance $S = \{1, 2, 2^2, \ldots, 2^i, \ldots, 2^{n-1}\}$ with $t \geqslant 2^n$ leads to such a worst case. Then, as the complexity of the merging of two sorted lists is the sum of the size of the input lists, the overall complexity of Algorithm 9.1, in the worst case, is $O(2^n)$.

3. Algorithm 9.2 discards elements that are too close to already kept elements (remember that elements are considered in nondecreasing order). We consider the list L_n at the end of Algorithm 9.3. Let $L_n = \{y_0, \ldots, y_{m-1}\}$. We want to establish an upper bound on m. For any i in $[1; m-1]$, the value y_i was kept by Algorithm 9.2 because it was greater than $(1+\delta)$ times the currently largest kept value, which was y_{i-1}. Furthermore, in all the calls to Algorithm 9.2, $\delta = \frac{\varepsilon}{2n}$. Therefore, as all integers in S are (strictly) positive:

$$t \geqslant y_{m-1} > (1+\delta)y_{m-2} > (1+\delta)^{m-2}y_0 \geqslant (1+\delta)^{m-2} = \left(1 + \frac{\varepsilon}{2n}\right)^{m-2}.$$

Therefore,

$$m \leqslant 2 + \frac{\log_2(t)}{\log_2(1 + \frac{\varepsilon}{2n})}.$$

We assume that $\varepsilon < 2n$ (as we are interested only in small values of epsilon). Then, as $\log_2(1+x) \geqslant x$ for $x \in [0;1]$, we obtain:

$$m \leqslant 2 + \frac{2n \log_2(t)}{\varepsilon}.$$

m is thus polynomial in $\log_2 t$, in $1/\varepsilon$ et in n, therefore, in the size of the input. Algorithm 9.3 has a complexity of $O(mn)$ because the complexity of each iteration of the loop is linear on the size of the list processed. The overall complexity is then $O\left(n \times \left(\frac{2n \log_2(t)}{\varepsilon} + 2\right)\right) = O\left(\frac{n^2 \log_2(t)}{\varepsilon}\right)$ and is polynomial in the size of the input and in $\frac{1}{\varepsilon}$.

4. As the complexity of Algorithm 9.3 is polynomial in the size of the input and in $\frac{1}{\varepsilon}$, to show that it is a fully polynomial-time approximation scheme, we have only to prove that its output is a $(1+\varepsilon)$-approximation of the optimal sum. In other words, we must show that $\max_{x \in L_n} x \geq \frac{\mathsf{Opt}}{(1+\varepsilon)}$, where Opt is the optimal sum.

   We denote by P_i the set of all the partial sums not greater than t that are obtained from the set of integers $\{x_1, \ldots, x_i\}$. (Then, $\mathsf{Opt} = \max_{x \in P_n} x$.)

   We prove by induction the following property: For any partial sum $x \in P_i$, there exists a partial sum $y \in L_i$, such that $x/(1+\delta)^i \leq y \leq x$.

   We initiate the reduction with the case $i = 1$. If x belongs to $P_i \setminus L_i$, then, by definition of Algorithm 9.2, there exists a partial sum $y \in L_i$ such that $x \leqslant (1+\delta)y$, and we have $y \leqslant x$ because Algorithm 9.2 works on nondecreasing lists. This proves the invariant for the case $i = 1$.

   We now assume that the invariant holds up to rank $(i-1)$ included, and we look at the case i. Let $x \in P_i$. As $P_i = P_{i-1} \cup (P_{i-1} + x_i)$, x can be decomposed only as follows: $x = x' + e$ where $x' \in P_{i-1}$ and where $e = 0$ or $e = x_i$. We then apply the invariant to x' as $x' \in P_{i-1}$. There exists some partial sum $y' \in L_{i-1}$ such that $x'/(1+\delta)^{i-1} \leq y' \leq x'$. We have two cases to consider depending whether $y' + e$ was kept or discarded by Algorithm 9.2 during the construction of L_i.

   (a) $y' + e$ was kept (i.e., $y' + e \in L_i$). Then, using the induction hypothesis recalled above, we get:

   $$\frac{x'+e}{(1+\delta)^i} \leqslant \frac{x'+e}{(1+\delta)^{i-1}} \leqslant \frac{x'}{(1+\delta)^{i-1}} + e \leqslant y' + e \leqslant x' + e.$$

   As $x' + e = x$ and as $y' + e \in L_i$, the invariant is satisfied.

   (b) $y' + e$ was discarded (i.e, $y' + e \notin l_i$). Therefore, there exists an element y'' of L_i such that $y'' \leq y' + e \leq (1+\delta)y''$. In addition to the definition of y'', we use the induction hypothesis to establish:

   $$\frac{x'+e}{(1+\delta)^i} \leqslant \frac{1}{1+\delta}\left(\frac{x'}{(1+\delta)^{i-1}} + e\right) \leqslant \frac{1}{1+\delta}(y'+e) \leqslant y''.$$

   As $y'' \leqslant y' + e \leqslant x' + e$ and as $x' + e = x$, the invariant is also satisfied in this case.

We then apply the invariant to the partial sum Opt: There exists an element $y^* \in L_n$ such that:

$$\mathsf{Opt} = \max_{x \in P_n} x \geqslant y^* \geqslant \frac{\mathsf{Opt}}{(1+\delta)^n} = \frac{\mathsf{Opt}}{\left(1+\frac{\varepsilon}{2n}\right)^n}.$$

As $\ln(1+x) \leqslant x$ for any $x \geqslant 0$, then $\left(1+\frac{\varepsilon}{2n}\right)^n \leqslant e^{\varepsilon/2}$. As for $|x| \leqslant 1$, $e^x \leqslant 1 + x + x^2$, and as $1 + \frac{x}{2} + (\frac{x}{2})^2 \leqslant 1 + x$ for $0 \geqslant x \geqslant 1$, then $\left(1+\frac{\varepsilon}{2n}\right)^n \leqslant 1+\varepsilon$. Therefore,

$$\max_{x \in L_n} x \geqslant y^* \geqslant \frac{\mathsf{Opt}}{(1+\delta)^n} = \frac{\mathsf{Opt}}{\left(1+\frac{\varepsilon}{2n}\right)^n} \geqslant \frac{\mathsf{Opt}}{1+\varepsilon}$$

and Algorithm 9.3 is an $(1+\varepsilon)$-approximation algorithm. Because of the algorithm complexity, this algorithm is an FPTAS.

Solution to Exercise 9.3: SET-COVER

1. The greedy algorithm has obviously a polynomial complexity. We have given, however, a reference to the NP-completeness of the associated decision problem. Therefore, one should not believe for a single second that this algorithm is optimal.

   We consider the following instance of SET-COVER. $X = \{1,2,3,4,5,6\}$, $S_1 = \{1,2,3\}$, $S_2 = \{4,5,6\}$, and $S_3 = \{1,2,5,6\}$. As all subsets of X are strict subsets, any set cover must contain at least two of the subsets, and thus $\{S_1, S_2\}$ is an optimal solution. Furthermore, S_1 is the sole subset containing 3, and S_2 is the sole subset containing 4. Therefore, any solution must contain S_1 and S_2. The greedy algorithm starts by picking S_3. Thus, it builds a solution containing the three subsets and is not optimal.

2. The set X contains n elements. We assume there exists a set cover of X made of k of the subsets $S_1, \ldots, S_m$ with $k > 1$ (we do not consider the trivial case where one of the subsets is equal to X). Therefore, there must exist at least one subset that contains at least $\frac{n}{k}$ elements. Then, S_{i_1}, the first subset chosen by the greedy algorithm, is such a set. Let n_1 be the number of elements of X not covered by S_{i_1}. Then:

$$n_1 = n - |S_{i_1}| \leqslant n - \frac{n}{k} = n\left(1 - \frac{1}{k}\right).$$

   We then apply the previous reasoning to $X \setminus S_{i_1}$ and to $S_1 \setminus S_{i_1}, \ldots, S_m \setminus S_{i_1}$. We know, by hypothesis, that there exists a k-cover of X. Thus, there exists a k-cover of $X \setminus S_{i_1}$. Thus, the second subset chosen by the greedy algorithm, S_{i_2}, contains at least $\frac{|X \setminus S_{i_1}|}{k} = \frac{n_1}{k}$ elements of

$X \setminus S_{i_1}$. Then,

$$n_2 = n_1 - |S_{i_2}| \leqslant n_1 - \frac{n_1}{k} = n_1 \left(1 - \frac{1}{k}\right) \leqslant n \left(1 - \frac{1}{k}\right)^2.$$

It immediately comes by induction that after m iterations of the greedy algorithm, the size n_m of the elements of X that are not yet covered satisfies

$$n_m \leqslant n \left(1 - \frac{1}{k}\right)^m.$$

We want to upper bound m. When m tends to infinity, the right-hand side of the above inequality tends to zero. Therefore, as soon as m is large enough for

$$n \left(1 - \frac{1}{k}\right)^m < 1,$$

the greedy algorithm has built a cover:

$$n \left(1 - \frac{1}{k}\right)^m < 1 \Leftrightarrow \ln(n) + m \ln\left(1 - \frac{1}{k}\right) < 0 \Leftrightarrow m > \frac{-\ln(n)}{\ln\left(1 - \frac{1}{k}\right)}.$$

Therefore, we take as upper bound for m:

$$m_{\lim} = 1 + \frac{-\ln(n)}{\ln\left(1 - \frac{1}{k}\right)}.$$

Then, when $x \in]0;1[$, we have $\ln(1-x) \leqslant -x$. This is equivalent to $\frac{-1}{\ln(1-x)} \leqslant \frac{1}{x}$. Therefore,

$$m_{\lim} \leqslant 1 + k \ln(n).$$

3. The greedy algorithm is an approximation algorithm with an approximation factor of $\ln(n) + 1$ (because $1 + k\ln(n) \leqslant k(\ln(n) + 1)$).

Solution to Exercise 9.4: VERTEX-COVER

1. Let OPT be an optimal vertex cover. By definition, the set of the edges that are incident to the vertices of OPT is E itself. Therefore, $\sum_{v \in \text{OPT}} \text{degree}(v) \geqslant |E|$ (this is an inequality because an edge can be incident to two vertices of OPT). Then,

$$w(\text{OPT}) = \sum_{v \in \text{OPT}} w(v) = c \sum_{v \in \text{OPT}} \text{degree}(v) \geqslant c|E|.$$

We now consider the whole set of vertices as a solution:

$$w(V) = \sum_{v \in V} w(v) = c \sum_{v \in V} \text{degree}(v) = 2c|E|$$

and thus $w(V) \leqslant 2w(\text{OPT})$.

2. We are looking for a weight function w' such that:

   - for any $v \in V$, $w'(v) \leqslant w(v)$;
   - there exists c' such that $w'(v) = c'.\text{degree}(v)$.

   Therefore, for any vertex v in V whose degree is nonnull, we must have $c'.\text{degree}(v) \leqslant w(v)$, which is equivalent to $c' \leqslant \frac{w(v)}{\text{degree}(v)}$. Furthermore, we have no constraints for the vertices whose degree is null. w' is obviously maximal when c' is. The solution to our problem is then to define c' as:

   $$c' = \min_{v \in V, \text{degree}(v) \neq 0} \frac{w(v)}{\text{degree}(v)}.$$

3. At each step $|S_t| \geqslant 1$ because w'_t is maximal. As $G_{t+1} \leftarrow G_t \setminus (D_t \cup S_t)$, then $V_{t+1} \subsetneq V_t$ and the "while" loop is executed at most $|V|$ times. Then, any of the steps inside the "while" loop can be completed in time $O(|V|)$ except the construction of the new graph whose complexity is at most $|V|^2$. The overall complexity of Algorithm 9.4 is polynomial in the size of G and thus in the problem size.

4. Let $e = (u, v)$ be any edge of $E = E_0$. There is an index i such that $e \in E_i$ and $e \notin E_{i+1}$. Therefore, either u, v, or both vertices belong to $(S_i \cup D_i)$. Without loss of generality, we suppose $u \in (S_i \cup D_i)$. As $(u, v) \in E_i$ by hypothesis, $\text{degree}_i(u) \geqslant 1$, and thus u does not belong to D_i. Therefore, u belongs to S_i, and thus u belongs to C and e is covered by C. C is thus a vertex cover.

5. Let K be the number of iterations of the "while" loop: $C = \cup_{i=0}^{K-1} S_i$. At each iteration we have $w_{t+1} \leftarrow w_t - w'_t$. Let v be a vertex in the cover C. Then, there exists an index $j \in [0; K-1]$ such that $v \in S_j$. Then, $w(v) = \sum_{k=0}^{j} w'_k(v)$. Therefore, $w(v) = \sum_{k=0}^{K-1} w'_k(v)$.

   We now consider a vertex u that is not in C: $u \in V \setminus C$. There is thus an index $j \in [0; K-1]$ such that $u \in D_j$. We then have $w(u) = \left(\sum_{i=0}^{j-1} w'_i(u)\right) + w_j(u) \geqslant \sum_{i=0}^{j-1} w'_i(u) = \sum_{i=0}^{K-1} w'_i(u)$.

6. For any $i \in [0; K-1]$, let OPT_i be an optimal cover of $H_i = (V_i, E_i, w'_i)$, i.e., H_i as the same structure than G_i, but the weight function is w'_i instead of w_i. A cover is stable when we take its restriction to a subgraph. Therefore, both $C \cap G_i$ and $C^* \cap G_i$ are covers of H_i. Then, according to Question 1, we have: $w'_i(C \cap G_i) \leqslant 2 \cdot w'_i(\text{OPT}_i)$. As $C^* \cap G_i$ is a cover of H_i and OPT_i is an optimal cover of H_i, then $w'_i(C^* \cap G_i) \geqslant w'_i(\text{OPT}_i)$. Therefore, $w'_i(C \cap G_i) \leqslant 2 \cdot w'_i(C^* \cap G_i)$.

7. We start by using the inequality established at Question 5:

$$w(C) = \sum_{v \in C} w(v) = \sum_{v \in C} \sum_{i=0}^{K-1} w_i'(v) = \sum_{i=0}^{K-1} \sum_{v \in C} w_i'(v) = \sum_{i=0}^{K-1} w_i'(C \cap G_i).$$

We then use the inequality established at the previous question:

$$w(C) = \sum_{i=0}^{K-1} w_i'(C \cap G_i) \leqslant 2 \cdot \sum_{i=0}^{K-1} w_i'(C^* \cap G_i) \leqslant 2w(C^*).$$

Therefore, Algorithm 9.4 is a 2-approximation for VERTEX-COVER.

Solution to Exercise 9.5: Scheduling independent tasks in parallel

1. Let $M_1, \ldots, M_p$ denote the respective completion time of the processors. The makespan of **greedy-online**, i.e., the time at which the overall computation completes, is then: $M = \max_{1 \leqslant i \leqslant p} M_i$. Let j be the index of one of the processors whose completion time defined the makespan: $M_j = M$. Then, let k be the index of the last task assigned to processor P_j. As **greedy-online** assigns a task to the least loaded processor at the time of the assignment, then all processors were working at the time of this assignment, that is, at time $M_j - a_k$. Therefore, for all $i \in [1; p]$, $M_i \geqslant M_j - a_k$. Let W be the sum of the sizes of the tasks: $W = \sum_{i=1}^{n} a_i$. Then, since processors are never left idle,

$$\begin{aligned} W = \sum_{i=1}^{p} M_i &= M_j + \sum_{1 \leqslant i \leqslant p, i \neq j} M_i \\ &\geqslant M_j + (p-1)(M_j - a_k) = p \cdot M_j - (p-1)a_k. \end{aligned}$$

Therefore,

$$M = M_j \leqslant \frac{W}{p} + \left(1 - \frac{1}{p}\right) a_k.$$

Let M^{OPT} be the optimal makespan. There are two obvious bounds to the makespan. The makespan cannot be smaller than any task (for all $i \in [1; n]$, $a_i \leqslant M^{\text{OPT}}$), and it cannot be smaller than the average load of a processor ($\frac{W}{p} \leqslant M^{\text{OPT}}$). Therefore,

$$M \leqslant \frac{W}{p} + \left(1 - \frac{1}{p}\right) a_k. \leqslant M^{\text{OPT}} + \left(1 - \frac{1}{p}\right) M^{\text{OPT}} = \left(2 - \frac{1}{p}\right) M^{\text{OPT}}$$

and $\lambda = 2 - \frac{1}{p}$.

To show that this approximation ratio is tight, we consider an instance including $p(p-1)$ tasks of size 1 and one task of size p. The optimal solution is to assign p tasks of size 1 to each of the first $p-1$ processors and the task of size p to the last processor, achieving a makespan of p. If, in the arbitrary task order considered by **greedy-offline**, the task of size p is scheduled last, at that time all processors were assigned $p-1$ tasks of size 1 and the makespan achieved is equal to $2p-1$. Hence, the approximation ratio in that case is equal to $\frac{2p-1}{p} = 2 - \frac{1}{p}$.

2. The **greedy-offline** algorithm considers the tasks in nonincreasing order: $a_1 \geqslant a_2 \geqslant \cdots \geqslant a_n$. Once again, we focus on the task a_k that completes last. We consider two cases. If $a_k \leqslant \frac{M^{\text{OPT}}}{3}$, we reuse the previous analysis and obtain:

$$M \leqslant \frac{W}{p} + \left(1 - \frac{1}{p}\right) a_k \leqslant M^{\text{OPT}} + \left(1 - \frac{1}{p}\right) \frac{M^{\text{OPT}}}{3} = \left(\frac{4}{3} - \frac{1}{3p}\right) M^{\text{OPT}}.$$

We now consider the case $a_k > \frac{M^{\text{OPT}}}{3}$. We can assume that $k = n$, i.e., that a_k is the last task. Indeed, let $M^{\text{OPT}'}$ be the optimal makespan for the instance $a_1, \ldots, a_k$, and let M' be the makespan achieved by **greedy-offline** on that instance. Then, we obviously have $M^{\text{OPT}'} \leqslant M^{\text{OPT}}$ and $M = M'$. Therefore, if we show that $M' \leqslant \left(\frac{4}{3} - \frac{1}{3p}\right) M^{\text{OPT}'}$, we will have established the desired result. Therefore, we assume that $k = n$. Because (i) tasks are sorted in nonincreasing order, (ii) a_k is the last task, and (iii) $a_k > \frac{M^{\text{OPT}}}{3}$, then all tasks have a size strictly greater than $\frac{M^{\text{OPT}}}{3}$ and each processor is assigned at most two tasks. Then, the schedule built by **greedy-offline** is optimal. We establish this result by contradiction assuming that **greedy-offline** is not optimal for that instance. Let P_j be the processor on which a_k is assigned under **greedy-offline**. In any optimal solution, no two of the first p tasks are assigned to the same processor; otherwise, that processor would be at least as loaded as P_j under **greedy-offline**, contradicting the hypothesis. Without loss of generality, we assume that the i-th task, for $i \in [1;p]$, is assigned to processor P_i under any optimal solution and under **greedy-optimal**. None of the processors $P_1, \ldots, P_{j-1}$ are assigned two tasks in an optimal solution because it would then be at least as loaded as P_j under **greedy-optimal**. No processor can be assigned three tasks under an optimal solution because each task has a size strictly greater than $\frac{M^{\text{OPT}}}{3}$. Then, each of the processors P_p through P_j is assigned two tasks under an optimal schedule. Then, P_j is assigned a second task under any optimal schedule, a task whose size is at least equal to a_k, as a_k is the smallest task. Hence, there is a contradiction, which concludes the proof.

To show that this approximation ratio is tight, we consider the following instance that includes $2p+1$ tasks: three tasks of size p and then two tasks each of the sizes $p+1, p+2, \ldots, 2p-1$. The optimal solution is to schedule on one processor the three tasks of size p, and then on each of the other processors two tasks whose sum of sizes is $3p$. The makespan is then equal to $3p$. We now focus on the assignment of the $2p$ largest tasks by **greedy-offline** (that is, all tasks except one of the tasks of size p). **greedy-offline** assigned exactly two of these tasks per processor, one task of size $2p-1$ with one of size p, one task of size $2p-2$ with one of size $p+1$, etc. Then, all processors have a load of $3p-1$. Whatever the processor the last task of size p is assigned to, the makespan will be $4p-1$ and the approximation ratio: $\frac{4p-1}{3p} = \frac{4}{3} - \frac{1}{3p}$.

Solution to Exercise 9.6: Point clustering

1. We consider Algorithm 9.6.

```
1 for i = 1 to k and S ≠ ∅ do
2 |  Randomly pick a point p_i ∈ S
3 |  C_i = {p' | d(p_i, p') ⩽ d}
4 |  P = P ∪ {C_i}
5 |_ S ← S \ C_i
6 return P
```

ALGORITHM 9.6: Partition a set $\mathcal{S}$ in k parts of diameters at most $2d$.

As we have assumed that the points of $\mathcal{S}$ belong to a metric system, the distances satisfy the triangle inequality, and the diameter of any part is at most equal to $2d$. Indeed, let us consider the subset C_i for any integer $i \in [1;k]$. We have $\forall q \in C_i, \forall r \in C_i, d(q,r) \leqslant d(q,p_i)+d(p_i,r) \leqslant 2d$. We still have to show that $\mathcal{P}$ is indeed a partition. For that, we must show that any point of $\mathcal{S}$ belongs to exactly one of the C_is. By construction, any point of $\mathcal{S}$ belongs to at most one of the C_is. Let us suppose that there exists a point q of $\mathcal{S}$ that does not belong to any of the C_is. Then, for any $i \in [1;k]$, by definition of C_i, $d(p_i,q) > d$ and $d(p_i,p_j) > d$ for any $j \in [1;k]$, $j \neq i$. Therefore, the set $\{p_1,\ldots,p_k,q\}$ is a set of $k+1$ distinct points of $\mathcal{S}$ so that any two of them are at a distance strictly greater than d from each other. This contradicts the hypothesis that there exists a k partition of diameter d. Indeed, in any partition of size k, at least one of the k parts contains at least two of the $k+1$ elements of the set $\{p_1,\ldots,p_k,q\}$, and its diameter is strictly greater than d.

2. Let $\mathcal{A}$ be the algorithm we must study in this question. We compare its output to that of Algorithm 9.6. At each of its iteration, Algorithm 9.6 randomly picks the seed p_i of the new part it builds. If, among the

possible choices of Algorithm 9.6, there are exactly the centers picked by Algorithm $\mathcal{A}$, then the latter algorithm is a 2-approximation algorithm because Algorithm 9.6 is one. Let us suppose that this choice is not available to Algorithm 9.6. Then, let i be the first step at which $\mathcal{A}$ picks a center q_i that does not belong to $\mathcal{S} \setminus \left(\cup_{j=1}^{i-1} C_j\right)$. (As the choice was available to Algorithm 9.6, we assume it picked the same centers as $\mathcal{A}$ at the first $i-1$ steps.) Therefore, there exists an index $l \in [1; i-1]$ such that $q_i \in C_l$ and thus $d(q_i, p_l) \leqslant d$ as $q_l = p_l$ by minimality of i. As $\mathcal{A}$ picks at each step the point that is the farthest from the already chosen centers, this means that all points in $\mathcal{S}$ are at a distance at most d from the chosen centers and that $\mathcal{A}$ has already defined a 2-approximation solution.

Solution to Exercise 9.7: k-center

1. We want to prove that, for a subset S of the vertices, S is a dominating set of G_i if and only if $\mathit{center}(S) \leqslant w(e_i)$.

$$\begin{array}{ll} S \text{ is a dominating set of } G_i & \Leftrightarrow \\ \forall v \in V \setminus S,\ \exists s \in S \text{ such that } (v,s) \in E_i & \Leftrightarrow \\ \forall v \in V \setminus S,\ \exists s \in S \text{ such that } w(v,s) \leqslant w(e_i) & \Leftrightarrow \\ \forall v \in V \setminus S,\ \mathit{connect}(v,S) \leqslant w(e_i) & \Leftrightarrow \\ \mathit{center}(S) = \max\limits_{v \in V \setminus S} \{\mathit{connect}(v,S)\} \leqslant w(e_i) & \end{array}$$

   Therefore, to minimize $\mathit{center}(S)$ with $|S| \leqslant k$ is equivalent to finding the smallest $w(e_i)$ (i.e., the minimum index i as the $w(e_i)$s are sorted in nondecreasing order) such that G_i has a dominating set S of size k.

2. We want to show that if I is an independent set of $H^{(2)}$ (i.e., $\forall x \in I,\ \forall y \in I,\ (x,y) \notin E_H^{(2)}$), then $|I| \leqslant \mathit{dom}(H)$. Let D be a dominating set of H. Then, any vertex of I is dominated by (i.e., is linked to or is equal to) a vertex of D in H, by definition of a dominating set. Furthermore, any vertex of D dominates at most one vertex of I. Let us assume that a vertex u of D dominates two distinct vertices v and w of I. If u is equal to either v or w, there is an edge between u and the other vertex, and the vertices v and w are at distance 1 in H. If u is distinct from both v and w, there is an edge between u and each of these vertices and v and w are at most at distance 2 in H. Whatever the case, v and w are at most at distance 2 from each other in H and thus are at distance 1 in $H^{(2)}$, which contradicts the hypothesis that I is an independent set of $H^{(2)}$. Therefore, any vertex of D dominates at most one vertex of I, and thus $|I| \leqslant \mathit{dom}(H)$.

3. (a) By definition of j, for any $i \in [1; j-1]$, $|M_i| > k$. On the other hand, for any i, according to Question 2, $|M_i| \leqslant \mathit{dom}(G_i)$. There-

fore, for any $i \in [1; j-1]$, $dom(G_i) > k$. According to Question 1, $OPT = \min_i\{w(e_i) \mid dom(G_i) \leqslant k\}$. Then, $OPT = \min_{i \geqslant j}\{w(e_i) \mid dom(G_i) \leqslant k\}$, and thus $OPT \geqslant w(e_j)$ because the edges are ordered by nondecreasing weights.

(b) We need to show that $center(M_j) \leqslant 2 \cdot OPT$. By definition:

$$center(M_j) = \max_{v \in V \setminus M_j}\{connect(v, M_j)\}.$$

We are going to estimate $connect(v, M_j)$ for any vertex of $V \setminus M_j$. Let v be any such vertex. As M_j is a maximal independent set of $G_j^{(2)}$, $M_j \cup \{v\}$ is not an independent set of $G_j^{(2)}$. Therefore, there exists a vertex s of M_j such that (v, s) is an edge of $G_j^{(2)}$. We then have two cases to consider:

- (v, s) is an edge of G_j. Then, $w(v, s) \leqslant w(e_j)$.
- (v, s) is not an edge of G_j. Then, there exists a vertex z such that G_j includes the edges (v, z) and (z, s). Using the triangle inequality we obtain: $w(v, s) \leqslant w(v, z) + w(z, s) \leqslant 2 \cdot w(e_j)$.

In both cases, $w(v, s) \leqslant 2 \cdot w(e_j)$ and thus $connect(v, M_j) \leqslant 2 \cdot w(e_j)$. Therefore, $center(M_j) = \max_{v \in V \setminus M_j}\{connect(v, M_j)\} \leqslant 2 \cdot w(e_j) \leqslant 2 \cdot OPT$ using the result of Question 3.

Finally, we must check that the proposed algorithm runs in polynomial time. Step 1 can be executed in time $O(m \times n^3)$ ($G_i^{(2)}$ is built from G_i in $O(n^3)$ using a product of adjacency matrices). Step 3 can be executed in $O(n^2)$ with a greedy algorithm, which picks a vertex in $V \setminus M_i$, discards this vertex and its neighbors from $G_i^{(2)}$, and iterates on the remaining of the graph. Step 4 can be executed in $O(m)$. Overall, Algorithm 9.5 runs in polynomial time and is a 2-approximation algorithm.

4. A graph on which the bound of 2 is reached is, for example, a wheel with $n + 1$ vertices (see Figure 9.2) where each edge incident to the center has a weight of 1 and any over edge has a weight of 2. When $k = 1$, the optimal solution is to take the center of the wheel. We then have $OPT = 1$. $G_n^{(2)}$ is the first square graph that is a clique. Therefore, Algorithm 9.5 returns $j = n$ and M_j contains a single vertex. If this vertex is not the center, the solution has a cost of 2.

5. We will show that if there existed such an approximation algorithm, it would solve an NP-complete problem in polynomial time. We build a reduction from DOMINATING SET. Let $(G = (V, E), k)$ be an instance of DOMINATING SET. We build a complete graph $G' = (V, E')$ with the following weight function:

$$cost(u, v) = \begin{cases} 1, & \text{if } (u, v) \in E \\ 2 & \text{otherwise.} \end{cases}$$

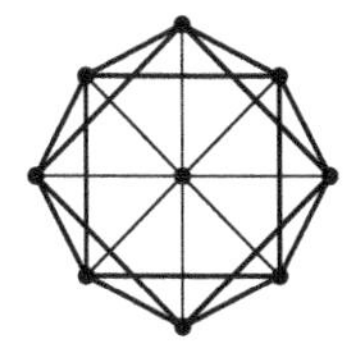

FIGURE 9.2: Wheel with $n = 8$: The edges of weight 1 are thin and the edges of weight 2 are thick. Not all thick edges are drawn.

One can check that G' satisfies the triangle inequality. Furthermore, the reduction satisfies the following conditions:

- If $dom(G) \leqslant k$, then G' has a k-center of cost 1;
- If $dom(G) > k$, then the optimal cost of a k-center of G' is 2.

Using a $(2-\varepsilon)$-approximation algorithm on the graph G' would produce a solution of cost 1 because it cannot use an edge of cost 2. We would then have, with such an approximation algorithm, a way to check in polynomial time whether there exists a dominating set of size k. If P $\neq$ NP, this is impossible; therefore, such an approximation algorithm does not exist if P $\neq$ NP.

Solution to Exercise 9.8: Knapsack

1. We consider the following instance I including two objects: $p_1 = a_1 = 1$, $p_2 = B - 1$ and $a_2 = B$, and we take $B = K + 2$. We then have $m^*(I) = B - 1 = K + 1$. The greedy algorithm considers first the first object, adds it to the knapsack, and stops there and its solution is valued $m_g(I) = 1$. Therefore, we have:

$$\frac{m^*(I)}{m_g(I)} = \frac{K+1}{1} > K.$$

2. Let j be the index of the first object not taken by the greedy algorithm. When j is considered, the value of knapsack and the size of its content are, respectively:

$$\overline{p}_j = \sum_{i=1}^{j-1} p_i \leqslant m_g(x) \qquad \text{and} \qquad \overline{a}_j = \sum_{i=1}^{j-1} a_i \leqslant B.$$

We first show that for an optimal solution $m^*(X) < \overline{p}_j + p_j$. Let $K_j = \{x_1, \ldots, x_{j-1}\}$ be the current solution of the greedy algorithm. Let x'_j be a task of size $a'_j = B - \overline{a}_j$ and of value $p'_j = \frac{B-\overline{a}_j}{a_j} p_j$. Note that x'_j has exactly the same value-to-size ratio as x_j. The optimal solution for the instance $X \cup \{x'_j\}$ is obviously $K_j \cup \{x'_j\}$ because this solution completely fills the knapsack while using only the objects of largest value-to-size ratio. Furthermore, we obviously have $m^*(X) \leqslant m^*(X \cup \{x'_j\})$ because any solution of instance X is a solution of instance $X \cup \{x'_j\}$. Therefore,

$$m^*(X) \leqslant m^*(X\{x'_j\}) = \overline{p}_j + p'_j < \overline{p}_j + p_j.$$

To complete the proof, we must consider two cases. If $p_j \leqslant \overline{p}_j$, then

$$m^*(X) < 2\overline{p}_j \leqslant 2m_g(X) \leqslant 2\max\{m_g(X), p_{max}\}.$$

Otherwise, $p_j > \overline{p}_j$ and then $p_{max} > \overline{p}_j$. In this case

$$m^*(X) < \overline{p}_j + p_j \leqslant \overline{p}_j + p_{max} < 2p_{max} \leqslant 2\max\{m_g(X), p_{max}\}.$$

From what precedes, taking the best solution among the one produced by the greedy algorithm and the one including the most valuable object defines a solution that is never more than two times less valuable than the optimal one. As the greedy algorithm obviously runs in polynomial time, the enhanced greedy algorithm we just defined is a 2-approximation algorithm.

3. For values of k and p for which $M^*(k,p)$ is undefined, we let $S^*(k,p) = 1 + \sum_{i=1}^{n} a_i$, i.e., we specify a size that exceeds the bound.

   We have the following limit conditions when our choice is limited to the first object:

   - $M^*(1,0) = \emptyset$ and $S^*(1,0) = 0$.
   - $M^*(1,p_1) = \{x_1\}$ and $S^*(1,p_1) = a_1$.
   - $M^*(1,p) =$ *undefined* and $S^*(1,p) = 1 + \sum_{i=1}^{n} a_i$ for any positive value $p \neq p_1$.

   To define the recursion for the general case ($k \geqslant 2$), we remark that the best subset of $\{x_1, \ldots, x_k\}$ of value p is either the best subset of $\{x_1, \ldots, x_{k-1}\}$ of value p or the best subset of $\{x_1, \ldots, x_{k-1}\}$ of value $p - p_k$ plus the object x_k. Therefore, if $p_k \leqslant p$, if $M^*(k-1, p-p_k)$ is defined, if $S^*(k-1, p-p_k) + a_k < S^*(k-1, p)$, and if $S^*(k-1, p-p_k) + a_k \leqslant B$, then $M^*(k,p) = M^*(k-1, p-p_k) \cup \{x_k\}$ and $M^*(k,p) = M^*(k-1,p)$ otherwise (even if the latter is not defined, in which case $M^*(k,p)$ is also not defined). From this property, we define Algorithm 9.7.

   Algorithm 9.7 builds a solution in time $O(n \sum_{i=1}^{n} p_i)$. The execution time is thus polynomial in the sum of objects values, which is exponential in the input size.

4. (a) The execution time is $O(n \sum_{i=1}^{n} p'_i)$ when running Algorithm 9.7 on the instance $\{(p'_i, a_i)\}_{i \in [1;n]}$. For any $i \in [1;n]$, using the definition of p'_i and of t, we obtain:

$$p'_i = \left\lfloor \frac{p_i}{2^t} \right\rfloor \leqslant \frac{p_i}{2^t} = \frac{p_i}{2^{\left\lfloor \log_2\left(\frac{r-1}{r}\frac{p_{max}}{n}\right)\right\rfloor}} \leqslant \frac{p_i}{2^{\left(\log_2\left(\frac{r-1}{r}\frac{p_{max}}{n}\right)\right)-1}}$$

```
1  for p = 1 to Σ_{i=1}^n p_i do
2  |   M*(1,p) ← undefined
3  |_  S*(1,p) ← 1 + Σ_{i=1}^n s_i
4  M*(1,0) ← ∅ and S*(1,0) ← 0
5  M*(1,p_1) ← {x_1} and S*(1,p_1) ← a_1
6  for k = 2 to n do
7  |   for p = 0 to Σ_{i=1}^n p_i do
8  |   |   if (p_k ⩽ p) and (M*(k − 1, p − p_k) ≠ undefined) and
   |   |   (S*(k − 1, p − p_k) + a_k < S*(k − 1, p)) and
   |   |   (S*(k − 1, p − p_k) + a_k ⩽ B) then
9  |   |   |   M*(k,p) ← M*(k − 1, p − p_k) ∪ {x_k}
10 |   |   |   S*(k,p) ← S*(k − 1, p − p_k) + a_k
11 |   |   else
12 |   |   |   M*(k,p) ← M*(k − 1, p)
13 |   |   |_  S*(k,p) ← S*(k − 1, p)
   |   |_
   |_
14 p* ← max{p | M*(n,p) ≠ undefined}
15 return M*(n,p*)
```

ALGORITHM 9.7: Another dynamic program to solve the knapsack problem for instance (p_1, a_1), ..., (p_n, a_n).

and, therefore,

$$p'_i \leqslant \frac{2\ r\ n\ p_i}{(r-1)p_{max}} \leqslant \frac{2\ r\ n}{(r-1)}$$

as $p_i \leqslant p_{max}$. The execution time of the algorithm on that instance is thus $O(\frac{r}{r-1}n^3)$.

Note that we have made no assumption to ensure that t is positive. In fact, when r tends to 1, t tends to $-\infty$.

(b) Let $\mathcal{I}$ be the set of the indices of the objects taken in a optimal solution for X, and $\mathcal{J}$ for an optimal solution for X'. Then,

$$\begin{aligned}
m^*(X) - n \cdot 2^t &= \left(\sum_{i\in\mathcal{I}} p_i\right) - n \cdot 2^t \\
&\leqslant \sum_{i\in\mathcal{I}} (p_i - 2^t) = 2^t \sum_{i\in\mathcal{I}} \left(\frac{p_i}{2^t} - 1\right) \\
&\leqslant 2^t \sum_{i\in\mathcal{I}} \left\lfloor \frac{p_i}{2^t} \right\rfloor \\
&\leqslant 2^t \sum_{j\in\mathcal{J}} \left\lfloor \frac{p_j}{2^t} \right\rfloor \\
&\leqslant 2^t \sum_{j\in\mathcal{J}} \frac{p_j}{2^t} = \sum_{j\in\mathcal{J}} p_j = m_{AS}(X, r).
\end{aligned}$$

Therefore, we have $m^*(X) - m_{AS}(X,r) \leqslant n \cdot 2^t$. Furthermore, as p_{max} is the largest value of an object in X, $n \cdot p_{max} \geqslant m^*(X) \geqslant p_{max}$ (assuming, without loss of generality, that no object has a size greater than B). Consequently,

$$\frac{m^*(X) - m_{AS}(X,r)}{m^*(X)} \leqslant \frac{n \cdot 2^t}{p_{max}},$$

which leads to

$$m^*(X) \leqslant \frac{p_{max}}{p_{max} - n \cdot 2^t} m_{AS}(X,r).$$

Using the definition of t, we can show that

$$\frac{p_{max}}{p_{max} - n \cdot 2^t} \leqslant r,$$

which completes the proof.

Solution to Exercise 9.9: Mixed integer linear program for replica placement

1. The constraints of the integer linear program are the following:

   (a) If $j \notin ancestors(i)$, we set $y_{i,j} = 0$, because the requests can be served only by nodes on the path from i to the root.

   (b) Every request is assigned a server: $\forall i \in \mathcal{C}, \sum_{j \in ancestors(i)} y_{i,j} = r_i$.

   (c) Server capacities are not exceeded: $\forall j \in \mathcal{N}$, $\sum_{i \in \mathcal{C}} y_{i,j} \leqslant W_j x_j$. Note that this ensures that if j is the server for one or more requests, there is indeed a replica located in node j.

   (d) Distance constraints are fulfilled:
   $\forall i \in \mathcal{C}, \forall j \in ancestors(i)$, $dist(i,j) y_{i,j} \leqslant d_i y_{i,j}$.

   The objective function is the cost of replicas, $\sum_{j \in \mathcal{N}} W_j x_j$.

2. We build an integer solution, keeping the same x_js and without breaking any constraint. In the following, for any variable y, $\lfloor y \rfloor$ is the integer part of y, and $\tilde{y}$ is the fractional part: $y = \lfloor y \rfloor + \tilde{y}$, and $\tilde{y} < 1$. We denote by $subtree(j)$ the tree rooted in node j.

   Let us consider a client $i \in \mathcal{C}$ such that $\exists j \in \mathcal{N} \mid \tilde{y}_{i,j} > 0$, i.e., $y_{i,j}$ is not an integer. We consider j_1 being the closest server to i not serving an integer number of requests of client i, and more generally j_k, $(1 \leqslant k \leqslant K)$ the servers on the path from i to the root, such that $\tilde{y}_{i,j_k} > 0$. We want to move bits of requests in order to obtain an integer value for y_{i,j_1}. This elementary transformation is called $trans(i, j_1)$. We consider the two following cases.

(a) If $\sum_{i' \in subtree(j_1) \cap \mathcal{C}} y_{i',j_1} \leqslant W_{j_1} - (1 - \tilde{y}_{i,j_1})$, there is enough space at server j_1 to fulfill an integer number of requests from client i. Because the total number of requests of client i is an integer, $\sum_{k=1}^{K} \tilde{y}_{i,j_k}$ is a nonnull integer. Thus, $\sum_{k=2}^{K} \tilde{y}_{i,j_k} \geqslant 1 - \tilde{y}_{i,j_1}$, and we can move down $1 - \tilde{y}_{i,j_1}$ bits of requests from servers j_k $(2 \leqslant k \leqslant K)$ to j_1. No constraints will be violated because there is enough space on the server. The move is done by changing the values of the y_{i,j_k}s. After such a transformation, y_{i,j_1} is an integer variable.

(b) Otherwise, if server j_1 is already too full in order to add a fraction of requests from client i, we need to exchange some requests with other clients. First, if there is some free space on the server, we start by filling completely server j_1 with fractions of requests of client i from servers j_k $(2 \leqslant k \leqslant K)$. We know there are such requests; otherwise, y_{i,j_1} would be an integer. This transformation is similar to the one done in the first case. We now have $\sum_{i' \in subtree(j_1) \cap \mathcal{C}} y_{i',j_1} = W_{j_1}$. Let us denote by i_t, for $1 \leqslant t \leqslant T$, the clients $i_t \in subtree(j_1) \cap \mathcal{C} \setminus \{i\}$ such that $\tilde{y}_{i_t,j_1} > 0$. Because W_{j_1} is an integer and $\tilde{y}_{i,j_1} > 0$, we have $\sum_{t=1}^{T} \tilde{y}_{i_t,j_1} \geqslant 1 - \tilde{y}_{i,j_1}$ and also $\sum_{k=2}^{K} \tilde{y}_{i,j_k} \geqslant 1 - \tilde{y}_{i,j_1}$. We can select in both sets $1 - \tilde{y}_{i,j_1}$ bits of requests that will be exchanged, i.e., bits of requests from client i_t initially treated by j_1 will be moved on some servers j_k, which are in $ancestors(j_1)$, and the corresponding amount of requests of i will be moved back on server j_1. In this case, we may break a distance constraint because it is not certain that clients i_ts can be served higher than j_1 in order to respect their distance constraint. However, we will see that in the general transformation process, we prevent such cases to happen. Note that all other constraints are still fulfilled.

Once $trans(i, j_1)$ has been done, y_{i,j_1} is an integer, and because only noninteger bits of requests have been moved, we have not assigned any integer part of the solution and have decreased at least by one the number of noninteger variables in the solution.

Let us detail now the complete transformation algorithm in order to obtain an integer solution. Particular attention must be paid to respect the distance constraints at all time. We consider each server in a bottom-up order, so that we are sure that each time we perform an elementary transformation, the server is the first one on the way from the client to the root having a noninteger number of requests. In fact, when transforming server j, each server in $subtree(j)$ has already been transformed and, thus, has no fraction numbers of requests.

In order to transform server j, we look at the set $\mathcal{C}'$ of clients having a noninteger number of requests processed at j. If the set is empty, there is nothing to transform at j. Otherwise, we perform the elementary

```
1 for j ∈ N taken in a bottom-up traversal order do
2     finish=1
3     while (finish==1) do
4         C' = {i' ∈ C ∩ subtree(j) | ỹ_{i',j} > 0}
5         if C' == ∅ then
6             finish=0
7         else
8             i = min_{i'∈C'} {d_{i'} − dist(i', j)}
9             trans(i, j)
```

ALGORITHM 9.8: Algorithm to transform a rational into an integer solution.

transformation with the client i that minimizes $(d_{i'} - dist(i', j))$, for $i' \in \mathcal{C}'$. This ensures that when we perform an elementary transformation as in the second case above, the distance constraint will be respected for all clients i_t, because we are moving their requests into servers at a distance of at most $d = d_i - dist(i, j)$ from j, and their own distance constraint allows them to be processed at a distance $d_{i_t} - dist(i_t, j) \geqslant d$. Figure 9.3 illustrates this phase of the algorithm; the algorithm being formally presented by Algorithm 9.8.

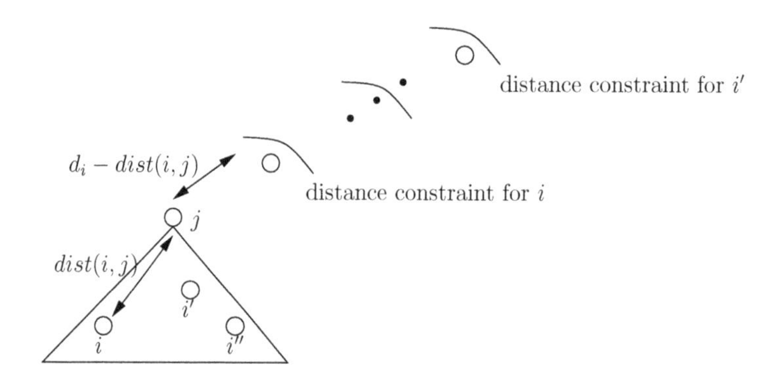

FIGURE 9.3: Illustration of the transformation algorithm.

At the end of the while loop, server j is processing only integer numbers of requests, and, thus, we will not modify its assignment of requests any more in the following.

The constraints are all respected in all steps of the transformation, and we do not add or remove any replica, so the solution has exactly the same cost as the initial LP-based solution, and the transformed solution is fully integer. Moreover, this transformation algorithm works in polynomial time, in the worst case in $|\mathcal{N}| + |\mathcal{C}|^2$, but most of the time it is much faster because the transformations do not concern all clients simultaneously, only a few of them.

3. If the x_js are also rational numbers, there is no hope in finding an optimal solution in polynomial time because the problem is NP-complete! The interested reader will refer to the case study of Chapter 12.

Solution to Exercise 9.10: A randomized algorithm for independent set

Let X be the random variable that counts the number of vertices that survive the first step. Because each vertex survives the first step with probability $\frac{1}{d}$, we have $E[X] = \frac{n}{d}$, where $E[X]$ denotes the expectation of X. Let Y be the random variable that counts the number of edges that survive the first step. Because an edge survives if and only if its two adjacent vertices survive, which happens with probability $\frac{1}{d^2}$, we have $E[Y] = \frac{m}{d^2}$ (as the expectation of the sum over all edges is equal to the sum of the expectations).

The second step removes all Y edges and at most Y vertices. The algorithm outputs an independent set of size at least $X - Y$, where $E[X-Y] = \frac{n}{d} - \frac{m}{d^2} = \frac{n^2}{2m} - \frac{n^2}{4m} = \frac{n^2}{4m}$.

We conclude by using the following result: If Z is a discrete random variable such that $E[Z] = \mu$, then $\mathbb{P}(X \geqslant \mu) > 0$, and there exists at least one instance in the sample space for which the value of Z is at least μ. We can take the ceiling function because the size of an independent set is an integer.

Solution to Exercise 9.11: Branch-and-bound applied to MAX-SAT

1. The search tree is depicted in Figure 9.4. The number associated with each leaf of the tree (i.e., a solution) indicates the number of clauses satisfied by this solution. Therefore, the formula cannot be satisfied, and the maximum number of clauses that can be satisfied is 9 (over 10 clauses).

2. Once a partial assignment is found, the idea is to check how many clauses cannot be satisfied and to cut the branches once a solution with more clauses is found. The algorithm explores the tree with a depth-first search strategy and obtains the first solution with $x_i = 1$ for all i, for which n clauses cannot be satisfied. In the example, $n = 1$ since 9 clauses are satisfied by $x_1 = x_2 = x_3 = x_4 = 1$. Pursuing the depth-first search, we cut branches that cannot lead to a better solution, i.e., partial

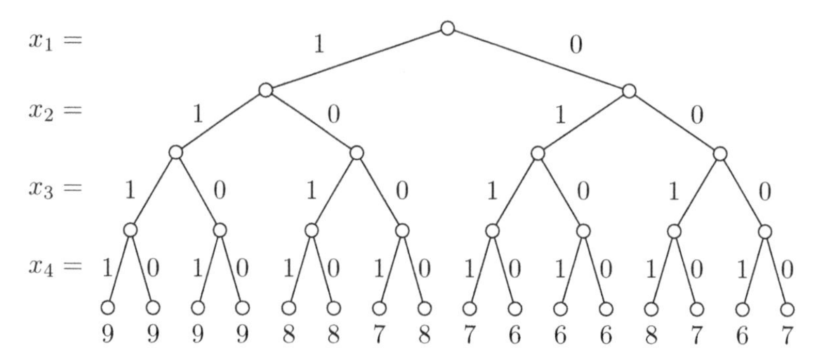

FIGURE 9.4: Search tree.

assignments with at least n clauses that cannot be satisfied. If a better solution is found, we update n before continuing the tree exploration. In the example, for the partial assignment $x_1 = x_2 = x_3 = 1$, clause $\overline{x_1} \vee \overline{x_3}$ cannot be satisfied by any choice of the free variable x_4, and, therefore, we do not explore the branch $x_4 = 0$. Then, for the partial assignment $x_1 = x_2 = 1$, all clauses can be satisfied; we explore the branch with $x_3 = 0$, but then clause x_3 cannot be satisfied and we do not go further. Similarly, we stop the exploration at $x_1 = 1$ and $x_2 = 0$ because clause $\overline{x_1} \vee x_2$ cannot be satisfied, and then we stop at $x_1 = 0$ because clause x_1 cannot be satisfied.

3. With a breadth-first search strategy, the idea is to explore first the branch with the smallest number of clauses that cannot be satisfied.

4. The exploration is done in a different order when the variables are ordered differently, leading to a modified algorithm complexity. We let the reader draw the corresponding trees and apply both the depth-first search and the breadth-first search strategies.

9.4 Bibliographical notes

Many exercises in this chapter are quite classical and can be found in several textbooks. Exercise 9.9 (mixed integer linear program for replica placement) is inspired from [11]. Exercise 9.11 (branch-and-bound applied to MAX-SAT) is inspired from the book by Hromkovic [51].

Part III

Reasoning on problem complexity

Chapter 10

Reasoning to assess a problem complexity

The third part of the book builds upon the previous chapters, and its objective is to show the reader how to establish complexity results in their field, by providing hints on how to assess the complexity of a new problem.

This chapter is organized as follows: In Section 10.1, we first review the basic reasoning to assess a problem complexity. Then, in Section 10.2, we recall some very classical problems for which there exist polynomial-time optimal algorithms. Finally, in Section 10.3, we exhibit a set of useful NP-complete problems that may be helpful when looking for a reduction.

The following chapters provide a comprehensive set of case studies, which illustrate the different techniques that can be used to determine the complexity of a new problem. Following the reasoning of Chapter 8, some practical solutions are proposed to tackle NP-complete problem instances.

10.1 Basic reasoning

In our opinion, the basic reasoning to assess a problem complexity is the following. First, we try to design an algorithm to solve the problem in polynomial time (see Section 10.1.1). If the complexity of the algorithm remains exponential despite our best efforts, we acknowledge the combinatorial nature of the problem and try to prove its NP-completeness (see Section 10.1.2). In this latter case, some practical solutions are then investigated to (approximately) solve the NP-complete problem.

10.1.1 Polynomial instances

When dealing with a new problem, the most natural approach is to try to solve it with a *greedy* algorithm, following the guidelines of Chapter 3. Counterexamples may be exhibited to prove the nonoptimality of the greedy choice. While a counterexample is found, we refine the greedy choice in order to take better local decisions, if possible. If no counterexample can be exhibited, we may try to prove the optimality of this greedy algorithm.

If there is no obvious greedy choice to make, or if the attempts to design a greedy algorithm lead only to nonoptimal algorithms, the next step is to try to design a *dynamic-programming* algorithm to solve the problem, as in Chapter 4. This means identifying subproblems whose optimal solutions can be used to build an optimal solution to the original problem. While searching for subproblems, one may resort to the *divide-and-conquer* approach of Chapter 2, where identical problems of smaller size are solved, and the solution to the original problem can be reconstructed from these partial solutions. Note that a particular care should be taken to identify the complexity of the dynamic-programming algorithm, which may well turn out to be exponential.

If the previous attempts to solve the problem have been unsuccessful or have resulted in exponential-time algorithms, another way to approach the problem is to fix the target value to be achieved and search for a solution to the decision problem that needs to match this target, rather than solve the optimization problem. Again, a greedy or dynamic-programming algorithm can be used to solve the decision problem. It may become easier to take local choices because the target value is known. In case of success, one can resort to a *binary search* to solve the original optimization problem. Then, particular care must be taken about the number of iterations required by the binary search. This technique will lead to a polynomial-time algorithm only if the domain of the target value is well identified.

Finally, if none of these techniques work, thanks to the knowledge gained on the problem by trying to solve it with polynomial algorithms, we can change gear and try to establish the NP-completeness following the approach of Section 10.1.2.

10.1.2 NP-complete instances

In order to prove the NP-completeness of the problem, the first step consists of proving that the problem belongs to the class NP. Even though this step is often trivial, it should not be neglected, in order to be sure that the problem is well defined.

Then, it is necessary to understand the combinatorial nature of the problem. In other words, why does the greedy approach not work? This knowledge is usually acquired while trying to solve greedily the problem and can be refined by running several examples and identifying in which cases a greedy algorithm is likely to fail.

The next step is to decide from which NP-complete problem the reduction should be done. In theory, all NP-complete problems are equivalent candidates by definition, but in practice the reduction is much easier if it involves a problem "similar" to our problem.

One can refer to the set of NP-complete problems provided in Section 10.3 and identify a problem that is close enough to the problem of open complexity. For problems with numbers, the reduction often comes from a partition problem, and toy examples may exhibit a partition problem, hence, leading to the proof. For graph problems, there is a whole set of different problems, and, depending upon the structure of the graph, one may resort to one or the other. If none of the classical problems seem to lead to an easy reduction, further investigation should be done; a whole set of NP-complete problems can be found in [38]. Note that while some reductions are straightforward, it is not unusual to think about the problem for several months before the appropriate NP-complete problem for the reduction can be identified, and the reduction may have some tricky parts. Our advice: Do not give up.

It also may be the case that it is not obvious whether to search for a polynomial-time algorithm or for an NP-completeness proof, because the problem complexity seems to lie in between. In such cases, one should alternate between phases searching for a polynomial algorithm, following the guidelines of Section 10.1.1, and phases trying to establish the NP-completeness of the problem, following the guidelines of this section. During each phase, some new insights about the complexity of the problem may be gained and help decide whether the problem is NP-complete.

Once the problem has been successfully proved NP-complete, the research around this problem does not necessarily come to an end. It might be interesting, from a theoretical point of view, to discuss approximation results. From the NP-completeness proof, it is sometimes possible to derive some inapproximability result. Also, the algorithms that were designed while investigating the combinatorial nature of the problem may well turn out to be good approximation algorithms (see Section 8.1). Some of those algorithms also may be adapted to provide a bound on the solution (for instance, a lower bound for a minimization problem) and, hence, they may prove useful to assess the performance of heuristic algorithms that would solve the problem. More generally, all techniques discussed in Chapter 8 may be considered.

10.2 Set of problems with polynomial-time algorithms

There are a few algorithmic kernels that are used time and again in the design of new algorithms. We briefly review some of them here.

Sorting. The problem of sorting a set of n objects on which a total order exists can obviously be solved in polynomial time. A naive solution is to find the maximum object, to remove it from the set, and then to iterate. This scheme has a complexity of $\Theta(n^2)$. Using decision trees, one can easily prove that the running time of any sorting algorithm is $\Omega(n \log(n))$ in the worst case. (This result holds as long as the algorithm has no knowledge on the objects other than the results of the comparisons that it performs.) Heapsort and mergesort are algorithms that sort n objects in time $O(n \log(n))$ and, thus, are asymptotically optimal.

Shortest paths. When dealing with a graph G where edges are labeled with weights, a common question is what is the shortest path between a node u and a node v? The length of the path is then defined as the sum of the weights of its constitutive edges. Many variants of this problem exist, depending on whether the weights are all nonnegative or can take negative values, whether edges are directed, and whether one wishes to know the answer for a given pair of graph vertices or for all possible pairs. Let V denote the set of the graph vertices and E the set of the graph edges.

- The Bellman–Ford algorithm solves in time $O(|V||E|)$ the single-source (all destinations) shortest-paths problem for undirected graphs whose edges can have negative weights.
- Dijkstra's algorithm solves in time $O(|V|^2)$ the single-source (all destinations) shortest-paths problem for directed graphs whose edges have nonnegative weights. (The running time of this algorithm can be lowered to $O(|V| \log(|V|) + |E|)$.)
- The Floyd–Warshall algorithm is a dynamic-programming algorithm that solves the all-pairs shortest-paths problem on directed graphs with negative edges in time $\Theta(|V|^3)$. Obviously, when dealing with graphs including negative-weight edges, we assume that the graph does not contain any negative-weight cycles. Indeed, shortest paths are undefined in the presence of such cycles.

Maximum bipartite matching. A graph $G = (V, E)$ is a bipartite graph if its set V of vertices can be partitioned in two subsets S and T such that any edge in G connects a vertex of S to a vertex of T (see also Section 3.3.1). A matching is a subset M of the edges such that at most one edge in M is incident to any given vertex. Finding a maximum matching, that is, a matching of maximum cardinality, in a bipartite graph can be solved in time $O(|V||E|)$ using the Ford–Fulkerson method.

10.3 Set of NP-complete problems

First, the reader needs to identify the nature of the problem whose complexity must be assessed. Usually, the combinatorial nature of the problem comes from numbers (as, for instance, task weights, processor speeds, and so on) or from graphs (if there are no numbers that render the problem combinatorial). Occasionally, the problem may not belong to any of these two main categories that are discussed in Sections 10.3.1 and 10.3.2. One may then resort to different kinds of NP-complete problems, as, for instance, the mother-source

3-SAT problem (see Section 6.4), even though 3-SAT may look even more different from the original problem than classical number and graph problems.

10.3.1 Numbers

For problems with numbers, most reductions can be done from 2-PARTITION, 3-PARTITION, or a variant of these problems. If the reduction does not come easily from a partition problem because the problem is slightly too complex, we often found the three-dimensional matching and permutation sums problems very useful.

First, we recall the 2-PARTITION problem and some of its most useful variants.

DEFINITION 10.1 (2-PARTITION). Given n integers $a_1, \ldots, a_n$, is there a subset I of $\{1, \ldots, n\}$ such that $\sum_{i \in I} a_i = \sum_{i \notin I} a_i$?

DEFINITION 10.2 (2-PARTITION-EQUAL). Given $2n$ integers $a_1, \ldots, a_{2n}$, is there a subset I of $\{1, \ldots, 2n\}$ such that $|I| = n$ and $\sum_{i \in I} a_i = \sum_{i \notin I} a_i$?

DEFINITION 10.3 (2-PARTITION-EVEN-ODD). Given $2n$ integers $a_1, \ldots,$ a_{2n}, is there a subset I of $\{1, \ldots, 2n\}$ such that $\sum_{i \in I} a_i = \sum_{i \notin I} a_i$, and for $1 \leqslant j \leqslant n$, either $a_{2j-1} \in I$ and $a_{2j} \notin I$, or $a_{2j} \in I$ and $a_{2j-1} \notin I$?

DEFINITION 10.4 (THREE-2-PARTITION). Given n integers $a_1, \ldots, a_n$, are there three subsets I_1, I_2, and I_3 that realize a partition of $\{1, \ldots, n\}$, and such that $\sum_{i \in I_1} a_i = \sum_{i \in I_2} a_i = \sum_{i \in I_3} a_i$?

It is easy to see that all these partition problems can be solved in pseudo-polynomial time. THREE-2-PARTITION is a bit misleading; it is a problem very similar to 2-PARTITION, and despite its name, it is very different from 3-PARTITION, which is NP-complete in the strong sense. Therefore, it is better, whenever possible, to perform the reduction from 3-PARTITION, hence, proving that the problem is NP-complete in the strong sense.

DEFINITION 10.5 (3-PARTITION). Given an integer B, $3n$ integers a_1, $\ldots$, a_{3n}, can we partition the $3n$ integers into n sets, each of sum B? We can assume that $\sum_{i=1}^{3n} a_i = nB$ (otherwise, there is no solution) and that $B/4 < a_i < B/2$ (so that one needs exactly three elements to obtain a sum B).

We do not list all NP-complete problems involving numbers in this section but only those that we found the most useful to perform reductions. Numerical 3-Dimensional Matching (N3DM) is one of them.

DEFINITION 10.6 (N3DM). Given three disjoint sets W, X, and Y each containing n elements, given, for each element $a \in W \cup X \cup Y$, an integer $s(a)$, and given a bound B, can $W \cup X \cup Y$ be partitioned into n disjoint sets $A_1, \ldots, A_n$ such that each A_i contains exactly one element from each of W, X, and Y and such that, for $1 \leqslant i \leqslant n$, $\sum_{a \in A_i} s(a) = B$?

In fact, the following particular instance of N3DM has been proved to be strongly NP-complete by Yu, Hoogeveen, and Lenstra [111]. Instead of matching arbitrary elements, we have to find only two permutations of $[1..n]$, which may greatly ease the reduction.

DEFINITION 10.7 (RN3DM)**.** Given an integer vector $A = (A[1], \ldots, A[n])$ of size n, do there exist two permutations λ_1 and λ_2 of $\{1, 2, \ldots, n\}$ such that $\forall 1 \leqslant i \leqslant n, \quad \lambda_1(i) + \lambda_2(i) = A[i]$? We can assume that $2 \leqslant A[i] \leqslant 2n$ for all i and that $\sum_{i=1}^{n} A[i] = n(n+1)$ (otherwise, there is no solution).

10.3.2 Graphs

We list below the most usual NP-complete problems in graphs. Note that these problems are all NP-complete in the strong sense because they do not involve numbers.

DEFINITION 10.8 (CLIQUE)**.** Let $G = (V, E)$ be a graph and k be an integer such that $1 \leqslant k \leqslant |V|$. Does there exist a clique of size k (i.e., a complete subgraph of G with k vertices)?

DEFINITION 10.9 (VERTEX-COVER)**.** Let $G = (V, E)$ be a graph and k be an integer such that $1 \leqslant k \leqslant |V|$. Do there exist k vertices $v_{i_1}, \ldots, v_{i_k}$ such that any edge $e \in E$ is incident to at least one of the v_{i_j}, for $1 \leqslant j \leqslant k$?

DEFINITION 10.10 (HC – Hamiltonian Cycle)**.** Given a graph $G = (V, E)$, is there a circuit that goes through each vertex once and only once?

DEFINITION 10.11 (HP – Hamiltonian Path)**.** Given a graph $G = (V, E)$ and two vertices $u, v \in V$, is there a path from u to v that goes through each vertex once and only once?

DEFINITION 10.12 (TSP – Traveling Salesman Problem)**.** Given a complete graph $G = (V, E)$, a cost function $w : E \to \mathbb{N}$, and an integer k, is there a cycle $\mathcal{C}$ going through each vertex once and only once, with $\sum_{e \in \mathcal{C}} w(e) \leqslant k$?

DEFINITION 10.13 (COLOR)**.** Given a graph $G = (V, E)$ and an integer k ($1 \leqslant k \leqslant |V|$), can we color G with at most k colors?

The disjoint connecting paths problem is slightly less usual, but it turns out to be very useful for some problems. In particular, we point out that the problem has been shown to be NP-complete even when restricted to two paths in the case of directed graphs (2DCP) [35].

DEFINITION 10.14 (DCP – Disjoint Connecting Paths)**.** Given a graph $G = (V, E)$ and a collection of $k+1$ disjoint vertex pairs (x_1, y_1), (x_2, y_2), ..., $(x_{k+1}, y_{k+1}) \in V \times V$, does G contain $k+1$ mutually vertex-disjoint paths, one connecting x_i and y_i for each i, $1 \leqslant i \leqslant k+1$? The number of nodes in the graph is $n = |V|$, and we have $1 \leqslant k \leqslant n$.

DEFINITION 10.15 (2DCP). Given a *directed* graph $G = (V, E)$ and two disjoint vertex pairs (x_1, y_1), $(x_2, y_2) \in V \times V$, does G contain two mutually vertex-disjoint paths, one going from x_1 to y_1 and the other going from x_2 to y_2?

Chapter 11

Chains-on-chains partitioning

This case study is devoted to the chains-on-chains partitioning (CCP) problem. Given an array of n elements $a_1, a_2, \ldots, a_n$, the problem is to partition the array into p intervals whose element sums are well balanced. This problem has been extensively studied in the literature because it has various applications. In particular, it amounts to load-balancing n computations whose ordering must be preserved (hence, the restriction to intervals) onto p processors. Then, each a_i corresponds to the execution time of the i-th task, and the sum of the elements in interval $\mathcal{I}_k$ is the load of the processor to which $\mathcal{I}_k$ is assigned. Several algorithms and heuristics have been proposed to solve this load-balancing problem, including [19, 47, 52, 53, 81]. We refer the reader to the survey paper by Pinar and Aykanat [86] for a detailed overview and comparison of the literature.

In this case study, we first discuss, in Section 11.1, the classical version of the problem with identical processors, and we propose optimal efficient algorithms to solve the problem. Then, in Section 11.2, we study variants of the problem, such as taking into account communication costs or considering a chain of heterogeneous processors [87]. Finally, in Section 11.3, we assess the complexity of the problem when extending it to a clique of heterogeneous resources [13]. We conclude in Section 11.4.

11.1 Optimal algorithms for homogeneous resources

DEFINITION 11.1 (CCP). Given an array of n elements $a_1, a_2, \ldots, a_n$, the chains-on-chains partitioning (CCP) problem consists of partitioning the array into p consecutive intervals $\mathcal{I}_1, \mathcal{I}_2, \ldots, \mathcal{I}_p$, where $\mathcal{I}_k = [d_k, e_k]$ and $d_1 = 1$, $e_p = n$, $d_k \leqslant e_k$, and $d_{k+1} = e_k + 1$ for $1 \leqslant k \leqslant p - 1$. The objective is to minimize

$$\max_{1 \leqslant k \leqslant p} \sum_{i \in \mathcal{I}_k} a_i = \max_{1 \leqslant k \leqslant p} \sum_{i=d_k}^{e_k} a_i.$$

We invite the reader to design several polynomial-time algorithms to solve this problem in the next sections.

11.1.1 Dynamic-programming algorithm

We first solve the CCP problem with a dynamic-programming algorithm. Design such an the algorithm and give its complexity.

(*Hint:* The subproblem can be the optimal partitioning of a subset of the n elements into $k \leqslant p$ intervals.)

Solution. We define the cost of a partitioning as the maximum interval sum. For $1 \leqslant i \leqslant n$ and $1 \leqslant k \leqslant p$, let $g(i,k)$ be the optimal cost when partitioning $[a_1, \ldots, a_i]$ into k intervals. We want to find $g(n,p)$, with initial values $g(i,1) = \sum_{j=1}^{i} a_j$ for $1 \leqslant i \leqslant n$, and $g(i,k) = +\infty$ for $1 \leqslant i < k \leqslant p$. For $k \geqslant 2$, the recursion writes:

$$g(i,k) = \min_{1 \leqslant s < i} \left(\max \left(g(s, k-1), \sum_{j=s+1}^{i} a_j \right) \right) .$$

With the above equation, we simply try all possible splits into $[1,s]$ (with $k-1$ intervals) and the single interval $[s+1,i]$.

We precompute all values $f(s,i) = \sum_{j=s+1}^{i} a_j$, for $2 \leqslant i \leqslant n$ and $1 \leqslant s \leqslant i-1$, in time $O(n^2)$. Then, there are $n \times p$ values of $g(i,k)$ to compute, each in $O(n)$ (minimum taken over up to n values). Therefore, the complexity is $O(n^2 \times p)$.

11.1.2 Binary search algorithm

Another way to solve the CCP problem is to perform a binary search on the value of the objective function. Give such an algorithm and its complexity.

(*Hint:* Be careful to account properly for the complexity of the binary search.)

Solution. Given a bound M, can we partition $[a_1, a_2, \ldots, a_n]$ into p intervals $\mathcal{I}_1, \mathcal{I}_2, \ldots, \mathcal{I}_p$ whose sums are not greater than M? We answer this question with a simple greedy algorithm. For $\mathcal{I}_1$, we start with a_1 and include the next elements until the sum exceeds M. Formally, e_1 is defined by $\sum_{i=1}^{e_1} a_i \leqslant M$ and $\sum_{i=1}^{e_1+1} a_i > M$. For $\mathcal{I}_2$, we start from a_{e_1+1} and repeat the procedure. We have a solution if and only if we reach the last element within p steps. Checking for a solution with M has a cost $O(n)$.

Obviously, a lower bound for M is $L = \max\left(\max_i(a_i), \frac{1}{p}\sum_{i=1}^{n} a_i\right)$. An upper bound is $U = \sum_{i=1}^{n} a_i$. The number of iterations in the binary search for M is bounded by $\log(U-L)$, which is polynomial in the problem size.

11.1.3 Improved algorithms

In this section, we investigate how to improve the complexity of the dynamic-programming algorithm introduced in Section 11.1.1. The idea is to exploit the monotonicity of $g(i,k)$, which is nondecreasing in i for $i \geqslant k$ (see [81]).

First, we prove that $g(i, k)$ (the optimal cost when partitioning $[a_1, \ldots, a_i]$ into k intervals) is nondecreasing in i for $i \geqslant k$. The optimal partition for $g(i + 1, k)$ can be used for $[a_1, \ldots, a_i]$. Indeed, we just need to remove a_{i+1} from the last interval. If the last interval is containing only a_{i+1}, we divide any interval with more than two elements into two disjoint intervals (because $k < i + 1$, such an interval exists). Therefore, the cost of this new solution is not greater than $g(i + 1, k)$, and it is not lower than $g(i, k)$, because it is a partition of $[a_1, \ldots, a_i]$. Finally, $g(i, k) \leqslant g(i + 1, k)$, which concludes the proof.

Let us now define the *balance* of two integers i, k, with $1 \leqslant i \leqslant n$ and $2 \leqslant k \leqslant p$. The balance $b_{i,k}$ is such that:

- $1 \leqslant b_{i,k} \leqslant i$,
- $\sum_{j=b_{i,k}+1}^{i} a_j < g(b_{i,k}, k - 1)$, and
- $\sum_{j=b_{i,k}}^{i} a_j \geqslant g(b_{i,k} - 1, k - 1)$.

Intuitively, for $s \geqslant b_{i,k}$, the maximum is dictated by $g(s, k - 1)$, while it is dictated by $\sum_{j=s+1}^{i} a_j$ otherwise. It is easy to prove that the balance point exists and is unique, by monotonicity of the functions.

The next step is to use the balance $b_{i,k}$ to compute $g(i, k)$. We prove indeed that

$$g(i, k) = \min \left(g(b_{i,k}, k - 1), \sum_{j=b_{i,k}}^{i} a_j \right) .$$

The proof is done by contradiction, first for the case $\sum_{j=b_{i,k}}^{i} a_j \leqslant g(b_{i,k}, k-1)$, and then for the symmetrical case, exploiting the properties of the balance.

From these observations, we compute the optimal position of the separator s when computing $g(i, k)$, which we denote by $s_{i,k}$. We have $s_{i,k} \geqslant s_{i-1,k}$, and recursively,

$$s_{i,k} = \operatorname{argmin}_{s_{i-1,k} \leqslant s < i} \left(\max \left(g(s, k - 1), \sum_{j=s+1}^{i} a_j \right) \right) ,$$

with

$$g(i, k) = \max \left(g(s_{i,k}, k - 1), \sum_{j=s_{i,k}+1}^{i} a_j \right) .$$

The initialization writes $s_{1,k} = 1$, $s_{i,1} = i$, and $g(i, 1) = \sum_{k=1}^{i} a_j$. Therefore, the balance allows us to compute all $g(i, k)$ entries for $1 \leqslant i \leqslant n$ in only one pass over the elements (i.e., in $O(n)$), which reduces the complexity of the algorithm to $O(n \times p)$ (instead of $O(n^2 \times p)$).

The complexity can be further reduced to $O(p(n-p))$ by observing that there are no empty intervals, and, therefore, the pass over the elements is restrained to an interval of $n-p$ elements. The initialization becomes $g(i,i) = \max_{1 \leqslant j \leqslant i}(a_j)$; if $k = i$, we keep i intervals with one element in each interval.

11.2 Variants of the problem

In this section, we study two variants of the CCP problem. First, we add communication costs in Section 11.2.1, and then in Section 11.2.2 we consider that processors are heterogeneous but that their order is predefined (i.e., the target platform is a chain of processors).

11.2.1 Communication costs

If we want to account for communication costs, we first need to define the communication model. Our aim is to have a realistic model that remains tractable. For this problem, we assume that there is an amount of communication to be transferred from one processor to another, if they are processing contiguous intervals: for $1 \leqslant i \leqslant n$, δ_i is the size of the output of the i-th element. It is often assumed that intraprocessor communication time is negligible, even though it would be easy to account for this time by lumping it with the execution times, i.e., the size of the elements. We define below the chains-on-chains partitioning problem with communications (CCPC).

DEFINITION 11.2 (CCPC). Given an array of n elements $a_1, a_2, \ldots, a_n$, and an array of communication costs $\delta_1, \delta_2, \ldots, \delta_n$, the chains-on-chains partitioning problem with communications (CCPC) consists of partitioning the array of n elements into p consecutive intervals $\mathcal{I}_1, \mathcal{I}_2, \ldots, \mathcal{I}_p$, where $\mathcal{I}_k = [d_k, e_k]$ and $d_1 = 1$, $e_p = n$, $d_k \leqslant e_k$, and $d_{k+1} = e_k + 1$ for $1 \leqslant k \leqslant p-1$. The objective is to minimize

$$\max_{1 \leqslant k \leqslant p} \left(\sum_{i \in \mathcal{I}_k} a_i + comm(\mathcal{I}_k) \right) = \max_{1 \leqslant k \leqslant p} \left(\sum_{i=d_k}^{e_k} a_i + \delta_{e_k} \right).$$

We assume that the communication cost for interval $\mathcal{I}_k$, $comm(\mathcal{I}_k)$, is equal to the size of the output communication of the interval, i.e., δ_{e_k}. Is it possible to extend the previous algorithms for the CCPC problem? Do communication costs impact their complexity?

Solution. For the dynamic-programming algorithm, we just need to refine the values $f(s,i)$ to account for communication costs: $f(s,i) = \sum_{j=s+1}^{i} a_j + \delta_i$. Also, the initial values become $g(i,1) = \sum_{j=1}^{i} a_j + \delta_i$. The complexity remains identical.

For the binary search, the previous algorithm builds a first interval as soon as the bound M is exceeded. With communication costs, it may happen that the bound is exceeded because of a large output size, while it would be fine to include one more element in the interval. Formally, e_1 is now defined as the maximum value e such that $\sum_{i=1}^{e} a_i + \delta_e \leqslant M$. It can still be obtained in $O(n)$. We now start with the last interval and with $e = n$ and stop as soon as we find a valid interval. The complexity remains the same.

However, it is not possible to adapt easily the improved algorithms because they heavily rely on the *balance*, which cannot be defined with communication costs. The classical dynamic-programming algorithm must be used.

11.2.2 Chain of heterogeneous resources

In this section, we consider the CCP problem in which the target platform is a *chain* of heterogeneous processors. The goal is still to partition the n elements into p intervals, but the element sums must now match p prescribed values (the processor speeds) as closely as possible. Let $s_1, s_2, \dots, s_p$ denote these values. The order of the processors is known, i.e., interval $\mathcal{I}_k$ is mapped onto the k-th processor, and the sum of its elements must match s_k. This heterogeneous version of CCP is called heterogeneous chains-on-chains partitioning problem (CCPH) and is discussed extensively in [87].

DEFINITION 11.3 (CCPH). Given an array of n elements $a_1, a_2, \dots, a_n$, and an array of p values $s_1, s_2, \dots, s_p$, the heterogeneous chains-on-chains partitioning problem (CCPH) consists of partitioning the array of n elements into p consecutive intervals $\mathcal{I}_1, \mathcal{I}_2, \dots, \mathcal{I}_p$, where $\mathcal{I}_k = [d_k, e_k]$ and $d_1 = 1$, $e_p = n$, $d_k \leqslant e_k$, and $d_{k+1} = e_k + 1$ for $1 \leqslant k \leqslant p-1$. The objective is to minimize

$$\max_{1 \leqslant k \leqslant p} \frac{\sum_{i \in \mathcal{I}_k} a_i}{s_k} \quad .$$

We ask whether it is possible to extend previous algorithms for the CCPH problem.

Solution. Here again, the extension of the first two algorithms is quite straightforward, because we know at each step which processor we are targeting, and, therefore, we just need to divide the sum of the a_i values by the corresponding s_k. We also must slightly modify the algorithms by observing that it may now be the case that a processor is left behind (if its speed s_k is too slow). To compute $g(i, k)$, we take the minimum for $1 \leqslant s \leqslant i$ instead of $1 \leqslant s < i$. However, the complexity of the algorithms remains the same.

In this case, it is even possible to adapt the improved algorithm; the properties of the balance remain true, by dividing the sums as explained above. However, the complexity is in $O(n \times p)$, and the last optimization is not possible because a slow processor may be discarded.

11.3 Extension to a clique of heterogeneous resources

The advent of heterogeneous clusters leads to the following generalization of the CCPH problem. The goal is still to partition the n elements into p intervals whose element sums match p prescribed values (the processor speeds $s_1, s_2, \ldots, s_p$) as closely as possible. But, now, we search not only for a partition of $[1..n]$ into p intervals $\mathcal{I}_k = [d_k, e_k]$ but also for a permutation σ of $\{1, 2, \ldots, p\}$, with the objective to minimize

$$\max_{1 \leqslant k \leqslant p} \frac{\sum_{i \in \mathcal{I}_k} a_i}{s_{\sigma(k)}}.$$

Another way to express the problem is that intervals are now weighted by the s_i values, while we had $s_i = 1$ for the homogeneous version CCP. This problem is called CPH (heterogeneous chains partitioning). Can we extend the efficient algorithms described above to solve CPH? In fact, the problem seems combinatorial because of the search over all possible permutations to weight the intervals.

Indeed, we prove the NP-completeness of (the decision problem associated with) CPH in Section 11.3.1 before discussing practical solutions to solve the problem in Sections 11.3.2 and 11.3.3.

DEFINITION 11.4 (CPH). Given an array of n elements $a_1, a_2, \ldots, a_n$, and an array of p values $s_1, s_2, \ldots, s_p$, the heterogeneous chain partitioning problem (CPH) consists of finding a partition of $[1..n]$ into p intervals $\mathcal{I}_1, \mathcal{I}_2, \ldots, \mathcal{I}_p$, with $\mathcal{I}_k = [d_k, e_k]$ and $d_1 = 1$, $e_p = n$, $d_k \leqslant e_k$, and $d_{k+1} = e_k + 1$ for $1 \leqslant k \leqslant p-1$, and a permutation σ of $\{1, 2, \ldots, p\}$. The objective is to minimize

$$\max_{1 \leqslant k \leqslant p} \frac{\sum_{i \in \mathcal{I}_k} a_i}{s_{\sigma(k)}}.$$

11.3.1 NP-completeness

THEOREM 11.1. *The decision problem associated with the CPH optimization problem is NP-complete.*

We consider the associated decision problem: Given a bound K, can we find a partition and a permutation such that $\max_{1 \leqslant k \leqslant p} \frac{\sum_{i \in \mathcal{I}_k} a_i}{s_{\sigma(k)}} \leqslant K$?

Because of the intervals, it seems difficult to have a straightforward reduction from 2-PARTITION for this problem. Actually, the proof is quite involved, and it provides a nice example of reduction from NUMERICAL MATCHING WITH TARGET SUMS (NMWTS) [38].

We first explain the reasoning that leads us to the proof before providing the formal proof. The first challenge is to create a repetitive pattern on which we use the initial problem: NMWTS. In order to do so, we introduce n large tasks such that they will be each mapped alone on a dedicated processor (the tasks of weight D below). Then, in order to add some combinatorial freedom in the choices made by the optimal solution, we add two processors per interval, and a set of M small tasks of weight 1, so that the interval must be split somewhere between two such tasks. This way, we force that the appropriate processors must be used for each interval, and, by tuning the parameters, we obtain a matching problem.

Another proof, based on 3-PARTITION, can be found in [87]. The idea is still to create a repetitive pattern with n large tasks, and all other tasks have a weight 1. The processor speeds correspond to the integers of 3-PARTITION, and we enforce that three processors must be assigned to each pattern.

Proof. The CPH problem clearly belongs to the class NP. Given a solution, it is easy to verify in polynomial time that the partition into p intervals is valid and that the maximum sum of the elements in a given interval divided by the corresponding s value does not exceed the bound K.

To establish the completeness, we use a reduction from NMWTS, which is NP-complete in the strong sense [38]. We consider an instance $\mathcal{I}_1$ of NMWTS. Given $3m$ numbers $x_1, x_2, \ldots, x_m$, $y_1, y_2, \ldots, y_m$, and $z_1, z_2, \ldots, z_m$, do there exist two permutations σ_1 and σ_2 of $\{1, 2, \ldots, m\}$, such that $x_i + y_{\sigma_1(i)} = z_{\sigma_2(i)}$ for $1 \leqslant i \leqslant m$? Because NMWTS is NP-complete in the strong sense, we can encode the $3m$ numbers in unary and assume that the size of $\mathcal{I}_1$ is $O(m \times M)$, where $M = \max_i\{x_i, y_i, z_i\}$. We also assume that $\sum_{i=1}^m x_i + \sum_{i=1}^m y_i = \sum_{i=1}^m z_i$; otherwise, $\mathcal{I}_1$ cannot have a solution.

We build the following instance $\mathcal{I}_2$ of CPH (we use the formulation in terms of task weights and processor speeds, which is more intuitive):

- We define $n = (M+3)m$ tasks, whose weights are outlined below:

$$A_1, \underbrace{1, 1, \cdots, 1}_{M}, C, D, A_2, \underbrace{1, 1, \cdots, 1}_{M}, C, D, \cdots, A_m, \underbrace{1, 1, \cdots, 1}_{M}, C, D.$$

  Here, $B = 2M$, $C = 5M$, $D = 7M$, and $A_i = B + x_i$ for $1 \leqslant i \leqslant m$. To define the a_is formally for $1 \leqslant i \leqslant n$, let $N = M + 3$. We have for $1 \leqslant i \leqslant m$:

$$\begin{cases} a_{(i-1)N+1} = A_i = B + x_i \\ a_{(i-1)N+j} = 1 \text{ for } 2 \leqslant j \leqslant M+1 \\ a_{iN-1} = C, \qquad a_{iN} = D. \end{cases}$$

- For the number of processors (and intervals), we choose $p = 3m$. As for the speeds, we let s_i be the speed of processor P_i where, for $1 \leqslant i \leqslant m$:

$$s_i = B + z_i, \qquad s_{m+i} = C + M - y_i, \qquad s_{2m+i} = D.$$

Finally, we ask whether there exists a solution matching the bound $K = 1$. Clearly, the size of $\mathcal{I}_2$ is polynomial in the size of $\mathcal{I}_1$. We now show that instance $\mathcal{I}_1$ has a solution if and only if instance $\mathcal{I}_2$ does.

Suppose first that $\mathcal{I}_1$ has a solution, with permutations σ_1 and σ_2 such that $x_i + y_{\sigma_1(i)} = z_{\sigma_2(i)}$. For $1 \leqslant i \leqslant m$,

- we map each task A_i and the following $y_{\sigma_1(i)}$ tasks of weight 1 onto processor $P_{\sigma_2(i)}$;
- we map the following $M - y_{\sigma_1(i)}$ tasks of weight 1 and the next task, of weight C, onto processor $P_{m+\sigma_1(i)}$;
- we map the next task, of weight D, onto processor P_{2m+i}.

We do have a valid partition of all the tasks into $p = 3m$ intervals. For $1 \leqslant i \leqslant m$, the load and speed of processor P_i are indeed equal:

- The load of $P_{\sigma_2(i)}$ is $A_i + y_{\sigma_1(i)} = B + x_i + y_{\sigma_1(i)}$ and its speed is $B + z_{\sigma_2(i)}$;
- The load of $P_{m+\sigma_1(i)}$ is $M - y_{\sigma_1(i)} + C$, which is equal to its speed;
- The load and speed of P_{2m+i} are both D.

The mapping does achieve the bound $K = 1$, hence, a solution to $\mathcal{I}_1$.

Suppose now that $\mathcal{I}_2$ has a solution, i.e., a mapping matching the bound $K = 1$. We first observe that $s_i < s_{m+j} < s_{2m+k} = D$ for $1 \leqslant i, j, k \leqslant m$. Indeed $s_i = B + z_i \leqslant B + M = 3M$, $5M \leqslant s_{m+j} = C + M - y_j \leqslant 6M$, and $D = 7M$. Hence, each of the m tasks of weight D must be assigned to a processor of speed D, and it is the only task assigned to this processor. These m singleton assignments divide the set of tasks into m intervals, namely, the set of tasks before the first task of weight D, and the $m - 1$ sets of tasks lying between two consecutive tasks of weight D. The total weight of each of these m intervals is $A_i + M + C > B + M + C = 10M$, while the largest speed of the $2m$ remaining processors is $6M$. Therefore, each of them must be assigned to at least two processors. However, there remains only $2m$ available processors, hence, each interval is assigned exactly two processors.

Consider such an interval $A_i \; 111 \cdots 1 \; C$ with M tasks of weight 1, and let P_{i_1} and P_{i_2} be the two processors assigned to this interval. Tasks A_i and C are not assigned to the same processor (otherwise, the whole interval would be assigned to the same processor). So, P_{i_1} receives task A_i and h_i tasks of weight 1, while P_{i_2} receives $M - h_i$ tasks of weight 1 and task C. The weight of P_{i_2} is $M - h_i + C \geqslant C = 5M$, while $s_i \leqslant 3M$ for $1 \leqslant i \leqslant m$. Hence, P_{i_1} must be some P_i, $1 \leqslant i \leqslant m$, while P_{i_2} must be some P_{m+j}, $1 \leqslant j \leqslant m$. Because this holds true on each interval, this defines two permutations $\sigma_2(i)$ and $\sigma_1(i)$ such that $P_{i_1} = P_{\sigma_2(i)}$ and $P_{i_2} = P_{\sigma_1(i)}$. Because the bound $K = 1$ is achieved, we have:

- $A_i + h_i = B + x_i + h_i \leqslant B + z_{\sigma_2(i)}$;
- $M - h_i + C \leqslant C + M - y_{\sigma_1(i)}$.

Therefore, $y_{\sigma_1(i)} \leqslant h_i$, $x_i + h_i \leqslant z_{\sigma_2(i)}$, and $\sum_{i=1}^m x_i + \sum_{i=1}^m y_i \leqslant \sum_{i=1}^m x_i + \sum_{i=1}^m h_i \leqslant \sum_{i=1}^m z_i$. By hypothesis, $\sum_{i=1}^m x_i + \sum_{i=1}^m y_i = \sum_{i=1}^m z_i$, hence, all inequalities are tight, and in particular $\sum_{i=1}^m x_i + \sum_{i=1}^m h_i = \sum_{i=1}^m z_i$.

We can deduce that $\sum_{i=1}^{m} y_i = \sum_{i=1}^{m} h_i = \sum_{i=1}^{m} z_i - \sum_{i=1}^{m} x_i$ and, because $y_{\sigma_1(i)} \leqslant h_i$ for all i, we have $y_{\sigma_1(i)} = h_i$ for all i. Similarly, we deduce that $x_i + h_i = z_{\sigma_2(i)}$ for all i, and, therefore, $x_i + y_{\sigma_1(i)} = z_{\sigma_2(i)}$.

Altogether, we have found a solution for $\mathcal{I}_1$, which concludes the proof. □

11.3.2 Practical solutions

The complexity study shows that we need to resort to heuristics in order to have practical solutions to the CPH problem. It turns out that the use of the optimal CCPH algorithms can help in deriving efficient heuristics. The idea is to fix arbitrarily an order for the processors and then to run the CCPH optimal algorithm. It turns out that this heuristic is efficient in practice, provided that the heterogeneity of the processors is moderate (see the experiments reported in [87]).

We provide below a simple approximation algorithm for CPH. In the following, $s_{\max} = \max_{1 \leqslant k \leqslant p} s_k$, and $s_{\min} = \min_{1 \leqslant k \leqslant p} s_k$.

THEOREM 11.2. *There exists a $\frac{s_{\max}}{s_{\min}}$ approximation algorithm for CPH.*

Proof. Let M_{opt} be an optimal solution to CPH. First, we compute the optimal CCP solution, assuming that all processor speeds are equal to $s_{\max}$, and we denote this solution by M_{max}. Because the speeds available in the optimal solution are all smaller than $s_{\max}$, we necessarily have $M_{opt} \geqslant M_{max}$; otherwise, we could find a better CCP solution to the problem with speeds equal to $s_{\max}$.

The approximation algorithm simply reuses the intervals created in the solution M_{max}, but the speeds of the processors are now the original speeds. The interval I that realizes the maximum is using a processor P_j of speed s_j, and the solution is $M_{algo} = \sum_{i \in I} a_i / s_j$. Because I is an interval obtained from M_{max}, we have $\sum_{i \in I} a_i / s_{\max} \leqslant M_{max}$. It follows that

$$M_{algo} = \frac{\sum_{i \in I} a_i}{s_j} \leqslant M_{max} \times \frac{s_{\max}}{s_j} \leqslant M_{opt} \times \frac{s_{\max}}{s_{\min}} ,$$

which concludes the proof. □

The problem of improving this approximation ratio is left open, and, in particular, the problem of establishing a better approximation ratio for the more sophisticated heuristic algorithm based on CCPH is open.

11.3.3 Integer linear program

We present here an integer linear program to compute the optimal solution of CPH. We denote by I_u the interval of elements handled by the processor of speed s_u, for $1 \leqslant u \leqslant p$. First, we need to define a few variables:

- For $1 \leqslant i \leqslant n$ and $1 \leqslant u \leqslant p$, $x_{i,u}$ is a Boolean variable equal to 1 if $a_i \in I_u$, and 0 otherwise;
- For $1 \leqslant u \leqslant p$, d_u is an integer variable that denotes the first element of I_u; similarly, e_u denotes the last element of I_u ($I_u = [d_u, e_u]$); of course, $1 \leqslant d_u \leqslant e_u \leqslant n$;
- K is the objective variable.

The objective function is to minimize K, and the following constraints need to be enforced:

- Each stage is in an interval: $\forall 1 \leqslant i \leqslant n$, $\sum_{1 \leqslant u \leqslant p} x_{i,u} = 1$;
- If $a_i \in I_u$, then necessarily $d_u \leqslant i \leqslant e_u$; this constraint writes $\forall 1 \leqslant i \leqslant n, \forall 1 \leqslant u \leqslant p$, $d_u \leqslant i \times x_{i,u} + n \times (1 - x_{i,u})$ and $e_u \geqslant i \times x_{i,u}$;
- If $a_i \in I_u$ and $a_{i+1} \in I_v$, with $u \neq v$, then necessarily $d_v \geqslant i + 1$ and $e_u \leqslant i$; this constraint writes
$$\forall 1 \leqslant i < n,\ \forall 1 \leqslant u, v \leqslant p, u \neq v,\ \ d_v \geqslant (i+1) \times (x_{i,u} + x_{i+1,v} - 1)$$
$$\text{and } e_u \leqslant i \times (x_{i,u} + x_{i+1,v} - 1) + n \times (2 - x_{i,u} - x_{i+1,v});$$
- There remains to compute the sum on each interval and to constrain it by K: $\forall 1 \leqslant u \leqslant p$, $\sum_{i=1}^{n} \frac{a_i}{s_u} x_{i,u} \leqslant K$.

We have $(n + 2) \times p + 1$ variables and $O(n \times p^2)$ constraints. All variables are Boolean or integer, except K, which is rational.

Note that it also is possible to extend this integer linear programming formulation to account for heterogeneous communication costs, hence, tackling the most complex combination of variants introduced in this case study. The formulation can be found in [13].

11.4 Conclusion

In this case study, we have thoroughly studied the chains-on-chains partitioning problem. Optimal solutions can be found for different variants of the problem, with dynamic-programming algorithms or a combination of binary search and greedy techniques. More efficient algorithms can be derived for easier problem instances.

The problem becomes NP-hard when we relax the chains-on-chains problem to a chain partitioning problem, in which the target platform is no longer a chain of processors but a clique of heterogeneous processors. We propose practical solutions to overcome the combinatorial nature of the problem: an efficient polynomial-time heuristic, an approximation algorithm, and an integer linear programming formulation.

Chains-on-chains partitioning is widely used in the context of streaming applications, when a linear chain application must be mapped onto a platform, with the objective to minimize the throughput of the application, i.e., the rate at which two consecutive data sets can be processed. The computation requirement of each pipeline stage corresponds to an element of the previous problem definitions, a_i, with $1 \leqslant i \leqslant n$. We also can consider communication costs, δ_i, as in the definition of CCPC.

We often restrict solutions to be interval mappings, hence, enforcing the chain partitioning constraint. Therefore, the goal is to find a partition of the pipeline stages into intervals, and a one-to-one mapping of these intervals onto processors, such that the period of the application, i.e., the longest cycle-time to operate a stage, is minimized. Note that the period is the inverse of the throughput that can be achieved, and it corresponds to the objective of CPH. The problem is, therefore, NP-hard, even without communication costs. However, when processors and bandwidth links are homogeneous, the problem is CCPC (with communication costs), and it can be solved in polynomial time.

Chapter 12

Replica placement in tree networks

This case study deals with the general problem of replica placement in tree networks. Informally, there are clients issuing requests to be satisfied by servers. The clients are known (both their position in the tree and their number of requests), while the number and location of the servers are to be determined. A client is a leaf node of the tree, and its requests can be served by one or several internal nodes. Initially, there are no replicas; when a node is equipped with a replica, it can process a number of requests up to its capacity limit. Nodes equipped with a replica, also called servers, can serve only clients located in their subtree (so that the root, if equipped with a replica, can serve any client); this restriction is usually adopted to enforce the hierarchical nature of the target application platforms where a node has knowledge only of its parent and children in the tree.

The rule of the game is to assign replicas to nodes so that some optimization function is minimized. Typically, this optimization function is the total utilization cost of the servers. If all the nodes are identical, this reduces to minimizing the number of replicas. If the nodes are heterogeneous, it is natural to assign a cost proportional to their capacity (so that one replica on a node capable of handling 200 requests is equivalent to two replicas on nodes of capacity 100 each).

We point out that the distribution tree (clients and nodes) is fixed in our approach. This key assumption is quite natural for a broad spectrum of applications, as, for instance, video on demand service delivery [24, 55, 56, 72, 105, 109]. The root server has the original copy of the database but cannot serve all clients directly, so a distribution tree is deployed to provide a hierarchical and distributed access to replicas of the original data. On the contrary, in other more decentralized applications (e.g., allocating Web mirrors in distributed networks), a two-step approach is used [57, 60, 74, 91, 102, 104]. First, determine a "good" distribution tree in an arbitrary interconnection graph, and then determine a "good" placement of replicas among the tree nodes. Both steps are interdependent, and the problem is much more complex, due to the combinatorial solution space (the number of candidate distribution trees may well be exponential).

We first detail the framework in Section 12.1, and, in particular, we motivate the different policies that can be enforced for replica placement; we investigate different access policies and compare them. Then, we provide exhaustive complexity results for the different policies in Section 12.2. Unsurprisingly, all problem instances become NP-hard with heterogeneous servers, but we exhibit variants of the problem in Section 12.3, for which the problem becomes NP-hard even in the homogeneous case. Either we must guarantee a quality of service (QoS), or we do not focus on the cost of the solution, but rather we try to minimize the power consumption of a given replica placement. Finally, we conclude in Section 12.4.

12.1 Access policies

First, we motivate and define various access policies in Section 12.1.1. Then, we study the impact of the policies that we retain for this case study on the existence of a solution (see Section 12.1.2) and on the cost of a solution (see Section 12.1.3).

12.1.1 Motivation

Given a distribution tree, with a set of client requests, and an access policy, the problem is to place replicas in the tree, following the rules of the access policy, so that the cost is minimized. In this section, we discuss several access policies.

Consider first that a client can be served by any internal node in the tree. We call this policy *One-to-any*. The problem turns out to be immediately NP-hard, even in the homogeneous case, from a straightforward reduction from 2-PARTITION (see Definition 10.1, p. 245): Consider an instance with n clients, the i-th client has a_i requests (the a_is are from the instance of 2-PARTITION), and the node capacity is $\frac{\sum_{i=1}^{n} a_i}{2}$. It is easy to see that there is a solution with two servers if and only if there is a 2-PARTITION of the a_is.

Why restrict to one server per client? The *One-to-many* policy is similar to the previous one, except that the processing of a given client's requests can be split among several servers. The previous reduction does not work anymore. Indeed, the problem becomes trivially polynomial in the homogeneous case: We fill the servers up to their capacities until there are no more requests to process. For the heterogeneous case, however, the problem remains NP-hard, again from a reduction from 2-PARTITION. This time, the server capacities are set to a_i (from the instance of 2-PARTITION), and the total number of requests is $\frac{\sum_{i=1}^{n} a_i}{2}$. It is easy to see that there is a solution of cost $\frac{\sum_{i=1}^{n} a_i}{2}$ if and only if there is a 2-PARTITION of the a_is. Otherwise, there is some server capacity that is unused, and the cost is greater.

From what precedes, both policies that we introduced so far have little theoretical interest. Moreover, as motivated earlier, it is often assumed that a server can serve only clients located in its subtree, so the practical interest of the previous policies is limited as well. Rather, in most papers from the literature, all requests of a client are served by the closest replica, i.e., the first replica found in the unique path from the client to the root in the distribution tree. This *Closest* policy is simple and natural but may be unduly restrictive, leading to a waste of resources. We discuss below two policies that are less constraining.

In the *Upwards* policy, we keep the restriction that all requests from a given client are processed by the same replica, but we allow client requests to "traverse" servers so as to be processed by other replicas located higher in the path (closer to the root). The trade-off to explore is that the *Closest* policy assigns replicas at proximity of the clients but may need to allocate too many of them if some local subtree issues a great number of requests. The *Upwards* policy will ensure a better resource usage, load-balancing the process of requests on a larger scale; the possible drawback is that requests will be served by remote servers, likely to take a longer time to process them. Taking quality of service constraints into account would typically be more important for the *Upwards* policy.

In the *Multiple* policy, we further relax access constraints and grant the possibility for a client to be assigned several replicas. With this policy, the processing of a given client's requests will be split among several servers located in the tree path from the client to the root, similarly to the above-described *One-to-many* policy. Obviously, this policy is the most flexible, and likely to achieve the best resource usage, among policies that restrict to the fact that a server can serve only clients located in its subtree. The only drawback is the (modest) additional complexity induced by the fact that requests must now be tagged with the replica server identifier in addition to the client identifier.

The comparison among the three latter access policies (*Closest*, *Upwards*, and *Multiple*) is done below in a framework with identical node capacities; thus, the problem amounts at minimizing the number of servers.

12.1.2 Impact of the policies on the existence of a solution

We first show the impact of the policies on a very simple instance of the problem. In this example, there are two nodes, B being the unique child of A, the tree root (Figure 12.1). Each node can process $W = 1$ request.

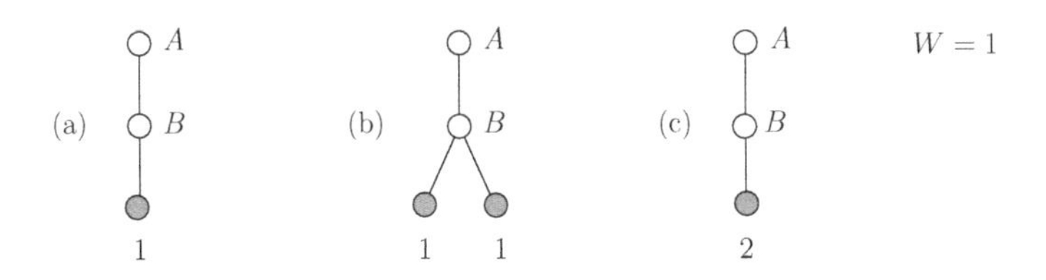

FIGURE 12.1: Solution existence.

If B has one client child making 1 request, the problem has a solution with all three policies, placing a replica on B or on A indifferently (Figure 12.1(a)). If B has two client children, each making 1 request, the problem has no more solution with *Closest*. However, we have a solution with both *Upwards* and *Multiple* if we place replicas on both nodes. Each server will process the request of one of the clients (Figure 12.1(b)). Finally, if B has only one client child making 2 requests, only *Multiple* has a solution because we need to process one request on B and the other on A, thus requesting multiple servers (Figure 12.1(c)).

12.1.3 Impact of the policies on the cost of a solution

Recall that the cost of a solution is the number of replicas that are placed in the tree, because we consider that all servers are identical. We first compare *Closest* and *Upwards*, and then we compare *Upwards* and *Multiple*. By transitivity, *Multiple* is better than *Closest*.

We construct an instance of the problem where the *Upwards* policy is arbitrarily better than the *Closest* policy. We consider the tree network of Figure 12.2, where there are $2n + 2$ internal nodes of capacity $W = n$, and $2n+1$ clients, each of them making one request. With the *Upwards* policy, we place three replicas in A, B, and S_{2n}. All requests can be satisfied with these three replicas. When considering the *Closest* policy, first we need to place a replica in A to cover its client. Then, we can decide to place a replica on B or not. If we place a replica on B, then this replica is handling n requests, but there remain n other requests from the clients in its subtree that cannot be processed by B. Thus, we need to add n replicas among $S_1, \ldots, S_{2n}$. Otherwise, $n - 1$ requests of the $2n$ clients in the subtree of B can be processed by A in addition to its own client. We need to add $n + 1$ extra replicas among $S_1, \ldots, S_{2n}$. In both cases, we are placing $n + 2$ replicas, instead of the 3 replicas needed with the *Upwards* policy, thus, a performance factor of $\frac{n+2}{3}$. This proves that *Upwards* can be arbitrary better than *Closest*, even in the simple case with homogeneous servers.

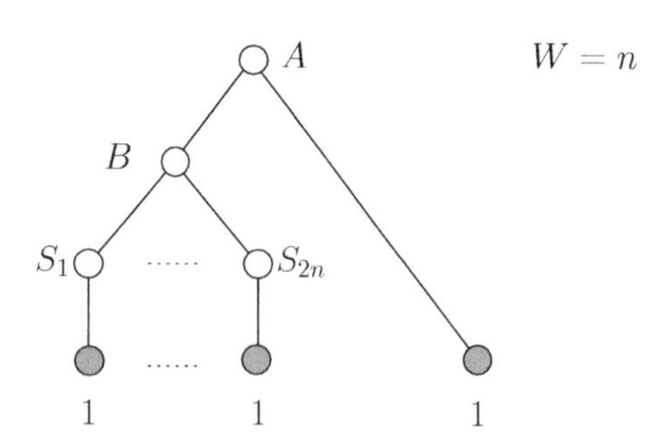

FIGURE 12.2: Solution cost: *Upwards* versus *Closest*.

The second comparison is between *Multiple* and *Upwards*. We build an instance of the replica placement problem where both access policies have a solution, but the solution of *Multiple* is arbitrarily better than the solution of *Upwards*. Consider the instance represented in Figure 12.3, with $3 + n$ nodes of capacity $W = 4n$. The root A has $n + 2$ children nodes: B, C, and $S_1, ..., S_n$. Node B has two client children, one with $2n - 1$ requests and the other with $4n$ requests. Node C has two client children, one with $2n$ requests and the other with $2n + 1$ requests. Each node S_i $(1 \leqslant i \leqslant n)$ has a unique child, a client with 2 requests.

- The *Multiple* policy assigns three replicas, one each to A, B, and C. B handles the $4n$ requests of its second client, while the other client is served by A. C handles $2n$ requests from both of its clients, and the one remaining request is processed by A. Server A, therefore, processes $(2n - 1) + 1 = 2n$ requests coming up from B and C. Requests coming from the n remaining nodes sum up to $2n$; thus, A is able to process all of them.

- For the *Upwards* policy, we need to assign replicas everywhere. Indeed, with this policy, C cannot handle more than $2n + 1$ requests because it is unable to process requests from both of its children, and, thus, A has $(2n - 1) + 2n$ requests coming from B and C. It cannot handle any of the $2n$ remaining requests, and each remaining node S_i $(1 \leqslant i \leqslant n)$ must process requests coming from its own client. This leads to a total of $n + 3$ replicas.

The performance factor is $\frac{n+3}{3}$, which can be arbitrarily big when n becomes large. This proves that *Multiple* can be arbitrary better than *Upwards*, even in the simple case with homogeneous servers.

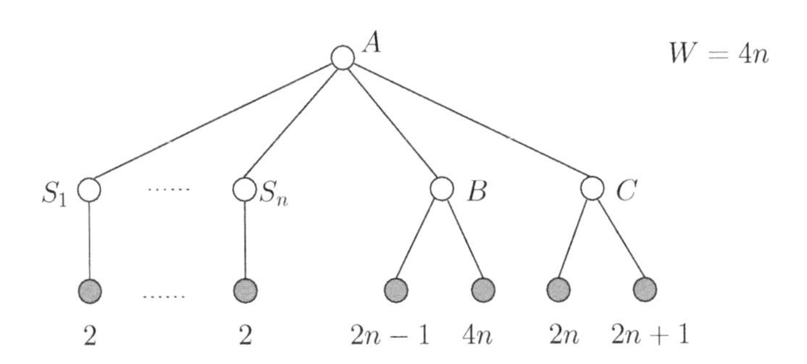

FIGURE 12.3: Solution cost: *Multiple* versus *Upwards*.

12.2 Complexity results

One major goal of this case study is to assess the impact of the access policy on the problem with homogeneous versus heterogeneous servers, as explained below. First, we formally define the different problem instances in Section 12.2.1. Then, we tackle the homogeneous instances in Section 12.2.2 and the heterogeneous ones in Section 12.2.3. We conclude this complexity study with an integer linear programming formulation (see Section 12.2.4) that allows us to solve the combinatorial instances of the problem.

12.2.1 Definitions

We consider a tree $\mathcal{T}$ whose set of nodes is $\mathcal{C} \cup \mathcal{N}$, where the clients $\mathcal{C}$ are leaves of the tree. Each client $i \in \mathcal{C}$ has r_i requests; each node $j \in \mathcal{N}$ has processing capacity W_j and storage cost $sc_j = W_j$. We need to decide which node is equipped with a replica and, thus, becomes a server ($j \in \mathcal{R}$, where $\mathcal{R}$ is the set of replicas). This problem comes in two flavors, either with homogeneous nodes ($W_j = W$ for all $j \in \mathcal{N}$) (MinNb, the goal is to minimize the number of servers $|\mathcal{R}|$) or with heterogeneous nodes, i.e., servers with different capacities/costs (MinCost, the goal is to minimize the total storage cost $\sum_{j \in \mathcal{R}} sc_j$). The comparison among policies was done in the homogeneous case and holds true for the heterogeneous case.

In the single server version of the problem (*Closest* and *Upwards* policies), we need to find a server $server(i)$ for each client $i \in \mathcal{C}$. $\mathcal{R}$ is the set of replicas, i.e., the servers chosen among the nodes in $\mathcal{N}$. The constraint is that server capacities cannot be exceeded; this translates into

$$\sum_{i \in \mathcal{C}, server(i)=j} r_i \leqslant W_j \quad \text{for all } j \in \mathcal{R}.$$

The objective is to find a valid solution of minimal storage cost $\sum_{j \in \mathcal{R}} W_j$.

In the *Multiple* policy with multiple servers per client, for any client $i \in \mathcal{C}$ and any node $j \in \mathcal{N}$, $r_{i,j}$ is the number of requests from i that are processed by j ($r_{i,j} = 0$ if $j \notin \mathcal{R}$, and $\sum_{j \in \mathcal{N}} r_{i,j} = r_i$ for all $i \in \mathcal{C}$). The capacity constraint now writes

$$\sum_{i \in \mathcal{C}} r_{i,j} \leqslant W_j \quad \text{for all } j \in \mathcal{R},$$

while the objective function is the same as for the single server version.

The decision problems associated with the previous optimization problems are easy to formulate. Given a bound on the number of servers (homogeneous version) or on the total storage cost (heterogeneous version), is there a valid solution that meets the bound?

12.2.2 MinNb problem

In this section, we establish the complexity of the MinNb problem, i.e., minimizing the number of (identical) servers.

First, we provide a greedy algorithm to solve the problem for the *Closest* policy in polynomial time [110]. Then, we prove the NP-completeness of the *Upwards* case [11], which comes as a surprise because all previously known instances were shown to be polynomial. Finally, we provide a multipass greedy algorithm to show the polynomial complexity of the *Multiple* problem [11].

THEOREM 12.1. *The instance of the* MinNb *problem with the* Closest *policy can be solved in polynomial time.*

This problem can be solved by greedily adding nodes in the replica set. A node is chosen such that there is an optimal solution that contains it, and, hence, we are able to prove the optimality of this greedy algorithm.

Proof. We provide a greedy polynomial-time algorithm to solve the problem. First, we greedily compute the *workload* of all internal nodes, which is defined as the total number of requests in the subtree of a node. Thus, the workload on a leaf $i \in \mathcal{C}$ is r_i, and the workload on an internal node is the sum of the workloads of its children. This computation is done in $O(|\mathcal{N}|)$. If a tree node has a workload greater than W, we call it a *heavy* node; otherwise, it is a *light* node. If a light node has a heavy parent, we call it a *critical* node.

We remove from the tree all light nodes that are noncritical nodes, because we prove that there exists an optimal replica set that does not contain any light and noncritical nodes. Indeed, we can easily build a solution in which the server placed on a light and noncritical node $j \in \mathcal{N}$ is replaced by a server on the only ancestor of j that is critical, j_c. (If no ancestor of j is critical, then the root is not a heavy node, and an optimal solution is to place a single server at the root of the tree.) By definition, j_c is a light node, so it can process its entire subtree, and the workload of the ancestors of j_c has not increased. Note that if there were already a server at node j_c, the solution would not be optimal because we could simply remove the server at node j and obtain a better solution. Therefore, we need only to consider heavy and critical nodes when searching for the optimal replica set.

Then, at each step, we greedily add a replica in the solution by selecting a heavy node j with no heavy node child and adding its child j^* with maximum workload to the replica set. Indeed, we prove that if j is a heavy node whose children are all critical nodes (i.e., no heavy node child), then there exists an optimal replica set that contains the child of j that has the maximum workload. Indeed, because j is heavy, at least one of its children must have a replica; otherwise, j would have too many requests to handle. Let j' be one of these children in the optimal solution, and let j^* be the child that has the maximum workload. If there is no replica at node j^*, we replace the replica at j' with a replica at j^*, hence, processing more requests, and reducing only the workload of node j. Because j^* is a light node (by hypothesis), this replacement is always possible.

The greedy choice is always optimal, and, therefore, node j^* is added to the replica set and removed from the tree. Also, the workloads of all ancestors of j^* are then updated (removing the requests coming from j^*). Some nodes may become noncritical, and we remove them from the tree as well. We then find a solution to the subproblem that we still need to solve by iterating the greedy choice until the tree is empty or the root node becomes a light node, which means that all remaining requests can be processed by the root node, if there are any.

The number of iterations is equal to the number of replicas in the solution, and it is bounded by $|\mathcal{N}|$. Moreover, each iteration is clearly done in polynomial time, which completes the proof. □

THEOREM 12.2. *The instance of the* MinNb *problem with the* Upwards *policy is NP-complete in the strong sense.*

Because all requests of a client must be processed by the same server, the intuition leads directly to a partition problem. With two servers A and B, where B is the unique child of A, and n clients that are the children of B, the problem is a 2-PARTITION (see Definition 10.1, p. 245), where the a_is correspond to the number of requests of each child, and the server capacity is $W = \frac{\sum_{i=1}^{n} a_i}{2}$. We can use the same reasoning with a 3-PARTITION (see Definition 10.5, p. 245), hence, proving the NP-completeness in the strong sense.

Proof. The problem clearly belongs to the class NP. Given a solution, it is easy to verify in polynomial time that all requests are served and that no server capacity is exceeded. To establish the completeness in the strong sense, we use a reduction from 3-PARTITION (see Definition 10.5, p. 245). We consider an instance $\mathcal{I}_1$ of 3-PARTITION: Given $3m$ positive integers $a_1, a_2, \ldots, a_{3m}$ such that $B/4 < a_i < B/2$ for $1 \leqslant i \leqslant 3m$, and $\sum_{i=1}^{3m} a_i = mB$, can we partition these integers into m triples, each of sum B?

We build the following instance $\mathcal{I}_2$ of MinNb (Figure 12.4), with $3m$ clients c_i with $r_i = a_i$ for $1 \leqslant i \leqslant 3m$, and m internal nodes n_j $(1 \leqslant j \leqslant m)$ with $W = B$, such that the children of n_1 are all the $3m$ clients c_i, and for $1 \leqslant j \leqslant m-1$, the parent of n_j is n_{j+1} (hence, n_m is the root). Finally, we ask whether there exists a solution with total storage cost mB, i.e., with a replica located at each internal node. Clearly, the size of $\mathcal{I}_2$ is polynomial (and even linear) in the size of $\mathcal{I}_1$.

We now show that instance $\mathcal{I}_1$ has a solution if and only if instance $\mathcal{I}_2$ does. Suppose first that $\mathcal{I}_1$ has a solution. Let $(a_{k_1}, a_{k_2}, a_{k_3})$ be the triple in $\mathcal{I}_1$. We assign the three clients c_{k_1}, c_{k_2}, and c_{k_3} to server n_k. Because $a_{k_1} + a_{k_2} + a_{k_3} = B$, no server capacity is exceeded. Because the m triples partition the a_is, all requests are satisfied. We do have a solution to $\mathcal{I}_2$.

Suppose now that $\mathcal{I}_2$ has a solution. Let I_k be the set of clients served by node n_k if there is a replica located at n_k, then $\sum_{i \in I_k} a_i \leqslant B$. The total number of requests to be satisfied is $\sum_{i=1}^{3m} a_i = mB$, and there are at most m replicas of capacity B. Therefore, no set I_k can be empty, and $\sum_{i \in I_k} a_i = B$ for $1 \leqslant k \leqslant m$. Because $B/4 < a_i < B/2$, each I_k must be a triple. This leads to the desired solution of $\mathcal{I}_1$. □

THEOREM 12.3. *The instance of the* MinNb *problem with the* Multiple *policy can be solved in polynomial time.*

We outline below an optimal algorithm to solve the problem, and then we illustrate it in an example. The proof of optimality is quite technical, and we refer to [11] for the details that are omitted.

Algorithm for the *Multiple* policy

We propose a greedy algorithm to solve the MinNb problem. Recall that W is the total number of requests that a server can handle. This algorithm works in three passes. First, we select the nodes that will have a replica handling exactly W requests. Then, a second pass allows us to select some extra servers that are fulfilling the remaining requests. Finally, we need to decide for each server how many requests of each client it is processing. Each pass is a greedy algorithm.

We assume that each node i knows its parent $parent(i)$ and its children $children(i)$ in the tree. Also, we introduce a new variable that is the flow coming up in the tree (requests that are not already fulfilled by a server). It is denoted by
$mathitflow_i$ for the flow between i and $parent(i)$. Initially, $\forall i \in \mathcal{C}$,
$mathitflow_i = r_i$ and $\forall i \in \mathcal{N}$,
$mathitflow_i = -1$. Moreover, the set of replicas is empty in the beginning: $\mathcal{R} = \emptyset$.

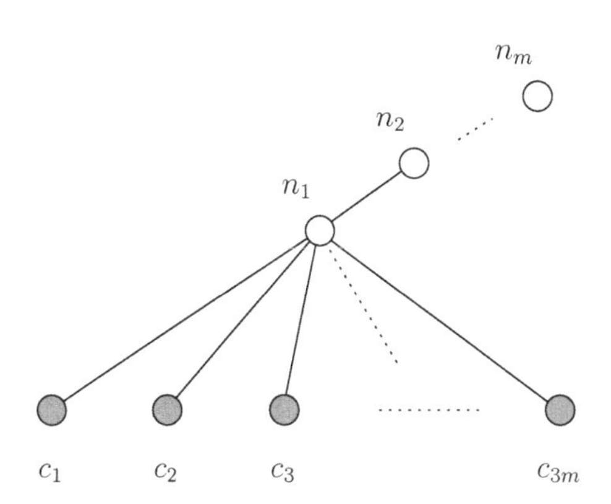

FIGURE 12.4: The platform used in the reduction for Theorem 12.2.

Pass 1. We greedily select in this step some nodes that will process W requests and that are as close to the leaves as possible, and we place a replica on such nodes. We call such a node a *saturated* node. Starting from the root of the tree r, the procedure goes down the tree recursively in order to compute the flows (the flow of an internal node is the sum of the flows of its children). When a flow exceeds W, we place a replica because the corresponding server will be fully used, and we remove the W processed requests from the flow going upwards.

At the end of this pass, if $\mathit{flow}_r = 0$ or ($\mathit{flow}_r \leqslant W$ and $r \notin \mathcal{R}$), we have an optimal solution because all replicas are fully used and all requests are satisfied by adding a replica in r if $\mathit{flow}_r \neq 0$. In this case, we skip pass 2 and go directly to pass 3. Otherwise, we need some extra replicas because some requests are not satisfied yet, and the root cannot satisfy all the remaining requests. To place these extra replicas, we go through pass 2.

Pass 2. In this pass, we need to select the nodes where to add replicas. To do so, while there are too many requests going up to the root, we select the node that can process the highest number of requests, and we place a replica there. The number of requests that a node $j \in \mathcal{N}$ can eventually process is the minimum of the flows between j and the root r, denoted uflow_j (for *u*seful *flow*). Indeed, some requests that are accounted for in the flow of node j might be processed by a server on the path between j and r, where a replica has been placed in pass 1. This is a key property of the greedy choice to prove the optimality of this algorithm, which is not trivial.

It may happen that this pass attempts to place replicas on all nodes, but this solution is not feasible because there are still some requests that are not processed going up to the root. In this case, the original problem instance had no solution. However, if we succeed to place replicas such that $\mathit{flow}_r = 0$, we have a set of replicas that succeeds to process all requests. We then go through pass 3 to assign requests to servers, i.e., to compute how many requests of each client should be processed by each server.

Pass 3. This pass is, in fact, straightforward, starting from the leaves and distributing the requests to the servers from the bottom until the top of the tree. We decide, for instance, to assign requests from clients starting to the left. Starting from the root of the tree r, the procedure goes down the tree recursively. Note that a server that was computing W requests in pass 1 may end up computing fewer requests if one of its descendants in the tree has earned a replica in pass 2. But, this does not impact the optimality of the result, because we keep the same number of replicas.

The sketch of proof below shows the equivalence between the solution built by this algorithm and any optimal solution, thus proving the optimality of the algorithm. First, we illustrate the step-by-step execution of the algorithm in an example.

Example. Figure 12.5(a) provides an example of a network on which we are placing replicas with the *Multiple* policy. The network is homogeneous, and we fix $W = 10$. Pass 1 of the algorithm is quite straightforward to unroll, and Figure 12.5(b) indicates the flow on each link and the saturated replicas are the black nodes.

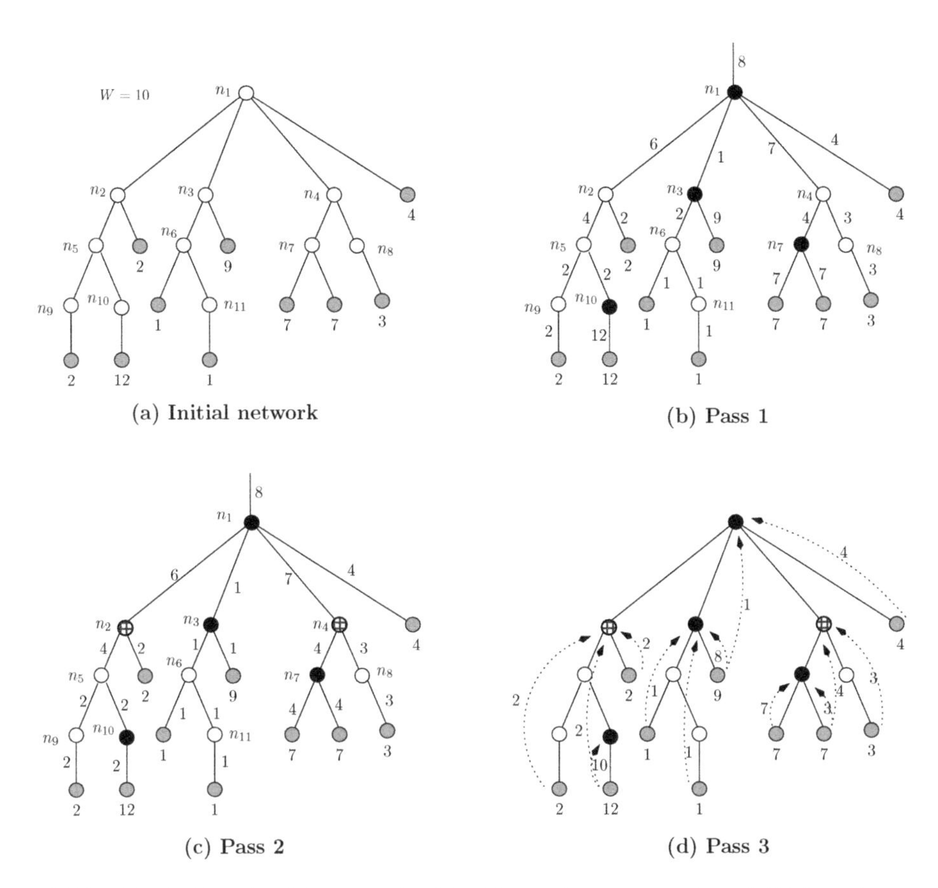

FIGURE 12.5: Algorithm for the MinNb problem with the *Multiple* policy.

During pass 2, we select the nodes of maximum useful flow. Figure 12.5(c) represents these useful flows at the beginning of the pass; we see that node n_4 is the one with the maximum useful flow (7), so we assign it a replica and update the useful flows. All the useful flows are then reduced down to 1 because there is only 1 request going through the root n_1. The first node of maximum useful flow 1 to be selected is n_2, which is set to be a replica of pass 2. The flow at the root is then 0, and it is the end of pass 2.

Finally, pass 3 assigns the servers to the clients and decides which requests are served by which replica (Figure 12.5(d)). For instance, the client with

12 requests shares its requests between n_{10} (10 requests) and n_2 (2 requests). Requests are assigned from the bottom of the tree up to the top. Note that the root n_1, even though it was a saturated replica of pass 1, has only 5 requests to proceed in the end.

Sketch of the proof of optimality

Let R_{opt} be an optimal solution to an instance of the problem. The core of the proof consists of transforming this solution into an equivalent canonical optimal solution R_{can}. We then show that the multipass algorithm is building this canonical solution, and thus it is producing an optimal solution.

First, given a distribution tree and an optimal solution R_{opt}, we define the flow flow_j of a node j by the number of requests going through this node up to its parents, accounting for the requests processed by node j. Also, we define the total flow tflow_j, which is summing all requests, including those processed by replicas. Then, we show that it is possible to change request assignments while keeping an optimal solution. The flows need to be recomputed after any such modification.

Let $j \in \mathcal{N} \cap R_{opt}$ be a server processing strictly fewer than W requests. We can change the request assignment between replicas of the optimal solution in such a way that node j processes exactly $\min(\mathit{tflow}_j, W)$ requests. We point out that we just change the flows of the solution, but at this point, we do not move, add, or delete replicas. Therefore, the solution is still optimal. We invite the reader to write formally the proof of this assertion, which can be found in [11].

We now introduce a new definition, completely independent from the optimal solution but related to the tree network. The *canonical flow* is obtained by distinguishing nodes that receive a flow greater than W from the other nodes. We compute the canonical flow *cflow* of the tree, independently of the replica placement, and define a subset of nodes that are *saturated*, SN. We also compute the number of saturated nodes in $\mathit{subtree}(k)$, denoted nsn_k, for any node $k \in \mathcal{C} \cup \mathcal{N}$ of the tree. For $i \in \mathcal{C}$, $\mathit{cflow}_i = r_i$ and $nsn_i = 0$, and we then compute recursively the canonical flows for nodes $j \in \mathcal{N}$. Let $f_j = \sum_{i \in children(j)} \mathit{cflow}_i$ and $x_j = \sum_{i \in children(j)} nsn_i$. If $f_j \geqslant W$, then $j \in SN$, $\mathit{cflow}_j = f_j - W$ and $nsn_j = x_j + 1$. Otherwise, j is not saturated, $\mathit{cflow}_j = f_j$ and $nsn_j = x_j$. We then have the following result:

LEMMA 12.1. *For all nodes* $j \in \mathcal{C} \cup \mathcal{N}$, $\mathit{tflow}_j \geqslant nsn_j \times W$.

This lemma is proved recursively on the tree. Indeed, we first note that a nonsaturated node always has a canonical flow being less than W: $\forall j \in \mathcal{N} \setminus SN$, $\mathit{cflow}_j < W$. Then, we show that $\mathit{cflow}_j = \mathit{tflow}_j - nsn_j \times W$. This property is true for the clients, and we can see that it remains true for a node if the property is true for all of its children (see [11]).

We are now ready to transform R_{opt} into a new optimal solution, R_{sat}, by redistributing the requests among the replicas and moving some replicas, in order to place a replica at each saturated node, and assigning W requests to this replica. Note that it is easy to show that it is always possible to move a replica to a node without a replica that is one of its ancestors in the tree, while keeping an optimal solution.

This transformation is done starting at the leaves of the tree and considering all nodes $j \in SN$. If there is already a replica on node j, we just need to change the assignment of requests, and Lemma 12.1 ensures that there are enough requests that can be moved, so that the workload of this saturated node is W. Otherwise, we need to move a replica of R_{opt} and place it at node j while keeping a valid solution. If there is a replica in $subtree(j)$ that is not in SN, it is straightforward to move it. Otherwise, we need to rearrange requests, and because of the properties of the flows, we can prove that we can assign W requests to node j and remove a replica that is, in the optimal solution, on the path from j to the root.

Once we have applied this procedure up to the root, we have an optimal solution R_{sat} in which all nodes of SN have a replica and are processing W requests. We will not change the assignment of these replicas anymore in the following. Note that nodes of SN correspond to nodes that are chosen by pass 1 of the greedy algorithm.

In a next step, we further modify the R_{sat} optimal solution in order to obtain what we call the *canonical solution* R_{can}. R_{can} is the solution that is built with the greedy algorithm in polynomial time, and the transformation is aiming at applying the greedy choice for the nonsaturated replicas of the optimal solution. To do so, we change the request assignment of the replicas that are not in SN. We "saturate" some of them as much as we can, and we integrate them into the subset of nodes SN, redefining the *cflow* accordingly. At the end of the process, $SN = R_{can}$ and we have not added replicas, i.e., $|R_{can}| = |R_{sat}| = |R_{opt}|$. This is the most technical part of the proof, and we omit it from this case study. The interested reader can try to finish the proof, using the useful flows introduced in the greedy algorithm. The detailed proof can be found in [11].

12.2.3 MinCost problem

All problems become NP-complete when dealing with resource heterogeneity. Note that previous NP-completeness results involved general graphs rather than trees, and the combinatorial nature of the problem came from the difficulty to extract a good replica tree out of an arbitrary communication graph [74, 91]. Here the tree is fixed, but the problem remains combinatorial due to resource heterogeneity [11].

THEOREM 12.4. *All three instances of the* MinCost *problem with heterogeneous nodes are NP-complete.*

Compared to the proof of the NP-completeness of the *Upwards* policy in the homogeneous case (see Theorem 12.2), we can now play on the heterogeneity of

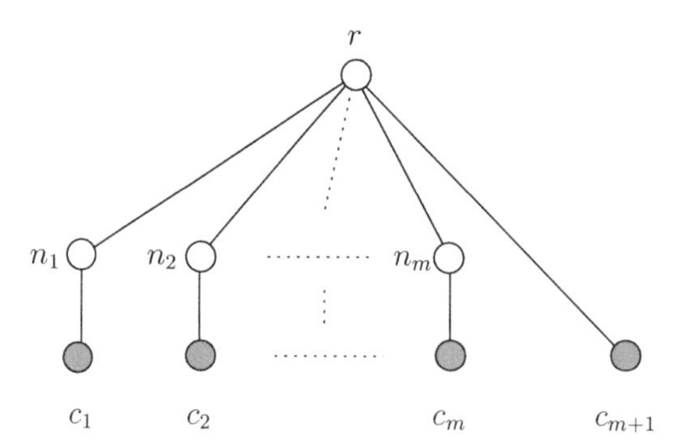

FIGURE 12.6: The platform used in the reduction for Theorem 12.4.

servers to enforce a partition. The reduction comes from 2-PARTITION (see Definition 10.1, p. 245), the root of the tree can process half of the a_is, and the other servers can process only one of the a_is. Because of the cost function, the cost of the solution will be minimized if the root server is processing exactly half of the a_is. Moreover, the reduction works for all policies. The reader is invited to search for this proof before reading the solution below.

Proof. Obviously, the NP-completeness of the *Upwards* policy is a consequence of Theorem 12.2. For the other two policies, the problem clearly belongs to the class NP; given a solution, it is easy to verify in polynomial time that all requests are served and that no server capacity is exceeded. To establish the completeness, we use a reduction from 2-PARTITION (see Definition 10.1, p. 245). We consider an instance $\mathcal{I}_1$ of 2-PARTITION: Given m positive integers $a_1, a_2, \ldots, a_m$, does there exist a subset $I \subset \{1, \ldots, m\}$ such that $\sum_{i \in I} a_i = \sum_{i \notin I} a_i$? Let $S = \sum_{i=1}^{m} a_i$. We build the following instance $\mathcal{I}_2$ of MINCOST (see Figure 12.6):

- $m+1$ clients c_i with $r_i = a_i$ for $1 \leqslant i \leqslant m$ and $r_{m+1} = 1$;
- m internal nodes n_j, $1 \leqslant j \leqslant m$, with $W_j = sc_j = a_j$; the only child of n_j is c_j;
- A root node r with $W_r = sc_r = S/2+1$, and r has $m+1$ children nodes: n_j for $1 \leqslant j \leqslant m$ and c_{m+1}.

Finally, we ask whether there exists a solution with total storage cost $S+1$. Clearly, the size of $\mathcal{I}_2$ is polynomial (and even linear) in the size of $\mathcal{I}_1$. We now show that instance $\mathcal{I}_1$ has a solution if and only if instance $\mathcal{I}_2$ does. The same reduction works for both policies, *Closest* and *Multiple*.

Suppose first that $\mathcal{I}_1$ has a solution. We assign a replica to each node n_i, $i \in I$, and one to the root r. Client c_i is served by n_i if $i \in \mathcal{I}$, and by the root r otherwise, i.e., if $i \notin \mathcal{I}$ or if $i = m+1$. The total storage cost is $\sum_{j \in I} W_j + W_r = S+1$. Because $W_r = S/2+1 = \sum_{i \notin I} r_i + r_{m+1}$, the capacity of the root is not exceeded. Note that the server allocation is compatible with both the *Closest* and *Multiple* policies. In both cases, we have a solution to $\mathcal{I}_2$.

Suppose now that $\mathcal{I}_2$ has a solution. Necessarily, there is a replica located at the root; otherwise, client c_{m+1} would not be served. Let I be the index set of nodes n_j, $1 \leqslant j \leqslant n$ that have been allocated a replica in the solution of $\mathcal{I}_2$. For $j \notin I$, there is no replica at node n_j, hence, all requests of client c_j are processed by the root, whose storage capacity is $S/2 + 1$. We derive that $\sum_{j \notin I} r_j \leqslant S/2$. Because the total storage capacity is $S + 1$, the total storage capacity of nodes in I is $S/2$. The proof is slightly different for the two server policies.

For the *Closest* policy, all requests from a client $c_j \in I$ are served by n_j, hence, $\sum_{j \in I} r_j \leqslant S/2$. Because $\sum_{j \in I} r_j + \sum_{j \notin I} r_j = S$, we derive that $\sum_{j \in I} r_j = \sum_{j \notin I} r_j = S/2$, hence, a solution to $\mathcal{I}_2$.

For the *Multiple* policy, consider a server $j \in I$. Let r'_j be the number of requests from client c_j served by n_j, and r''_j be the number of requests from c_j served by the root r (of course, $r_j = r'_j + r''_j$). All requests from a client c_j, $j \notin I$, are served by the root. Let $A = \sum_{j \in I} r'_j$, $B = \sum_{j \in I} r''_j$, and $C = \sum_{j \notin I} r_j$. The total storage cost is $A + B + S/2 + 1$, hence, $A + B \leqslant S/2$. We have seen that $C \leqslant S/2$. However, $A+B+C = S$. Therefore, $A+B = S/2$ and $C = S/2$. Thus, $B = 0$ because all the capacity of the server at the root is used to process the requests from c_{m+1} and from the c_js with $j \notin I$. Hence, $A = C = S/2$ and we have a solution to $\mathcal{I}_2$. □

12.2.4 Integer linear program

We also provide an expression of the optimization problem in terms of an integer linear program. We deal with the most general instance of the problem on a heterogeneous tree, including bounds on server capacities. We derive a formulation for each of the three server access policies, namely, *Closest*, *Upwards*, and *Multiple*.

We introduce some extra notations to express the tree hierarchy. If $i \in \mathcal{C}$ is a client, then *ancestors*$(i) \subseteq \mathcal{N}$ is the set of internal nodes that are *ancestors* of i in the tree, i.e., on the path between i and the root of the tree. Similarly, we define the ancestors of a node $j \in \mathcal{N}$ as *ancestors*(j). Finally, we denote by *subtree*(j) the tree rooted in node j.

12.2.4.1 Single server

We start with the single server policies, namely, the *Upwards* and *Closest* access policies. First, we define a few variables:

- For all $j \in \mathcal{N}$, x_j is a Boolean variable equal to 1 if j is a server (for one or several clients);

- For all $i \in \mathcal{C}$ and $j \in \mathcal{N}$, $y_{i,j}$ is a Boolean variable equal to 1 if $j = server(i)$; if $j \notin ancestors(i)$, we directly set $y_{i,j} = 0$.

The objective function is the total storage cost, namely, $\sum_{j\in\mathcal{N}} sc_j x_j$. We list below the constraints, common to the *Closest* and *Upwards* policies, that express server usage:

- Every client is assigned a server: $\forall i \in \mathcal{C}$, $\sum_{j\in ancestors(i)} y_{i,j} = 1$.
- The processing capacity of any server cannot be exceeded: $\forall j \in \mathcal{N}, \sum_{i\in\mathcal{C}} r_i y_{i,j} \leqslant W_j x_j$. Note that this constraint ensures that if j is the server of i, there is indeed a replica located at node j.

Altogether, we have fully characterized the linear program for the *Upwards* policy. We need additional constraints for the *Closest* policy, which is a particular case of the *Upwards* policy (thus, all constraints and equations remain valid). We need to express that if node j is the server of client i, then no ancestor of j can be the server of a client in the subtree rooted at j. Indeed, a client in this subtree would need to be served by j and not by one of its ancestors, according to the *Closest* policy. A direct way to write this constraint is $\forall i \in \mathcal{C}, \forall j \in ancestors(i), \forall i' \in \mathcal{C} \cap subtree(j), \forall j' \in ancestors(j), y_{i,j} \leqslant 1 - y_{i',j'}$. Indeed, if $y_{i,j} = 1$, meaning that $j = server(i)$, then any client i' in the subtree rooted in j must have its server in that subtree, not closer to the root than j. Hence, $y_{i',j'} = 0$ for any ancestor j' of j. There are $O(s^4)$ such constraints to write where $s = |\mathcal{C}| + |\mathcal{N}|$ is the problem size.

12.2.4.2 Multiple servers

We now proceed to the *Multiple* policy. We slightly change the definition of the variables: $y_{i,j}$ is now an integer variable equal to the number of requests from client i processed by node j.

The objective function is unchanged, as the total storage cost still writes $\sum_{j\in\mathcal{N}} sc_j x_j$, but the constraints must be slightly modified:

- Every request is assigned a server: $\forall i \in \mathcal{C}, \sum_{j\in ancestors(i)} y_{i,j} = r_i$.
- Server capacities: $\forall j \in \mathcal{N}, \sum_{i\in\mathcal{C}} y_{i,j} \leqslant W_j x_j$. Note that this ensures that if j is the server for one or more requests from i, there is indeed a replica located in node j.

Altogether, we have fully characterized the linear program for the *Multiple* policy.

12.2.4.3 An LP-based lower bound

The previous linear program contains Boolean or integer variables, because it does not make sense to assign half a request or to place one third of a replica on a node. Thus, it must be solved in integer values if we wish to obtain an exact solution to an instance of the problem, and there is no efficient algorithm to solve integer linear programs (unless P = NP). For each access policy, there is a large number of variables, and the problem cannot be solved for platforms of size $s > 50$, where $s = |\mathcal{N}| + |\mathcal{C}|$. Thus, we cannot use this approach for large-scale problems.

However, this formulation is extremely useful as it leads to an absolute lower bound; we can solve the integer linear program over the rationals. In this case, all constraints are relaxed, and we assume that all variables can take rational values. The optimal solution of the relaxed program can be obtained in polynomial time (in theory using the ellipsoid method [93], in practice using standard software packages [76, 39]), and the value of its objective function provides an absolute lower bound on the cost of any valid (integer) solution. For all practical values of the problem size, the rational linear program returns a solution in a few minutes. We tested up to several thousands of nodes and clients, and we always found a solution within 10 seconds. Of course, the relaxation makes the most sense for the *Multiple* policy, because several fractions of servers are assigned by the rational program.

However, we can obtain a more precise lower bound for trees with up to $s = 400$ nodes and clients by using a rational solution of the *Multiple* instance of the linear program with fewer integer variables. We treat the $y_{i,j}$ as rational variables and require only the x_j to be integer variables. These variables are set to 1 if and only if there is a replica on the corresponding node. Thus, forbidding to set $0 < x_j < 1$ allows us to get a realistic value of the cost of a solution of the problem. For instance, a server might be used at only 50% of its capacity; thus, setting $x = 0.5$ would be enough to ensure that all requests are processed. However, in this case, the cost of placing the replica at this node is halved, which is incorrect. While we can place a replica or not, it is impossible to place half of a replica.

In practice, this lower bound provides a drastic improvement over the unreachable lower bound provided by the fully rational linear program. The good news is that we can compute the refined lower bound for problem sizes up to $s = 400$, using GLPK (GNU Linear Programming Kit) [39]. In the next section, we show that this refined bound is an achievable bound, and we provide an exact solution to the *Multiple* instance of the problem, based on the solution of this mixed integer linear program.

12.2.4.4 An exact MIP-based solution for *Multiple*

THEOREM 12.5. *The solution of the linear program detailed in 12.2.4.2, when solved with all variables being rational except the x_is, is an achievable bound for the* Multiple *problem, and we can build an exact solution in polynomial time, based on the LP solution.*

Proof. Consider the solution of the LP program:

- $\forall i \in \mathcal{C},\ x_i \in \{0, 1\}$;
- $\forall i \in \mathcal{C}, \forall j \in \mathcal{N},\ y_{i,j} \in \mathbb{Q}$.

To prove that the lower bound obtained by this program is achievable, we are building an integer solution where the $y'_{i,j}$s are integer numbers, keeping the same x_is and without breaking any constraints.

In the following, for any variable y, $\lfloor y \rfloor$ is the integer part of y, and $\tilde{y}$ is the fractional part: $y = \lfloor y \rfloor + \tilde{y}$, and $\tilde{y} < 1$.

Let us consider a client $i \in \mathcal{C}$ such that $\exists j \in \mathcal{N} \mid \tilde{y}_{i,j} > 0$, i.e., $y_{i,j}$ is not an integer. We consider j_1 being the closest server to i not serving an integer number of requests of client i, and more generally, for $1 \leqslant k \leqslant K$, j_k is the k-th server on the path from i to the root, such that $\tilde{y}_{i,j_k} > 0$. Note that K is the number of servers that are not serving an integer number of requests of client i. We want to move bits of requests in order to obtain an integer value for y_{i,j_1}. This elementary transformation is called *trans*(i, j_1). We consider the two following cases.

First case:

$$\sum_{i' \in subtree(j_1) \cap \mathcal{C}} y_{i',j_1} \leqslant W_{j_1} - (1 - \tilde{y}_{i,j_1}) \, .$$

In this case, there is enough space at server j_1 to fulfill an integer number of requests from client i. Because the total number of requests of client i is an integer, $\sum_{k=1}^{K} \tilde{y}_{i,j_k}$ is a nonnull integer. Thus, $\sum_{k=2}^{K} \tilde{y}_{i,j_k} \geqslant 1 - \tilde{y}_{i,j_1}$, and we can move down $1 - \tilde{y}_{i,j_1}$ bits of requests from servers j_k $(2 \leqslant k \leqslant K)$ to j_1. No constraints will be violated because there is enough space on the server. The move is done by changing the values of y_{i,j_k}, for $1 \leqslant k \leqslant K$. After such a transformation, y_{i,j_1} is an integer variable.

Second case: If server j_1 is already too full in order to add a fraction of requests from client i, we need to exchange some requests with other clients. First, if there is some free space on the server, we start by filling completely server j_1 with fractions of requests of client i from servers j_k $(1 \leqslant k \leqslant K)$. We know there are such requests; otherwise, y_{i,j_1} would be an integer. This transformation is similar to the one done in the first case. We now have $\sum_{i' \in subtree(j_1) \cap \mathcal{C}} y_{i',j_1} = W_{j_1}$. Let us denote by i_t $(1 \leqslant t \leqslant T)$ the clients $i_t \in subtree(j_1) \cap \mathcal{C} \setminus \{i\}$ such that $\tilde{y}_{i_t,j_1} > 0$. Because W_{j_1} is an integer and $\tilde{y}_{i,j_1} > 0$, we have $\sum_{t=1}^{T} \tilde{y}_{i_t,j_1} \geqslant 1 - \tilde{y}_{i,j_1}$, and also $\sum_{k=2}^{K} \tilde{y}_{i,j_k} \geqslant 1 - \tilde{y}_{i,j_1}$. We can select in both sets $1 - \tilde{y}_{i,j_1}$ bits of requests that will be exchanged, i.e., bits of requests from client i_t initially treated by j_1 will be moved on some servers j_k that are in *ancestors*(j_1), and the corresponding amount of requests of i will be moved back on server j_1. Note that all constraints are still fulfilled.

Once *trans*(i, j_1) has been applied, y_{i,j_1} is an integer. Because only noninteger bits of requests have been moved, no variables with integral values were modified, and we have decreased at least by one the number of noninteger variables in the solution.

```
1 for j ∈ 𝒩 taken in a bottom-up traversal order do
2     finish=1
3     while (finish==1) do
4         C' = {i' ∈ C ∩ subtree(j) | ỹ_{i',j} > 0}
5         if C' == ∅ then
6             finish=0
7         else
8             Choose arbitrarily i ∈ C'
9             trans(i, j)
```

ALGORITHM 12.1: Building an integer solution for *Multiple* from a rational one.

Let us detail now the complete transformation algorithm, Algorithm 12.1, in order to obtain an integer solution. We consider each server in a bottom-up order, so that we are sure that each time we perform an elementary transformation, the server is the first one on the way from the client to the root having a noninteger number of requests. In fact, when transforming server j, each server in *subtree*(j) has already been transformed and thus has no fraction numbers of requests.

In order to transform server j, we look at the set $\mathcal{C}'$ of clients having a noninteger number of requests processed at j. If the set is empty, there is nothing to transform at j. Otherwise, we perform the elementary transformation with one client $i \in \mathcal{C}'$.

At the end of the "while" loop of Algorithm 12.1, server j is processing only integer numbers of requests, and thus we will not modify its assignment of requests anymore in the following.

The constraints are all respected at each step of the transformation, and we do not add or remove any replica, so the solution has exactly the same cost as the initial LP-based solution, and the transformed solution is fully integer. Moreover, this transformation algorithm works in polynomial time, in the worst case in $|\mathcal{N}| + |\mathcal{C}|^2$, but most of the time it is much faster because the transformations do not concern all clients simultaneously, only a few of them. □

12.3 Variants of the replica placement problem

In this section, we discuss two variants of the initial replica placement problem with homogeneous servers (MinNb) in order to exhibit problem instances for which the problem with the *Multiple* or *Closest* policy becomes NP-hard.

First, in Section 12.3.1, we introduce the notion of quality of service, in terms of distance between a client and its server. For the *Multiple* policy, the problem, which could be solved in polynomial time without QoS constraints, becomes NP-hard.

A second variant targets a power-aware replica placement problem (MinPower, see Section 12.3.2), where the objective is to minimize not the cost of the solution anymore but rather the power consumption. In this case, the instance of the problem with the *Closest* policy becomes NP-hard.

12.3.1 Enforcing a quality of service

We consider that a quality of service must be enforced for each client, in terms of distance between a client and its server. A distance is associated with each edge of the tree, and for each client, a maximum distance is fixed, which should not be exceeded. While it is possible to extend the *Closest* algorithm for MinNb with QoS (see [110]), the problem becomes NP-hard for the *Multiple* policy.

The reduction holds even when all distances in the tree are identical, and the QoS of a client $i \in \mathcal{C}$ is then defined as q_i, the maximum number of hops between i and $server(i)$.

THEOREM 12.6. *The instance of the* MinNb *problem with the* Multiple *policy and QoS constraints is NP-complete.*

The proof is more involved than the ones seen so far, but because we can play on the number of requests, it is natural to perform the reduction from a partition problem. In this case, we use the 2-PARTITION-EQUAL problem (see Definition 10.2, p. 245), which is a variant of 2-PARTITION where the a_is must be partitioned in two sets of identical cardinality, m. Clients with QoS equal to 1 allow us to enforce the placement of replicas anywhere in the tree. Then, we add nodes with two clients, and we must decide whether to place a replica on these nodes or not. A total of m replicas will be required. Therefore, this will lead to a 2-PARTITION-EQUAL.

Proof. The problem clearly belongs to the class NP. Given a solution, it is easy to verify in polynomial time that all requests are served, that all QoS constraints are satisfied, and that no server capacity is exceeded.

To establish the completeness, we use a reduction from 2-PARTITION-EQUAL (see Definition 10.2, p. 245). We consider an instance $\mathcal{I}_1$ of 2-PARTITION-EQUAL. Given $2m$ positive integers $a_1, a_2, \ldots, a_{2m}$, does there exist a subset $I \subset \{1, \ldots, 2m\}$ of cardinal m such that $\sum_{i \in I} a_i = \sum_{i \notin I} a_i$? Let $S = \sum_{i=1}^{2m} a_i$, $W = \frac{S}{2} + 1$, and $b_i = \frac{S}{2} - 2a_i$ for $1 \leqslant i \leqslant 2m$. We build the following instance $\mathcal{I}_2$ of our problem (Figure 12.7), with $5m-1$ clients c_i and $3m-1$ internal nodes n_j of identical capacity W:

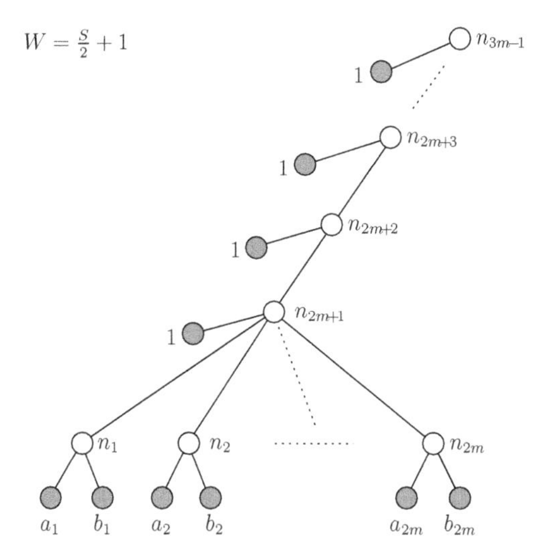

FIGURE 12.7: The platform used in the reduction for Theorem 12.6.

- Nodes:
  - For $1 \leqslant j \leqslant 2m$, the parent of node n_j is node n_{2m+1};
  - For $2m + 1 \leqslant j \leqslant 3m - 2$, the parent of node n_j is node n_{j+1};
  - Node n_{3m-1} is the root of the tree.

- Clients:
  - For $1 \leqslant i \leqslant 2m$, client c_i has $r_i = a_i$ requests of QoS $q_i = 2$, and its parent is node n_i;
  - For $2m + 1 \leqslant i \leqslant 4m$, client c_i has $r_i = b_{i-2m}$ requests of QoS $q_i = m$, and its parent is node n_{i-2m};
  - For $4m+1 \leqslant i \leqslant 5m-1$, client c_i has $r_i = 1$ request of QoS $q_i = 1$, and its parent is node n_{i-2m}.

Finally, we ask whether there exists a solution with $2m-1$ servers. Clearly, the size of $\mathcal{I}_2$ is polynomial (and even linear) in the size of $\mathcal{I}_1$. We now show that instance $\mathcal{I}_1$ has a solution if and only if instance $\mathcal{I}_2$ does.

Suppose first that $\mathcal{I}_1$ has a solution. We assign a replica to each node n_i, $i \in I$, (by hypothesis there are m of them) and one to each of the $m - 1$ top nodes n_{2m+1} to n_{3m-1}. All $m - 1$ clients with QoS 1 are served by their parent.

For $1 \leqslant i \leqslant 2m$, there are two cases: (i) If $i \in \mathcal{I}$, both clients c_i and c_{i+2m} are served by their parent n_i. Node n_i serves a total of $a_i + b_i = \frac{S}{2} - a_i \leqslant W$ requests. (ii) If $i \notin \mathcal{I}$, client c_i is served by node n_{2m+1} and client c_{i+2m} is served by one or several ancestors of n_{2m+1}, i.e., nodes n_{2m+2} to n_{3m-1}. Node n_{2m+1}, which also serves the unique request of client c_{2m+1}, serves a total of $\sum_{i \notin I} a_i + 1 = W$ requests. The $m - 2$ ancestors of n_{2m+1} receive the load $\sum_{i \notin I} b_i = mS - 2S$. They also serve $m-2$ clients with a single request, hence, a total load of $(m-2)S+m-2=(m-2)W$ requests to distribute among them. This is precisely the sum of their capacities, and any assignment will do the job.

Note that the allocation of requests to servers is compatible with all QoS constraints. All requests with QoS 1 are served by the parent node. All requests with QoS 2, i.e., with value a_i, are served either by the parent node (if $i \in I$) or by the grandparent node (if $i \notin I$). Altogether, we have a solution to $\mathcal{I}_2$.

Suppose now that $\mathcal{I}_2$ has a solution with $2m-1$ servers. Necessarily, there is a replica located in each of the top $m-1$ nodes n_{2m+1} to n_{3m-1}; otherwise, some request with QoS 1 would not be served satisfactorily. Each of these nodes serves one of these requests, hence, has remaining capacity $W-1=\frac{S}{2}$.

There remain m servers that are placed among nodes n_1 to n_{2m}. Let I be the set of indices of those m nodes that have not received a replica. Necessarily, requests a_i, with $i \in I$, are served by node n_{2m+1}, because of their QoS constraint. Hence, $\sum_{i\in I} a_i \leqslant \frac{S}{2}$. Next, all requests a_i and b_i, with $i \in I$, are served by nodes n_{2m+1} to n_{3m-1}, whose total remaining capacity is $(m-1)\frac{S}{2}$. There are $(\sum_{i\in I} a_i) + (m\frac{S}{2} - 2\sum_{i\in I} a_i)$ such requests, thus:

$$m\frac{S}{2} - \sum_{i\in I} a_i \leqslant (m-1)\frac{S}{2}.$$

From this equation, we derive that $\sum_{i\in I} a_i \geqslant \frac{S}{2}$. Finally, we have $\sum_{i\in I} a_i = \frac{S}{2}$, with $|I| = m$, hence, a solution to $\mathcal{I}_2$. □

12.3.2 Power-aware replica placement

In this section, we introduce a variant of the MinNb problem, MinPower, which aims at minimizing the power consumption of a replica placement. We prove the NP-completeness of this problem for the *Closest* policy, even with no static power, when processors can have an arbitrary number of modes.

We assume that servers may operate under a set $\mathcal{M} = \{W_1, \ldots, W_M\}$ of different speeds, or *modes*, depending upon the number of requests that they have to process per time unit. Modes are indexed according to increasing values, and $W_M = W$, the maximal capacity. If a server $j \in \mathcal{R}$ processes r_j requests, with $W_{i-1} < r_j \leqslant W_i$, then it is operated at mode W_i, and we let $mode(j) = i$. The power consumption of a server $j \in \mathcal{R}$ obeys the classical model:

$$\mathcal{P}(j) = \mathcal{P}^{(static)} + \left(W_{mode(j)}\right)^{\alpha},$$

where $\mathcal{P}^{(static)}$ is the static power consumption (constant part), while $W^{\alpha}_{mode(j)}$ is the dynamic part that depends upon the operated mode. Finally, $\alpha \in [2,3]$ is a rational constant that depends upon the model for power [6, 22, 23, 54, 90]. The total power consumption $\mathcal{P}(\mathcal{R})$ of the solution is the sum of the power consumption of all server nodes:

$$\mathcal{P}(\mathcal{R}) = \sum_{j \in \mathcal{R}} \mathcal{P}(j) = |\mathcal{R}| \times \mathcal{P}^{(static)} + \sum_{j \in \mathcal{R}} \left(W_{mode(j)}\right)^{\alpha}. \qquad (12.1)$$

Intuitively, this equation calls for balancing two conflicting terms: static power is minimized with few servers, while dynamic power is minimized with many servers operated in the slowest mode.

THEOREM 12.7. *The instance of the* MinPower *problem with the* Closest *policy is NP-complete.*

This proof, even though the reduction comes from the now well-known 2-PARTITION problem, is quite involved because of the α exponent in equation (12.1). The idea is to use all the capacity given by a mode. We enforce that the root of the tree must run at the highest mode, and a 2-PARTITION must exist so that the capacity is fully utilized.

Proof. We consider the associated decision problem. Given a total power consumption $\mathcal{P}$, is there a solution that does not consume more than $\mathcal{P}$?

First, the problem is clearly in NP. Given a solution, i.e., a set of servers, and the mode of each server, it is easy to check in polynomial time that no capacity constraint is exceeded and that the power consumption meets the bound.

To establish the completeness, we use a reduction from 2-PARTITION (see Definition 10.1, p. 245). We consider an instance $\mathcal{I}_1$ of 2-PARTITION. Given n strictly positive integers $a_1, a_2, \ldots, a_n$, does there exist a subset I of $\{1, \ldots, n\}$ such that $\sum_{i \in I} a_i = \sum_{i \notin I} a_i$? Let $S = \sum_{i=1}^{n} a_i$; we assume that S is even (otherwise, there is no solution).

We build an instance $\mathcal{I}_2$ of our problem where each server has $n+2$ modes. We assume that the a_i are sorted in increasing order, i.e., $a_1 \leqslant \cdots \leqslant a_n$. The modes are then, when sorted in increasing order:

- $W_1 = K$;
- $\forall 1 \leqslant i \leqslant n$, $W_{i+1} = K + a_i \times X$;
- $W_{n+2} = K + S \times X$;

where the values of K and X will be determined later.

We furthermore decide that there is no static power, and the power consumption for a server running at capacity W_i is, therefore, $\mathcal{P}_i = W_i^{\alpha}$, where α is the rational exponent used in the computation of the power, and $2 \leqslant \alpha \leqslant 3$. The idea is to have K large and X small, so that we have an upper bound on the power consumed by a server running at capacity W_{i+1}, for $1 \leqslant i \leqslant n$:

$$W_{i+1}^{\alpha} = (K + a_i \times X)^{\alpha} \leqslant K^{\alpha} + a_i + \frac{1}{n}. \qquad (12.2)$$

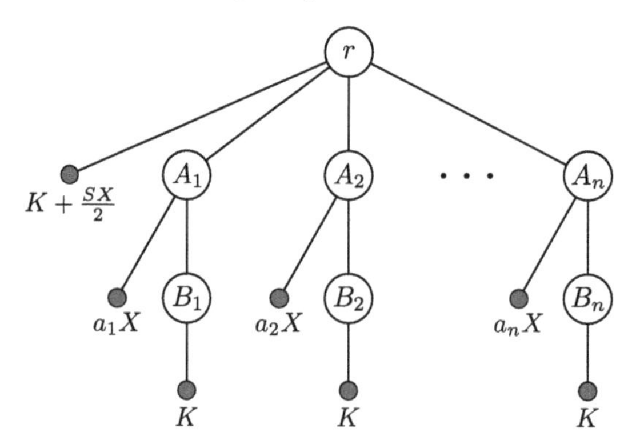

FIGURE 12.8: Illustration of the NP-completeness proof.

To ensure that equation (12.2) is satisfied, we set

$$X = \frac{1}{\alpha \times K^{\alpha-1}},$$

and then we have $(K + a_i \times X)^\alpha = K^\alpha(1 + \frac{a_i}{\alpha K^\alpha})^\alpha$, with $K > S$ and, therefore, $\frac{a_i}{\alpha K^\alpha} < 1$. We set $x_i = \frac{a_i}{\alpha K^\alpha}$, and we want to ensure that

$$(1 + x_i)^\alpha \leqslant 1 + \alpha \times x_i + \frac{1}{n \times K^\alpha}. \qquad (12.3)$$

To do so, we study the function

$$f(x) = (1 + x)^\alpha - (1 + \alpha \times x) - 5x^2,$$

and we show that $f(x) \leqslant 0$ for $x \leqslant \frac{1}{2}$ (thanks to the term in $-5x^2$). We have $f(0) = 0$, and $f'(x) = \alpha(1 + x)^{\alpha-1} - \alpha - 10x$. We have $f'(0) = 0$, and $f''(x) = \alpha(\alpha-1)(1+x)^{\alpha-2} - 10$. Because $\alpha \leqslant 3$, $\alpha(\alpha-1)(1+x)^{\alpha-2} \leqslant 6(1+x)$, and for $x \leqslant \frac{1}{2}$, $f''(x) < 0$. We deduce that $f'(x)$ is nonincreasing for $x \leqslant \frac{1}{2}$, and because $f'(0) = 0$, $f'(x)$ is nonpositive for $x \leqslant \frac{1}{2}$.

Finally, $f(x)$ is nonincreasing for $x \leqslant \frac{1}{2}$, and because $f(0) = 0$, we have $(1 + x)^\alpha < (1 + \alpha \times x) + 5x^2$ for $x \leqslant \frac{1}{2}$.

Equation (12.3), therefore, is satisfied if $5x_i^2 \leqslant \frac{1}{n \times K^\alpha}$, i.e., $K^\alpha \geqslant \frac{5a_i^2 \times n}{\alpha^2}$. This condition is satisfied for

$$K = n \times S^2,$$

and we then have $x_i < \frac{1}{2}$, which ensures that the previous reasoning was correct. Finally, with these values of K and X, equation (12.2) is satisfied.

Then, the distribution tree is the following: The root node r has one client with $K + \frac{S}{2} \times X$ requests and n children $A_1, \ldots, A_n$. Each node A_i has a client with $a_i \times X$ requests and a children node B_i that has K requests. Figure 12.8 illustrates the instance of the reduction.

Finally, we ask if we can find a placement of replicas with a maximum power consumption of

$$\mathcal{P}_{max} = (K + S \times X)^\alpha + n \times K^\alpha + \frac{S}{2} + \frac{n-1}{n}.$$

Clearly, the size of $\mathcal{I}_2$ is polynomial in the size of $\mathcal{I}_1$ because K and X are of polynomial size. We now show that $\mathcal{I}_1$ has a solution if and only if $\mathcal{I}_2$ does.

Let us assume first that $\mathcal{I}_1$ has a solution, I. The solution for $\mathcal{I}_2$ is then as follows: There is one server at the root, running at capacity W_{n+2}. Then, for $i \in I$, we place a server at node A_i running at capacity W_{1+i}, while for $i \notin I$, we place a server at node B_i running at capacity W_1. It is easy to check that all capacity constraints are satisfied for nodes A_i and B_i. At the root of the tree, there are $K+\frac{S}{2}\times X+\sum_{i\notin I} a_i \times X = K+S\times X$ requests. The total power consumption is then $\mathcal{P} = (K + S \times X)^\alpha + \sum_{i\in I}(K + a_i \times X)^\alpha + \sum_{i\notin I} K^\alpha$. Thanks to equation (12.2), $\mathcal{P} \leqslant (K + S \times X)^\alpha + \sum_{i\in I}\left(K^\alpha + a_i + \frac{1}{n}\right) + \sum_{i\notin I} K^\alpha$, and finally, $\mathcal{P} \leqslant (K+S\times X)^\alpha + n\times K^\alpha + \sum_{i\in I} a_i + \frac{n-1}{n}$. Because I is a solution to 2-PARTITION, we have $\mathcal{P} \leqslant \mathcal{P}_{max}$. Finally, $\mathcal{I}_2$ has a solution.

Suppose now that $\mathcal{I}_2$ has a solution. There is a server at the root node r, which runs at mode W_{n+2}, because this is the only way to handle its $K + \frac{S}{2} \times X$ requests. This server has a power consumption of $(K + S \times X)^\alpha$. Then, there cannot be more than n other servers. Indeed, if there were $n+1$ servers, running at the smallest mode W_1, their power consumption would be $(n+1)K^\alpha$, which is strictly greater than $n\times K^\alpha + \frac{S}{2} + 1$. Therefore, the power consumption would exceed $\mathcal{P}_{max}$. So, there are at most n extra servers.

Consider that there exists $i \in \{1, \ldots, n\}$ such that there is no server, neither on A_i nor on B_i. Then, the number of requests at node r is at least $2K$; however, $2K > W_{n+2}$, so the server cannot handle all of these requests. Therefore, for each $i \in \{1, \ldots, n\}$, there is exactly one server either on A_i or on B_i. We define the set I as the indices for which there is a server at node A_i in the solution. Now, we show that I is a solution to $\mathcal{I}_1$, the original instance of 2-PARTITION.

First, if we sum up the requests at the root node, we have

$$K + \frac{S}{2} \times X + \sum_{i\notin I} a_i \times X \leqslant K + S \times X.$$

Therefore, $\sum_{i\notin I} a_i \leqslant \frac{S}{2}$.

Now, if we consider the power consumption of the solution, we have

$$(K + S \times X)^\alpha + \sum_{i\in I}(K + a_i \times X)^\alpha + \sum_{i\notin I} K^\alpha \leqslant \mathcal{P}_{max}.$$

Let us assume that $\sum_{i\in I} a_i > \frac{S}{2}$. Because the a_is are integers, we have $\sum_{i\in I} a_i \geqslant \frac{S}{2} + 1$. It is easy to see that $(K + a_i \times X)^\alpha > K^\alpha + a_i$. Finally, $\sum_{i\in I}(K + a_i \times X)^\alpha + \sum_{i\notin I} K^\alpha \geqslant n \times K^\alpha + \sum_{i\in I} a_i \geqslant n \times K^\alpha + \frac{S}{2} + 1$. This implies that the total power consumption is greater than $\mathcal{P}_{max}$, which leads to a contradiction, and, therefore, $\sum_{i\in I} a_i \leqslant \frac{S}{2}$.

We conclude that $\sum_{i\notin I} a_i = \sum_{i\in I} a_i = \frac{S}{2}$, and so the solution I is a 2-PARTITION for instance $\mathcal{I}_1$. This concludes the proof. □

12.4 Conclusion

In this case study, we have introduced and analyzed three meaningful policies for the replica placement problem; the *Upwards* and *Multiple* policies are natural variants of the standard *Closest* approach.

On the theoretical side, we have fully assessed the complexity of the problem with each of these policies, both for homogeneous and heterogeneous platforms. The polynomial complexity of the *Multiple* policy in the homogeneous case is quite unexpected, and we have provided an elegant algorithm to compute the optimal cost for this policy. Not surprisingly, all three policies turn out to be NP-complete for heterogeneous nodes, which provides yet another example of the additional difficulties induced by resource heterogeneity.

When adding QoS constraints, the problem on homogeneous platforms with the *Multiple* policy becomes NP-complete, which illustrates the additional complexity induced by such constraints. We also have demonstrated the difficulty induced by power-aware replica placement, because the problem of minimizing the power consumption becomes NP-hard with identical servers for the *Closest* policy.

On the practical side, several polynomial-time heuristics for the *Closest*, *Upwards*, and *Multiple* policies can be found in [11], and their performance is compared using several problem instances, with or without QoS constraints. The impact of the policy is impressive. The number of trees that admit a solution is much higher with the *Upwards* and *Multiple* policies than with the *Closest* policy. In order to handle the power consumption problem, [12] proposes a pseudopolynomial algorithm capable of optimizing power consumption for a bounded cost. The algorithm is exponential in the total number of processor modes, and, therefore, it turns out to be of polynomial time with a fixed number of modes.

Chapter 13

Packet routing

In this case study, we deal with two famous routing problems, which have received considerable attention in the literature. The focus here is on approximation techniques rather than on NP-completeness proofs, which for once are not given in the text (but references are provided).

Informally, routing is the activity to move packets from source to destination. The underlying interconnection network is modeled as a directed graph $G = (V, E)$. There is a set $\mathcal{R}$ of routing requests to satisfy. Each request $R_i \in \mathcal{R}$ has a source s_i and a destination t_i, which are both vertices of V. In addition, each request has a number n_i of elementary data items, or *packets*, to be routed from s_i to t_i. For each packet of R_i, we must find a simple path in G, along whose edges the packet will "ripple down" from s_i to t_i.

We study the following two variants of the problem:

Maximum edge-disjoint path (MEDP) problem. Here the rule of the game is that all packets of a given request follow the same path. In addition, no two request paths are allowed to share any edge in the graph. In other words, each request is assigned a private path that is reserved for its own usage. It may well be the case that all requests cannot be accommodated simultaneously, so the natural objective is to maximize the number of accepted (routed) requests. In terms of graph theory, given a set of source/destination pairs, we aim at finding the maximum number of edge-disjoint paths between them.

Packet routing with variable-paths (PRVP). In this second variant, each packet can be routed along a distinct path. Two packet paths can share edges, but packets are routed sequentially. At any time step at most one packet can cross an edge of the graph (but several edges can be crossed by different packets simultaneously). Here the natural objective is to minimize the total execution time, i.e., the time needed to move each packet from its source to its destination. Technically, requests are less important than individual packets here, because two packets of the same request can follow two arbitrarily different paths, which may either intersect or be edge-disjoint. Requests can be viewed as different marks, or types, for the packets.

Both MEDP and PRVP, along with several extensions, have been extensively studied because they lie at the heart of many network problems. Comprehensive surveys are provided in [33] for MEDP and in [85] for PRVP. In this case study, we focus on proving the following results:

- A tight approximation factor for MEDP;
- An asymptotically optimal algorithm for PRVP.

13.1 MEDP: Maximum edge-disjoint paths

In this section, we study the MEDP problem, and we prove that some greedy algorithm provides the best possible approximation factor. We start with a precise formulation in terms of graph theory.

13.1.1 Problem statement

Given a directed graph $G = (V, E)$ and a set of terminal pairs $\mathcal{R} = \{R_i = (s_i, t_i)\}$, the goal of MEDP is to connect as many pairs as possible using edge-disjoint simple paths. In line with the previous discussion, each terminal pair in $\mathcal{R}$ is also called a request. A feasible solution is a subset $\mathcal{A}$ of accepted requests. In the solution, each $R_i \in \mathcal{A}$ must be assigned a simple path π_i from s_i to t_i in G, so that no two paths π_i and π_j, where $R_i \in \mathcal{A}$, $R_j \in \mathcal{A}$, and $i \neq j$ have an edge of the graph in common. Requests in $\mathcal{A}$ are said to be *accepted* while those in $\mathcal{R} \setminus \mathcal{A}$ are *rejected.*

The goal is to maximize $|\mathcal{A}|$, the cardinality of $\mathcal{A}$, i.e., the number of accepted requests. In the following, $\mathcal{A}^*$ will denote an optimal solution. We also let $n = |V|$ and $m = |E|$. The MEDP problem is NP-complete with only two requests [35] (note that Definition 10.15, p. 247, refers to a very similar problem, with vertex-disjoint paths). In the following, we analyze two greedy algorithms, and we show that the second one is, in essence, the best that we can hope for. Beforehand, we briefly mention two special classes of graphs, for which the optimal solution can be found in polynomial time.

Chains. When G is a chain, the problem should look familiar. Each request can be viewed as an interval, and the goal is to identify the maximum number of nonoverlapping intervals. We recognize the sports hall problem of Section 3.1. Recall that the optimal solution is to sort the intervals (requests) by ending time (termination vertex) and to select them greedily.

Stars. When G is a star, the problem can still be solved in polynomial time but with a more advanced graph algorithm. In a star graph, source vertices are connected to destination vertices by a single path of two edges that go through the same intermediate vertex, the center of the star. To accommodate for the case where a request starts from the central vertex, we simply add a new source vertex to the graph (and we add a new destination vertex if a request ends in the central vertex). Next, we build a bipartite graph from sources to destinations, and we add an edge between a source and a destination if and only if there is a request between them. There remains to be found a maximum cardinality matching in the bipartite graph [106] to derive the optimal solution.

13.1.2 Naive greedy algorithm

The simplest, or naive, greedy algorithm tries to accept the requests in any order. The algorithm accepts the first request R_1 if there exists a path from s_1 to t_1 in G. However, there may exist several paths. Because edges are the scarce resource in the problem, the algorithm looks for a shortest-path π_1 from s_1 to t_1 (if there exist many shortest paths, it picks one arbitrarily). Each time a request is accepted, we have to prune the graph and remove the edges that appear in the shortest path selected for that request, so as to enforce that only edge-independent paths will be used throughout the execution. We derive Algorithm 13.1:

```
1 A ← ∅
2 for i = 1 to |R| do
3     if there exists a path from s_i to t_i in G then
4         accept: A ← A ∪ {R_i}
5         route: π_i ← a shortest path from s_i to t_i in G
6         prune: remove all edges of π_i from G
```

ALGORITHM 13.1: Naive greedy algorithm.

The complexity of this naive greedy algorithm can easily be upper-bounded by $O(|\mathcal{R}|.|V|^2)$. There are many algorithms to find a shortest path from a source to a destination (in fact, all destinations) in a directed graph with nonnegative weights; for instance, Dijkstra's algorithm without advanced data structures has complexity $O(|V|^2)$ (see [27]).

THEOREM 13.1. *Algorithm 13.1 has approximation ratio $n-1$ for* MEDP *in graphs with n vertices, and this bound is tight.*

Proof. Given an instance of MEDP, let $\mathcal{A}$ be the solution returned by Algorithm 13.1, and let $\mathcal{A}^*$ be an optimal solution. We start with the tightness of the bound. Consider a chain with n vertices:

$$v_1 \rightarrow v_2 \rightarrow \cdots \rightarrow v_n \ .$$

There are n requests $R_1 = (v_1, v_n)$, and $R_i = (v_{i-1}, v_i)$ for $2 \leqslant i \leqslant n$. If the algorithm starts with R_1, no other request can be accepted, so that $|\mathcal{A}| = 1$. Obviously, the optimal solution is to accept the last $n - 1$ requests, so that $|\mathcal{A}^*| = n - 1$.

Now, we show that for any instance of MEDP on a graph G with n vertices, we have

$$|\mathcal{A}^*| \leqslant (n-1) \times |\mathcal{A}| \, .$$

For $R_i \in \mathcal{A}$, let π_i be the path assigned by Algorithm 13.1. Similarly, for $R_j \in \mathcal{A}^*$, let π_j^* be the path assigned by the optimal solution. Consider the execution of Algorithm 13.1 at each step. Each time it accepts a request R_i, we prune the optimal solution $\mathcal{A}^*$ by deleting R_i itself if it belongs to $\mathcal{A}^*$, and also all those requests $R_j \in \mathcal{A}^*$ whose paths π_j^* are not edge-disjoint with π_i. For instance, assume that Algorithm 13.1 rejects R_1 and then accepts R_2 and R_3, we do the following actions on $\mathcal{A}^*$:

- We leave it unchanged at step 1;
- We remove all requests whose paths intersect π_2 at step 2;
- We remove all remaining requests whose paths intersect π_3 at step 3.

At each step of the execution of Algorithm 13.1, there remains in $\mathcal{A}^*$ only requests R_j whose paths π_j^* are edge-disjoint with all the paths π_i of the requests R_i that have been previously accepted by Algorithm 13.1. Also, the set $\mathcal{A}^*$ must be empty at the end of the execution of Algorithm 13.1. Otherwise, if there remained a request, Algorithm 13.1 could (and greedily would) accept it. Therefore, we have proved the following result:

LEMMA 13.1. *To show that $|\mathcal{A}^*| \leqslant \lambda \times |\mathcal{A}|$ for some constant λ, it is sufficient to show that whenever the greedy algorithm accepts a request, no more than λ requests are deleted from $\mathcal{A}^*$.*

We now use Lemma 13.1 with $\lambda = n - 1$. Consider Algorithm 13.1 when it accepts a request $R_i = (s_i, t_i)$. The path π_i is a shortest (hence, simple) path from s_i to t_i in G , and it has at most $n - 1$ edges. There are two cases:

1. π_i has at most $n-2$ edges. Then, there are at most $n-2$ requests in $\mathcal{A}^*$ (one per edge of π_i) whose paths are not edge-disjoint from π_i. Adding R_i itself, there are at most $\lambda = n-1$ requests that are deleted from $\mathcal{A}^*$.

2. π_i has exactly $n - 1$ edges. Then, π_i goes through all vertices of G. Because π_i and π_i^* are both shortest paths from s_i to t_i, they cannot be edge-disjoint. In fact, we can show that they share the same first edge. Otherwise, let v_i be the first node visited by π_i and v_i^* the first node visited by π_i^*. Because π_i visits all nodes, it visits v_i^* at some point. But, replacing the beginning of π_i from s_i to v_i^* with the edge (s_i, v_i^*), which belongs to π_i^*, and is still available in the graph at that point of the execution, leads to a shorter path from s_i to t_i, a contradiction. Now, there are at most $n - 1$ requests in $\mathcal{A}^*$ (one per edge of π_i) whose paths are not edge-disjoint from π_i, and this set includes R_i.

Altogether, this concludes the proof of Theorem 13.1. □

13.1.3 Short-requests-first greedy algorithm

When executing the greedy algorithm, it would be quite desirable to give priority to *short* requests, i.e., requests that can be routed with the smallest number of edges in the graph. Note that we need to recompute the length of the shortest path from sources to destinations at each step, because we remove edges from the graph whenever accepting a request. This modification of Algorithm 13.1 leads to Algorithm 13.2:

```
1 A ← ∅
2 while there exists a request in R that can be routed do
3     choose: R_i ← a request in R whose shortest path from s_i to t_i
      in G has minimum length among all requests in R
4     accept: A ← A ∪ {R_i}
5     route: π_i ← a shortest path from s_i to t_i in G
6     prune: remove all edges of π_i from G
```

ALGORITHM 13.2: Short-requests-first greedy algorithm.

The complexity of this second greedy algorithm can easily be upper-bounded by $O(|\mathcal{R}|.|V|^3)$ using, for instance, the Floyd–Warshall algorithm (whose complexity is $O(|V|^3)$ to find a shortest path from all sources to all destinations in G (see [27]).

THEOREM 13.2. *Algorithm 13.2 has approximation ratio $\lceil \sqrt{m} \rceil$ for problem* MEDP *in graphs with m edges, and this bound is tight.*

Proof. Given an instance of MEDP, let $\mathcal{A}$ be the solution returned by Algorithm 13.2, and let $\mathcal{A}^*$ be an optimal solution. Again, we start with the tightness of the bound. Consider the following directed graph made of the following $q+1$ paths, each with q edges:

- a first path p_0 with q edges e_i, $1 \leqslant i \leqslant q$;
- q paths p_i, $1 \leqslant i \leqslant q$, such that:
  - the first edge of p_i is e_i;
  - the following $q-1$ edges of p_i are edges private to p_i.

The graph contains $m = q^2$ edges. There are $q+1$ requests R_i, one per path p_i, $0 \leqslant i \leqslant q$, whose source and destination are the first and final nodes of the path.

Because all paths have the same length, Algorithm 13.2 may well accept request R_0 first. No other request can be accepted, so that $|\mathcal{A}| = 1$. But, the optimal solution is to accept the other q requests, so that $|\mathcal{A}^*| = q$. The ratio is indeed $q = \sqrt{m}$ (which is an integer).

Now, we show that for any instance of MEDP on a graph G with m vertices, we have

$$|\mathcal{A}^*| \leqslant \lceil \sqrt{m}\, \rceil \times |\mathcal{A}| \, .$$

The proof is similar to that of Theorem 13.1; we use Lemma 13.1 but this time with the value $\lambda = \lceil \sqrt{m}\, \rceil$.

As before, for $R_i \in \mathcal{A}$, let π_i be the path chosen by Algorithm 13.2. Similarly, for $R_j \in \mathcal{A}^*$, let π_j^* be the path chosen by the optimal solution. Consider the path π_i chosen by Algorithm 13.2 when it accepts a request R_i. This path π_i is a shortest (hence, simple) path from s_i to t_i in G, and it has the fewest number of edges among all requests that can still be accepted at this point. There are two cases:

1. π_i has at most $\lceil \sqrt{m}\, \rceil - 1$ edges. Then, there are at most $\lceil \sqrt{m}\, \rceil - 1$ requests in $\mathcal{A}^*$ (one per edge of π_i) whose paths are not edge-disjoint from π_i. Adding R_i itself, there are at most $\lambda = \lceil \sqrt{m}\, \rceil$ requests that are deleted from $\mathcal{A}^*$.

2. π_i has at least $\lceil \sqrt{m}\, \rceil$ edges. Then, all remaining requests in $\mathcal{A}^*$ correspond to paths with at least as many edges, because they can all be accepted at this point of the execution of Algorithm 13.2, and because Algorithm 13.2 always chooses the shortest possible request. Hence, all the remaining requests in $\mathcal{A}^*$ (possibly including R_i itself) are edge-disjoint and have no fewer than $\lceil \sqrt{m}\, \rceil$ edges. With a total number of m edges in the graph (some of which may have been deleted), there cannot be more than $\lceil \sqrt{m}\, \rceil$ such requests. This shows that we never delete more than $\lceil \sqrt{m}\, \rceil$ requests in $\mathcal{A}^*$.

Altogether, this concludes the proof of Theorem 13.2

□

13.1.4 Inapproximability result

The performance ratio $\lceil \sqrt{m}\, \rceil$ of Algorithm 13.2 may appear disappointing. However, a surprising result is that we cannot do much better. We formally state this result in the following theorem, due to [46].

THEOREM 13.3. *Unless P = NP, for any $\varepsilon > 0$, there cannot exist an $m^{\frac{1}{2}-\varepsilon}$ approximation algorithm for* MEDP *in graphs with m edges.*

Proof. The proof is based on the NP-hardness of the instance of MEDP with two requests [35], which we call 2DIRPATH. Consider an arbitrary instance $\mathcal{I}_1$ of 2DIRPATH, i.e., a directed graph $G' = (V', E')$ and four different vertices u_1, v_1, u_2, v_2. We ask whether there exist two edge-disjoint paths in G', one from u_1 to v_1, and one from u_2 to v_2. Consider $\varepsilon > 0$ and let $k = \lceil |E'|^{\frac{1}{\varepsilon}} \rceil$. We build an instance $\mathcal{I}_2$ of MEDP by arranging a triangle mesh of $(k-1)(k-2)/2$ copies of G' and linking them by k chains of k edges each, as shown in Figure 13.1. The resulting graph G has $n = (k-1)(k-2)|V'|/2 + 2k$ vertices (corresponding to the copies of G' and to the additional $2k$ extremities of the paths), and $m = (k-1)(k-2)|E'|/2 + k^2 - 2k + 3$ edges (in addition to the edges of the copies of G', there are two input edges for each of these copies and an input edge for each of the t_is). There are k requests $R_i = (s_i, t_i)$ for $1 \leqslant i \leqslant k$.

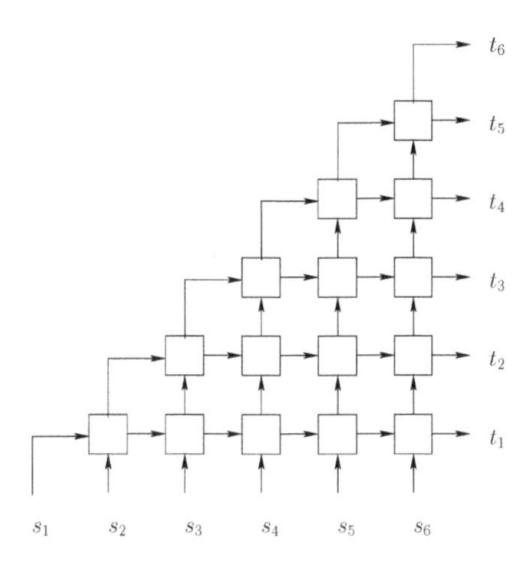

FIGURE 13.1: Construction of the instance $\mathcal{I}_2$ of MEDP (with $k = 6$).

The main idea is the following: If $\mathcal{I}_1$ has a solution, we can accept all k requests in $\mathcal{I}_2$ simply by traversing the copies of G' with its two independent paths. But, if $\mathcal{I}_1$ has no solution, then only one request can be accepted in $\mathcal{I}_2$: Once we have accepted any request, we cannot traverse the copies of G' that it includes with independent paths, which "blocks" all other requests and forbids their acceptance. In other words, the optimal solution to $\mathcal{I}_2$ has either k or one requests, depending upon whether $\mathcal{I}_1$ has a solution or not. If there would exist a polynomial approximation of ratio smaller than k, we would decide for $\mathcal{I}_1$, hence, solve the P = NP question. There remains to be checked that our choice of k is such that $m^{\frac{1}{2}-\varepsilon} < k$. We have $m = \Theta(k^2|E'|)$, where Θ denotes the order of magnitude. Then, $m^{\frac{1}{2}-\varepsilon} = \Theta(k^{1-2\varepsilon}|E'|^{\frac{1}{2}-\varepsilon})$. Since $k = \lceil |E'|^{\frac{1}{\varepsilon}} \rceil$, we get

$$m^{\frac{1}{2}-\varepsilon} = \Theta(k^{1-3\varepsilon/2-\varepsilon^2}).$$

We can assume that $|E'| \geqslant 2$ without loss of generality. Then, we have:

$$m = \frac{(k-1)(k-2)}{2}|E'| + (k^2 - 2k + 3) \leqslant \frac{k^2}{2}|E'| + k^2 \leqslant k^2|E'| \leqslant k^{2+\varepsilon}$$

by definition of k. Therefore,

$$m^{\frac{1}{2}-\varepsilon} \leqslant k^{1-\frac{3}{2}\varepsilon-\varepsilon^2} < k,$$

which leads to the result. □

13.2 PRVP: Packet routing with variable-paths

In this section, we study the PRVP problem, and we design a simple algorithm that is asymptotically optimal. The key technique is original and is based upon optimizing the throughput of the algorithm in steady-state mode. We start with a precise formulation of the problem and a quick overview of the main known complexity results.

13.2.1 Problem statement

Consider a directed graph $G = (V, E)$. To simplify notations, let $V = \{1, 2, \ldots, n\}$, and $(i \to j) \in E$ denote the edge from i to j (if it exists). In an instance of PRVP, we are given a set of node pairs $\mathcal{R} = \{(k, l)\}$; for each pair $(k, l) \in \mathcal{R}$ (each request), there is a number of packets $n_{k,l} > 0$ that need to be routed from source k to destination l. Such packets are called packets of type (k, l). See Figure 13.2 for an example with three requests (two of which have the same source).

The rules of the game are the following:

- Routing paths are not fixed a priori, and packets of the same type may follow different paths;
- A packet traverses an edge within one time step;
- At any time step, at most one packet traverses an edge.

So, not only do we have to find a path for each packet, but at any time step, we also have to decide which packet to route along each edge. We can view the PRVP problem as a *scheduling* problem, as the objective is to route all packets from source to destination in the fewest possible number of time steps. Indeed, the objective is to minimize the total execution time, or *makespan*. An example of possible actions at the first time step is shown in Figure 13.3. The scheduler has decided to route two black packets (of type (A, C)), one white packet (of type (A, D)), and two gray packets (of type (B, C)).

The NP-completeness of the PRVP problem is given in [85]. In fact, the proof of [85] shows that PRVP cannot be approximated with a factor $\frac{6}{5} - \varepsilon$ for any $\varepsilon > 0$ (unless P = NP). An approximation algorithm with some (complicated) constant factor is provided in [100], using an approach based on linear programming. Here, we present an algorithm due to Bertsimas and Gamarnik [16], which is also based upon a linear programming approach

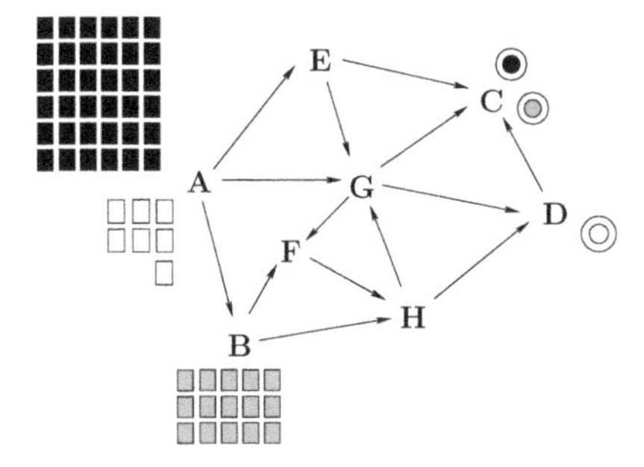

FIGURE 13.2: An instance of PRVP (for clarity, nodes are denoted with letters rather than numbers).

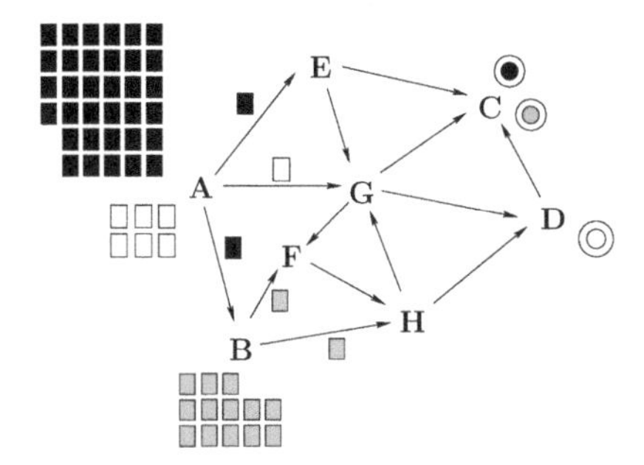

FIGURE 13.3: An example of possible actions at the first time step.

but which uses a very interesting relaxation technique. We show that the algorithm of [16] achieves a makespan $C \leqslant C^* + O(\sqrt{C^*})$, where C^* is the optimal makespan. As a consequence, the algorithm is asymptotically optimal when the total number of packets to be routed increases to infinity. The reader may want to refer to Section 8.3 for a background on linear programming techniques.

13.2.2 Bounding optimal makespan via linear programming

In this section, we derive a lower bound on the optimal makespan C^*. Consider the optimal solution, and let $x_{i,j}^{k,l}$ be the total number of packets of type (k, l) that traverse the edge (i, j):

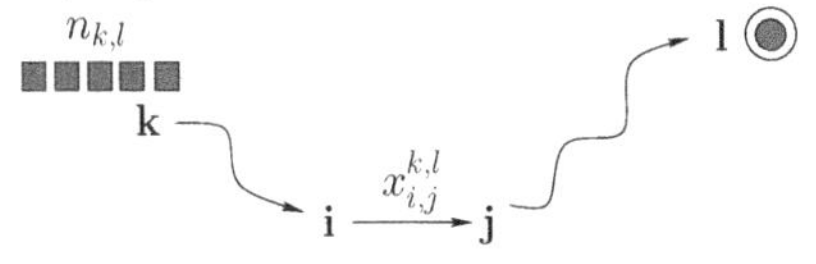

Let the *congestion* $C_{i,j}$ of an edge $(i \to j) \in E$ be the total number of packets crossing that edge:

$$C_{i,j} = \sum_{(k,l)\in\mathcal{R}} x_{i,j}^{k,l} .$$

Clearly, the optimal execution time is at least equal to the maximal congestion of any edge:

$$C^* \geqslant \max_{(i\to j)\in E} C_{i,j} .$$

This lower bound leads us to write the following linear program, which we state as a whole before explaining each equation with further details:

$$\left\{\begin{array}{lll} \multicolumn{3}{l}{\text{MINIMIZE } C_{\max} \text{ UNDER THE CONSTRAINTS}} \\ (13.1a) & \forall (k,l)\in\mathcal{R}, \forall (i\to j)\in E, & x_{i,j}^{k,l} \geqslant 0 \\ (13.1b) & \forall (k,l)\in\mathcal{R}, \forall (i\to k)\in E, & x_{i,k}^{k,l} = 0 \\ (13.1c) & \forall (k,l)\in\mathcal{R}, \forall (l\to j)\in E, & x_{l,j}^{k,l} = 0 \\ (13.1d) & \forall (k,l)\in\mathcal{R}, & \sum_{j:(k\to j)\in E} x_{k,j}^{k,l} = n_{k,l} \\ (13.1e) & \forall (k,l)\in\mathcal{R}, & \sum_{i:(i\to l)\in E} x_{i,l}^{k,l} = n_{k,l} \\ (13.1f) & \forall (k,l)\in\mathcal{R}, \forall i \neq k,l, & \sum_{h:(h\to i)\in E} x_{h,i}^{k,l} = \sum_{j:(i\to j)\in E} x_{i,j}^{k,l} \\ (13.1g) & \forall (i\to j)\in E, & C_{i,j} = \sum_{(k,l)\in\mathcal{R}} x_{i,j}^{k,l} \\ (13.1h) & \forall (i\to j)\in E, & C_{i,j} \leqslant C_{\max}. \end{array}\right. \qquad (13.1)$$

Here is an explanation of the equations (see also Figure 13.4):

- Equation (13.1a): All variables are nonnegative.
- Equation (13.1b): We can assume that a packet of type (k,l) never revisits its source node k.
- Equation (13.1c): We can assume that a packet of type (k,l) never leaves its destination node l.
- Equation (13.1d): All packets of type (k,l) eventually leave their source node k.
- Equation (13.1e): All packets of type (k,l) eventually reach their destination node l.
- Equation (13.1f): Conservation of packets (all types) in intermediate node i.
- Equation (13.1g): Definition of the congestion of edge $(i \to j)$.
- Equation (13.1h): Definition of $C_{\max}$ as the maximum of all edge congestions.

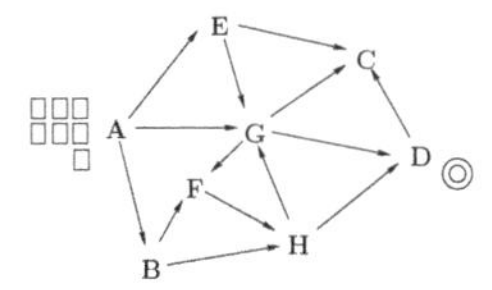

(a) Initial status, with seven packets of type (A, D).

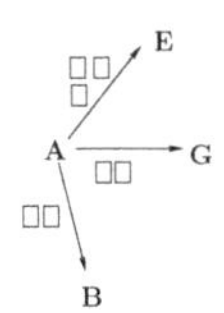

(b) Equation (13.1d): All seven packets eventually leave their source node A.

G
D
H

(c) Equation (13.1e): All seven packets eventually reach their destination node D.

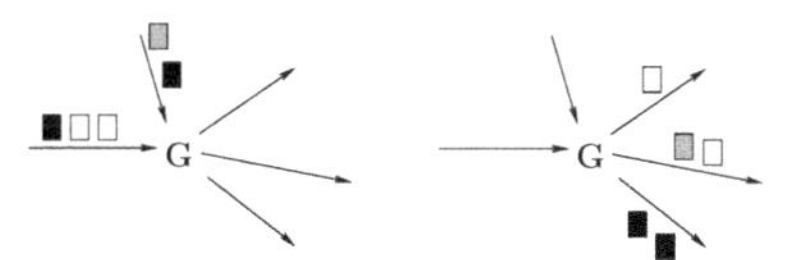

(d) Equation (13.1f): Conservation of packets (all types) in intermediate node G.

FIGURE 13.4: Illustrating the linear program.

Of course, for an actual execution, we need integer values for the variables $x_{i,j}^{k,l}$. But, we can solve the linear program with rational variables, hence, in polynomial time [93]. We then derive a rational value for the optimal solution of the linear program, which we still denote as $C_{\max}$. A fortiori, we have $C_{\max} \leqslant C^*$, where C^* is the optimal makespan. We will now use this value to derive the routing algorithm.

13.2.3 Routing algorithm

The routing algorithm is the following:

- Compute the optimal solution $C_{\max}$, $x_{i,j}^{k,l}$ of the linear program (equation (13.1)).
- Periodic schedule:
  - Define $\Omega = \lceil \sqrt{C_{\max}} \rceil$;
  - Use $P = \lceil \frac{C_{\max}}{\Omega} \rceil$ periods of length Ω;

– During each period, edge $(i \to j)$ forwards

$$a_{i,j}^{k,l} = \left\lfloor \frac{x_{i,j}^{k,l}\Omega}{C_{\max}} \right\rfloor$$

packets of type (k, l) (if enough such packets are available; otherwise, it forwards all available ones).

- Clean-up phase: Sequentially route each residual packet inside the network up to its destination.

Before discussing the performance of this algorithm, let us explain the intuition behind it. The value of Ω is not important by itself; what is important is the idea of organizing the algorithm into periods of length Ω. During each period, the same scheme of execution is repeated. Because we aim to reach an execution time that is close to $C_{\max}$, we try to route a number of packets of each type (k, l) in proportion; a total of $x_{i,j}^{k,l}$ must be routed in time $C_{\max}$, which means $x_{i,j}^{k,l}/C_{\max}$ per time step, which in turns means $\Omega \times x_{i,j}^{k,l}/C_{\max}$ in Ω time steps. And, we take the floor function to be on the safe side (and with an integer number of packets).

Now, consider an edge $(i \to j)$ at the beginning of a given period. There are two cases:

- There are at least than $a_{i,j}^{k,l}$ packets stored in node i; then, during the period, the algorithm routes $a_{i,j}^{k,l}$ of them across the edge.
- There are fewer than $a_{i,j}^{k,l}$ packets stored in node i; then, during the period, the algorithm routes all these packets across the edge.

Thus, despite the fact that the algorithm is divided into P periods, it is not periodic because it does not perform exactly the same operations within each period.

THEOREM 13.4. *The routing algorithm routes all the packets in time C_{ra}, where*

$$C_{\max} \leqslant C^* \leqslant C_{ra} \leqslant C_{\max} + O(\sqrt{C_{\max}}) \,.$$

Here, C^* denotes the optimal makespan.

As a consequence, we have that $\frac{C_{\text{ra}}}{C^*} \to 1$ as $\sum_{(k,l)\in\mathcal{R}} n_{k,l} \to +\infty$. The algorithm is asymptotically optimal when the total number of packets becomes large.

Proof. The proof that the algorithm is feasible is easy. It suffices to show that as many as $a_{i,j}^{k,l}$ packets of each type (k, l) can indeed be routed within a period. We derive

$$\sum_{(k,l)\in\mathcal{R}} a_{i,j}^{k,l} \leqslant \sum_{(k,l)\in\mathcal{R}} \frac{x_{i,j}^{k,l}\Omega}{C_{\max}} = \frac{C_{i,j}\Omega}{C_{\max}} \leqslant \Omega \,.$$

The tricky part is to bound the execution time. Here, the value of Ω comes into play. We show that at the end of the last period, i.e., at time

$$T = P \times \Omega = \left\lceil \frac{C_{\max}}{\Omega} \right\rceil \times \Omega \leqslant C_{\max} + \Omega\,, \tag{13.2}$$

only a small number of packets still resides in the network, and that the time to route them all up to their destination is negligible in front of $C_{\max}$.

Consider the $n_{k,l}$ packets of type $(k,l) \in \mathcal{R}$. During each period $[m\Omega, (m+1)\Omega[$, where $0 \leqslant m < P$, the source node k processes exactly $a_{i,j}^{k,l}$ packets, because they are all available there. Let $\delta(k)$ be the out degree of node k (the number of its outgoing edges). We have

$$\sum_{j:(k \to j) \in E} a_{k,j}^{k,l} \geqslant \sum_{j:(k \to j) \in E} \left(\frac{x_{k,j}^{k,l}\Omega}{C_{\max}} - 1 \right) \geqslant \frac{n_{k,l}\Omega}{C_{\max}} - \delta(k)$$

(see equation (13.1d)). Hence, the number of packets of type (k,l) that remain at node k at time T is at most

$$n_{k,l} - P \times \left(\frac{n_{k,l}\Omega}{C_{\max}} - \delta(k) \right) \leqslant n_{k,l} - \frac{C_{\max}}{\Omega} \times \left(\frac{n_{k,l}\Omega}{C_{\max}} - \delta(k) \right) = \frac{C_{\max}}{\Omega} \times \delta(k)\,.$$

Consider now another node $i \neq k$. We also will bound the number of packets of type (k,l) that are stored in i at time T. Initially, there is no packet of type (k,l) available in node i. Let $[m_0\Omega, (m_0+1)\Omega[$ be the first period during which one or more packets of type (k,l) arrive into node i. For $m \geqslant m_0$, let $f_i^{k,l}(m)$ be the number of packets of type (k,l) that arrive into node i, and $g_i^{k,l}(m)$ be the number of packets of type (k,l) that depart from node i, during the period $[m\Omega, (m+1)\Omega[$. We have

- Reception: For $m \geqslant m_0$, $f_i^{k,l}(m) \leqslant \sum_{h:(h \to i) \in E} a_{h,i}^{k,l} \leqslant \sum_{h:(h \to i) \in E} \frac{x_{h,i}^{k,l}\Omega}{C_{\max}}$.

- Emission: For $m \geqslant m_0 + 1$, the schedule sends $a_{i,j}^{k,l} \geqslant \frac{x_{i,j}^{k,l}\Omega}{C_{\max}} - 1$ packets along each outgoing edge j, if enough packets are available. Otherwise, it sends all available packets. So,

  $$f_i^{k,l}(m) - g_i^{k,l}(m) \leqslant \sum_{h:(h \to i) \in E} \frac{x_{h,i}^{k,l}\Omega}{C_{\max}} - \left(\sum_{j:(i \to j) \in E} \left(\frac{x_{i,j}^{k,l}\Omega}{C_{\max}} - 1 \right) \right) = \delta(i)$$

  (see equation (13.1f)). Hence, the number of packets of type (k,l) that reside in node i increases by at most $\delta(i)$ during each period.

Altogether, there remains at most $f_i^{k,l}(m_0) + P\delta(i) \leqslant \left(\sum_{h:(h \to i) \in E} \frac{x_{h,i}^{k,l}\Omega}{C_{\max}} \right) + P\delta(i)$ packets of type (k,l) in node i at time T.

We can now bound the total number of packets f_P that remain in the network at time T:

$$
\begin{aligned}
f_P &= \sum_{i \in V} \sum_{(k,l) \in \mathcal{R}} \left(\left(\sum_{h:(h \to i) \in E} \frac{x_{h,i}^{k,l} \Omega}{C_{\max}} \right) + P \times \delta(i) \right) \\
&\leqslant \left(\sum_{(i \to j) \in E} \frac{C_{i,j} \Omega}{C_{\max}} \right) + \left(\frac{C_{\max}}{\Omega} + 1 \right) 2 \times |\mathcal{R}| \times |E| \\
&\leqslant |E| \times \Omega + \left(\frac{C_{\max}}{\Omega} + 1 \right) 2 \times |\mathcal{R}| \times |E| \, .
\end{aligned}
$$

Because $\Omega = \lceil \sqrt{C_{\max}} \rceil$, we derive $f_P \leqslant (|E| + 2 \times |\mathcal{R}| \times |E|)(\sqrt{C_{\max}} + 1)$. Each of the remaining packets can be individually routed in time at most $|V|$, using any shortest (simple) path in the graph. The total duration of the clean-up phase is bounded by $f_P \times |V| = O(\sqrt{C_{\max}})$, which concludes the proof (recall that $|V|$ and $|E|$ are constants in the asymptotic analysis, and that $|\mathcal{R}| \leqslant |E|$). Note that the bound on the clean-up phase is very crude, as some packets could be routed in parallel, but it is enough to derive the result. □

13.2.4 Steady-state approach

We conclude this study with a few remarks on the approach of Bertsimas and Gamarnik [16]. Their approach can be summarized as follows:

- Use a bound $C_{\max}$ from the rational linear program.
- Concentrate on the steady-state operation of the schedule, divided into periods.
- Ensure that periods are long enough so that rounding down to integer numbers has a negligible impact.
- Ensure that periods are numerous enough so that the clean-up phase has a negligible impact.

This explains the trade-off achieved by choosing $\Omega = \sqrt{C_{\max}}$. In addition, using periods enables one to describe the schedule in compact form. There is no need to specify which packet is routed across which edge at which time step. Instead, only a small (and polynomial) number of values $a_{i,j}^{k,l}$ need to be specified to characterize the whole routing operation. Such a steady-state approach has successfully been applied to many other scheduling problems.

13.3 Conclusion

Routing problems are ubiquitous in computer science. This case study has dealt with two classical and important instances, MEDP and PRVP, and has illustrated very different techniques. For MEDP, we have followed the typical approach for an NP-complete problem: (i) design a greedy algorithm, (ii) prove that it is an approximation algorithm with bound B, and (iii) prove that the bound B cannot be improved (unless P = NP). For PRVP, the main idea is to relax the original problem and to concentrate on steady-state operation. In other words, the focus is on optimizing the throughput rather than the makespan. This technique is very powerful, as shown by the result of asymptotic optimality. Refer to [10] for further examples of the use of steady-state techniques to solve various scheduling problems.

Chapter 14

Matrix product, or tiling the unit square

In this case study, we deal with a simple geometric problem: How do we partition the unit square into p rectangles of given area $s_1, s_2, \ldots, s_p$ (such that $\sum_{i=1}^{p} s_i = 1$), so as to minimize the sum of the p semiperimeters of the rectangles? Note that there always exists a solution to this problem. For instance, we can tile the unit square into p horizontal slices of height $s_1, s_2, \ldots, s_p$. The difficulty is to minimize the objective function.

For an illustration, consider the following example with $p = 5$ rectangles $R_1, \ldots, R_5$ of areas $s_1 = 0.36$, $s_2 = 0.25$, $s_3 = s_4 = s_5 = 0.13$. A possible partition of the unit square is shown in Figure 14.1. The size of each rectangle is: $0.61 \times \frac{36}{61}$ for R_1; $0.61 \times \frac{25}{61}$ for R_2; and $0.39 \times \frac{1}{3}$ for R_3, R_4, and R_5. We compute that the sum of the semiperimeters is 4.39, while an absolute lower bound is $\sum_{i=1}^{p} 2\sqrt{s_i} \approx 4.36$ (obtained when all rectangles are squares, which is not achievable in this example). Here, the partition turns out to be excellent with respect to the objective function.

The geometric interpretation for the sum of the semiperimeters is nice. It is the length of the lines drawn to make the partition, plus two lines corresponding to the right and bottom edges of the unit square.

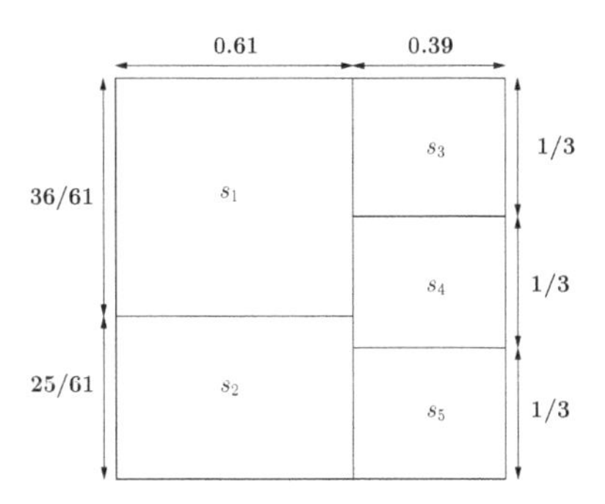

FIGURE 14.1: A simple example with $p = 5$ rectangles.

Before addressing the complexity of this geometric problem, we start (Section 14.1) by describing the original motivation for its study. The reader will not be surprised that the problem arises from the design of some parallel algorithm. However, we are not speaking of just yet another parallel algorithm; we target the product of two square matrices, which is the main building block of many computational kernels in numerical linear algebra. Then, in Section 14.2, we formally state the optimization problem, and we establish its NP-completeness. This is the key result of this case study because the reduction is quite involved. We outline the main ideas for each step of the proof. Section 14.3 is devoted to the design of approximation algorithms that are based on partitions into columns of rectangles. Finally, in Section 14.4, we briefly survey some related optimization problems.

14.1 Problem motivation

The motivation for this work is the design of parallel matrix product algorithms targeted to heterogeneous platforms, such as heterogeneous clusters of workstations or collections of such clusters. In the following, we describe a parallel matrix product algorithm with identical processors before moving to the case of different-speed processors.

With identical processors, the reference algorithm [18] works as follows: Let $C = A \times B$ the product to be computed, where A and B are square matrices of size $n \times n$. Assume that there are $p = q^2$ processors. Assume that q divides n and let $n = r \times q$. We assign an $r \times r$ block of each matrix to each processor according to a two-dimensional grid/torus topology of size $q \times q$. We denote the three matrix blocks assigned to processor $P_{i,j}$ by $\widehat{A_{i,j}}$, $\widehat{B_{i,j}}$, and $\widehat{C_{i,j}}$, as depicted in Figure 14.2 for matrix A.

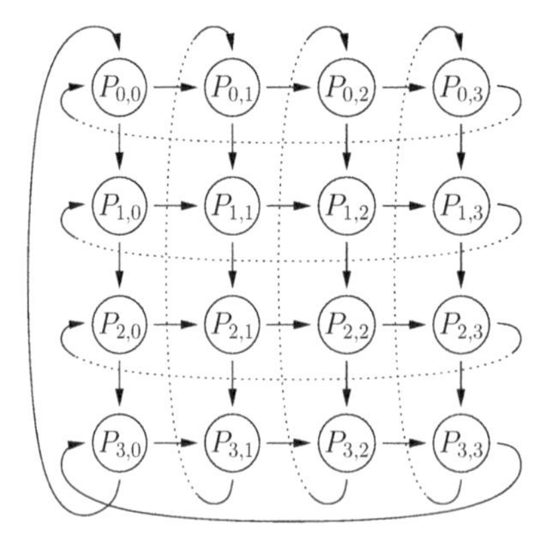

FIGURE 14.2: Block distribution of an $n \times n$ matrix ($n = 24$) on a grid/torus of $p = q^2$ processors ($q = 4$).

While the standard algorithm for matrix multiplication is often written as a sequence of inner product computations, operations can be ordered in many different ways. One option is to write the algorithm as a series of outer products, which amounts to simply switching the order of the loops from i, j, k to k, i, j:

for $k = 0$ **to** $n - 1$ **do**
  for $i = 0$ **to** $n - 1$ **do**
    for $j = 0$ **to** $n - 1$ **do**
      $C_{i,j} \leftarrow C_{i,j} + A_{i,k} \times B_{k,j}$

Here, we assume that all elements of matrix C are initialized to zero. It turns out that this so-called outer-product algorithm [1, 36, 68] leads to a particularly simple and elegant parallelization on a two-dimensional processor grid. The algorithm proceeds in n steps, that is, n iterations of the outer loop. At each step k, $C_{i,j}$ is updated using $A_{i,k}$ and $B_{k,j}$. Recall that all three matrices are partitioned in q^2 blocks of size $r \times r$, as on the right side of Figure 14.2. The algorithm above can be written in terms of matrix blocks and of matrix multiplications, and it proceeds in q steps as follows:

for $k = 0$ **to** $q - 1$ **do**
  for $i = 0$ **to** $q - 1$ **do**
    for $j = 0$ **to** $q - 1$ **do**
      $\widehat{C_{i,j}} \leftarrow \widehat{C_{i,j}} + \widehat{A_{i,k}} \times \widehat{B_{k,j}}$

Now, consider the execution of this algorithm on a two-dimensional grid of $p = q^2$ processors. Processor $P_{i,j}$ holds block $\widehat{C_{i,j}}$ and is responsible for updating this block at each step of the above algorithm. To perform this update at step k, processor $P_{i,j}$ needs blocks $\widehat{A_{i,k}}$ and $\widehat{B_{k,j}}$. At step $k = j$, $P_{i,j}$ already happens to hold $\widehat{A_{i,k}}$. For all other steps, $P_{i,j}$ must receive $\widehat{A_{i,k}}$ from the processor that holds it, that is, $P_{i,k}$. This is true for all $P_{i,j}$, $j \neq k$, processors. Therefore, at step k, processor $P_{i,k}$ must broadcast its block of matrix A to all processors $P_{i,j}$, $j \neq k$, that is, all processors that are on processor $P_{i,k}$'s processor row. This is true for all i. Similarly, blocks of matrix B must be broadcast at step k by $P_{k,j}$ to all processors on its processor column, for all j. The resulting communication pattern is illustrated in Figure 14.3. The figure shows which blocks of matrices A and B are sent to which processors at step $k = 1$ of the algorithm in the case of a 4×4 grid. For instance, block $\widehat{A_{2,1}}$, which is held by processor $P_{2,1}$, is sent to processors $P_{2,0}$, $P_{2,2}$, and $P_{2,3}$.

How can we extend this outer product algorithm to deal with heterogeneous platforms? The idea is to keep the same framework; at each step, one pivot column and one pivot row are communicated to all processors, and independent updates take place. However, with different-speed processors, we cannot

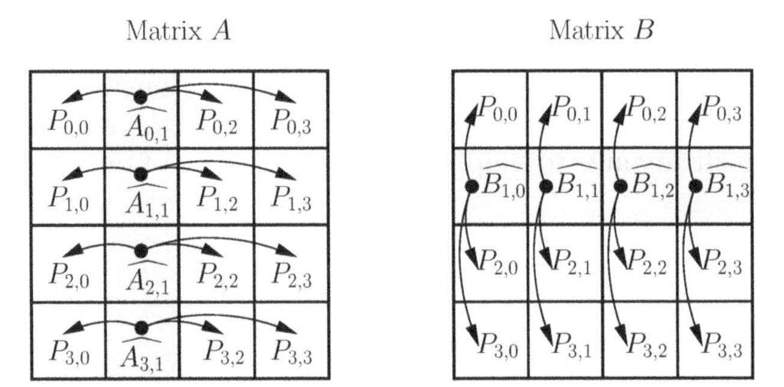

FIGURE 14.3: Communications of blocks of matrices A and B at step $k = 1$ of the outer-product matrix multiplication algorithm on a 4×4 processor grid.

distribute same-size rectangles from the C matrix to the processors. Intuitively, we want to balance the computing load so that each processor receives an amount of work in accordance to its computing power. In fact, a C rectangle block requires an amount of arithmetic operations that is proportional to its area, so that the areas of the blocks assigned to the processors should be proportional to their speeds. To parallelize the matrix product $C = AB$, we have to tile the C matrix into p nonoverlapping rectangles, each rectangle being assigned to one processor. If the speed of a processor is twice as large as that of another one, it should be assigned a rectangle of twice the area.

Figure 14.4 shows an example with 13 different-speed computing resources. For this example, we picked a prime number of processors to emphasize the fact that for solving the general problem it is not necessary that the processors be arranged along a two-dimensional grid. At each step of the outer-product algorithm, there is a horizontal broadcast of a column of A and a vertical broadcast of a row of B. These broadcasts involve different numbers of source and destination processors, depending upon where the column or row is located. The figure details the vertical broadcast of a row of B, which is partially held by four processors. For instance, the processor holding the top right block of the matrix is involved in receiving data from three source processors, while the processor holding the bottom left block of the matrix is involved in receiving data from only one source processor.

Given p processors with speeds $s_1, \ldots, s_p$, normalized so that $\sum_{i=1}^{p} s_i = 1$, it is always possible to partition the unit square in p rectangles with areas $s_1, s_2, \ldots, s_p$. From now on we always reason on the unit square, knowing that we can then multiply rectangle dimensions by the matrix size n and round to appropriate integer values to obtain actual matrix blocks. One possibility for partitioning the unit square is simply to use a unidimensional distribution in block rows. But, the question is: What is the best partitioning among the ones that achieve perfect load balancing?

To formalize this question, one must define an objective function. Given that we consider only partitioning schemes that balance the load perfectly,

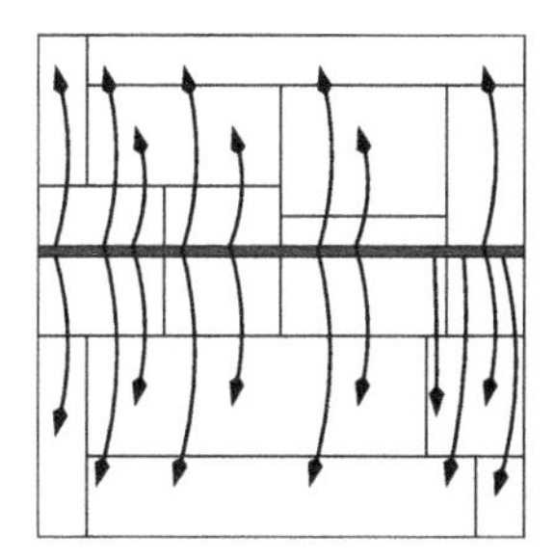

FIGURE 14.4: Distributing the matrices across 13 heterogeneous processors, and vertical broadcast of a row of matrix B during a step of the outer-product algorithm (the horizontal broadcast of a column of matrix A at the same step is not shown).

we can distinguish them by the amount of communication these partitioning schemes induce for our outer-product matrix multiplication algorithm. At each step (see Figure 14.4), each processor is involved in communicating an amount of data proportional to the semiperimeter of the rectangular matrix block associated with that processor (except for the right and bottom boundaries of the whole square). This leads to a very intuitive geometric interpretation of our problem. While the areas of the rectangular blocks are fixed, one can adjust their shapes to lead to the lowest amount of total communication volume. At last we have found the geometric interpretation of our problem: How can one partition a unit square in p rectangles with prescribed areas $s_1, s_2, \ldots, s_p$, such that $\sum_{i=1}^{p} s_i = 1$, so that the sum $\hat{C}$ of the semiperimeters of the rectangles is minimized? Indeed, $\hat{C} - 2$ (for the boundaries) is equal to the total communication volume of the algorithm.

14.2 NP-completeness

We now formally state the problem. We have to determine p rectangles R_i, of prescribed areas s_i, $1 \leqslant i \leqslant p$, where $\sum_{i=1}^{p} s_i = 1$. The shape of each R_i is the degree of freedom; we want to tile the unit square so as to solve the following optimization problem:

DEFINITION 14.1. PERISUM(s): Given p real positive numbers $s_1, \ldots, s_p$ such that $\sum_{i=1}^{p} s_i = 1$, find a partition of the unit square into p rectangles R_i of area s_i and of size $h_i \times v_i$, so that $\hat{C} = \sum_{i=1}^{p}(h_i + v_i)$ is minimized.

There is an obvious lower bound for PERISUM(s):

LEMMA 14.1. *For all solutions of* PERISUM(s), $\hat{C} \geqslant 2\sum_{i=1}^{p} \sqrt{s_i}$.

Proof. The semiperimeter of each rectangle R_i is always larger than $2\sqrt{s_i}$, the value when it is a square. Of course, tiling the unit square into p squares of area s_i is not always possible, so the lower bound for PERISUM(s) is not always tight. □

The decision problem associated with the optimization problem PERISUM(s) is the following.

DEFINITION 14.2. PERISUM(s, K): Given p real positive numbers $s_1, \ldots, s_p$ such that $\sum_{i=1}^{p} s_i = 1$ and a positive real bound K, is there a partition of the unit square into p rectangles R_i of area s_i and of size $h_i \times v_i$, so that $\sum_{i=1}^{p}(h_i + v_i) \leqslant K$?

THEOREM 14.1. PERISUM(s, K) *is NP-complete.*

Proof. The problem obviously belongs to NP. The proof of completeness is both lengthy and technical, so we start by outlining the main ideas:

- First, we polynomially reduce the decision problem PERISUM(s, K) to a geometric problem (which we name ASP for *All-Square Partition*), which amounts to checking if there exists a partition of the unit square into squares of given areas.

- Then, we prove the NP-completeness of ASP using a polynomial reduction to the 2-PARTITION-EQUAL problem (2PEQ for short), which is NP-complete [38].

The idea of partitioning into squares rather than into arbitrary instances comes from the lower bound: If such a partition exists, then it is optimal. Formally, we state the following lemma.

LEMMA 14.2. *2PEQ* $\xrightarrow{pr}$ *ASP* $\xrightarrow{pr}$ PERISUM*(s,K), where* $\xrightarrow{pr}$ *denotes polynomial reduction (see Section 6.3.1), and where 2PEQ and ASP are defined as follows.*

DEFINITION 14.3 (2PEQ - 2-PARTITION-EQUAL)**.** Given a set of p integers $\mathcal{A} = \{a_1, \ldots, a_p\}$, is there a partition of $\{1, \ldots, p\}$ into two subsets $\mathcal{A}_1$ and $\mathcal{A}_2$ such that

$$\sum_{i \in \mathcal{A}_1} a_i = \sum_{i \in \mathcal{A}_2} a_i \text{ and } \operatorname{card}(\mathcal{A}_1) = \operatorname{card}(\mathcal{A}_2) \quad ?$$

(recall that this variant of 2-PARTITION has been discussed in Section 7.6).

DEFINITION 14.4 (ASP - All-Square Partition)**.** Given a set $\mathcal{L} = \{l_1, \ldots, l_p\}$ of p real positive numbers such that $\sum_{i=1}^{p} l_i^2 = 1$, is there a partition of the unit square into p squares S_1, ..., S_p, where S_i is of width l_i?

Because 2PEQ is known to be NP-Complete [38], Lemma 14.2 will complete the proof of Theorem 14.1. We start by proving the easy part of Lemma 14.2, i.e., ASP $\xrightarrow{pr}$ PERISUM(s, K). Let $\mathcal{L} = \{l_1, \ldots, l_p\}$ be a set of p real positive numbers such that $\sum_{i=1}^{p} l_i^2 = 1$. Solving ASP is equivalent to solving PERISUM(s, K) with $K = \sum_{i=1}^{p} 2l_i$, and for $1 \leqslant i \leqslant p$, $s_i = l_i^2$. Therefore, ASP $\xrightarrow{pr}$ PERISUM(s, K).

Next, we consider an arbitrary instance of the 2-PARTITION-EQUAL (2PEQ) problem, i.e., a set $\mathcal{A} = \{a_1, \ldots, a_n\}$ of n integers. We assume that $n \geqslant 400$ without loss of generality. We have to transform this instance polynomially into an instance of the ASP problem that has a solution if and only if the original instance of 2PEQ has a solution. This transformation is sketched as follows:

- Move from set $\mathcal{A}$ to set $\mathcal{B}$: The idea is to build an equivalent instance of this problem using a set $\mathcal{B} = \{b_1, b_2, \ldots, b_n\}$, where $b_i = 2(a_i + 2n \max_k a_k)$. The goal is to have $b_i > \frac{2}{3} \max_k b_k$ for all i. Under a few technical assumptions, we show that there exists a solution to the initial 2PEQ problem if and only if $\mathcal{B}$ can be partitioned into two subsets of same sum (but not necessarily of same cardinal).

- Build from $\mathcal{B}$ an instance of ASP, as illustrated in Figures 14.5 and 14.6. The instance is made of three kinds of squares: 14 large squares (represented in Figure 14.5 and denoted as $A_{i,j}$), n squares of size $b_i \times b_i$ (denoted as A_{b_i} in Figure 14.6), and a polynomial number of other squares (denoted as $A_{\overline{b_i},j}$ in Figure 14.6). We show that the only possible configuration is the one shown in Figure 14.5. In this configuration, there are two nonadjacent $M \times S$ rectangular zones (where $M = \frac{4}{3} \max_k b_k$ and $S = \sum_i \frac{b_i}{2}$), which are partitioned as shown in Figure 14.6. Because of the condition $b_i > \frac{2}{3} \max_k b_k = \frac{M}{2}$, necessarily the A_{b_i} squares of size $b_i \times b_i$ are adjacent and never on top of each other. Therefore, for each rectangle, the sum of the b_is is not larger than S, and in fact it is equal to S because all the A_{b_i} rectangles have to fit inside the nonrectangular zones. Intuitively, the large squares are introduced to create the two nonadjacent rectangular zones of area $M \times S$; the n squares A_{b_i} must be aligned within these two zones, and the other squares are here to fill up the holes in the two rectangular zones.

Formally, recall that we have defined $b_i = 2(a_i + 2n \max_k a_k)$, $S = \frac{\sum_i b_i}{2}$, $M = \frac{4}{3} \max_k b_k$, and that $n \geqslant 400$. Then, $S \geqslant \frac{1}{2} n \frac{M}{2} \geqslant 100M$, and $\frac{M}{2} < b_i \leqslant \frac{3M}{4}$ for all i. The reason for introducing M is to fill up the two nonadjacent rectangular zones of area $M \times S$ with a polynomial number of rectangles. We will tile the n rectangles $\overline{R_i}$ of size $b_i \times (M - b_i)$ (see Figure 14.6) into a minimal number of squares $KS(i)$, following the procedure of Kenyon [59]. Here, KS stands for *Kenyon's squares*. To get a logarithmic number of squares $KS(i)$, the rectangle $\overline{R_i}$ must not be too elongated, which is ensured by the inequality

$M - b_i < b_i \leqslant 3(M - b_i)$. We obtain from [59] that $KS(i) \leqslant 3 + C \log b_i$, where C is a universal constant. In the following, for $1 \leqslant i \leqslant n$, we let $w(\overline{b_i}, j)$, $1 \leqslant j \leqslant KS(i)$, denote the widths of the $KS(i)$ squares obtained by the procedure in [59] to tile the rectangle $\overline{R_i}$ of size $b_i \times (M - b_i)$.

Altogether, we build the following instance of the ASP problem: Is there a partition of the unit square into $14 + n + \sum_{i=1}^{n} KS(i)$ squares of respective widths $\frac{(13S+11M)}{l}(\times 1)$, $\frac{(7S+6M)}{l}(\times 3)$, $\frac{(4S+3M)}{l}(\times 2)$, $\frac{(3S+3M)}{l}(\times 2)$, $\frac{(3S+2M)}{l}$ $(\times 2)$, $\frac{(2S+2M)}{l}(\times 4)$, $\frac{b_i}{l}$ (for all i), $\frac{w(\overline{b_i},j)}{l}$ (for all i and for all $j \leqslant KS(i)$), where $l = (20S + 17M)$?

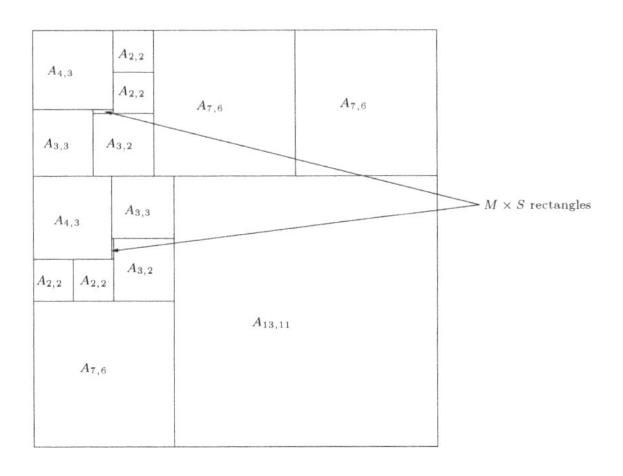

FIGURE 14.5: General position of the squares.

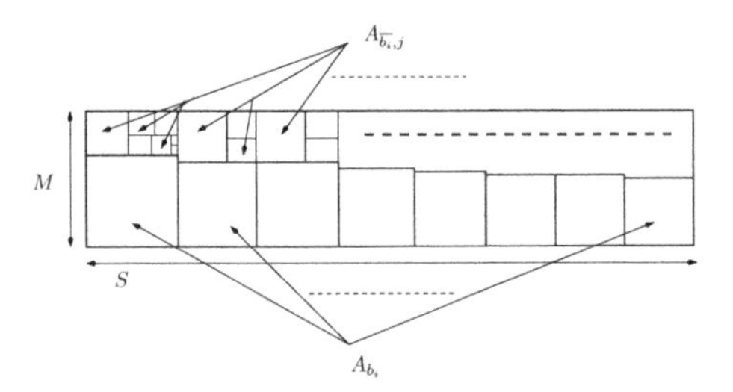

FIGURE 14.6: Zoom in on the $M \times S$ rectangle areas.

For convenience, we can consider the (equivalent) scaled problem: Is there a partition of the $(20S+17M) \times (20S+17M)$ square into $14+n+\sum_{i=1}^{n} KS(i)$

squares of respective width

$$\begin{cases} 13S + 11M & (\times 1), \\ 7S + 6M & (\times 3), \\ 4S + 3M & (\times 2), \\ 3S + 3M & (\times 2), \\ 3S + 2M & (\times 2), \\ 2S + 2M & (\times 4), \\ b_i & (\forall i,\ 1 \leqslant i \leqslant n), \\ w(\overline{b_i}, j) & (\forall i,\ 1 \leqslant i \leqslant n;\ \forall j,\ 1 \leqslant j \leqslant KS(i)) \ ? \end{cases}$$

Let $A_{i,j}$ denote a square of width $(iS+jM)$, A_{b_i} denote a square of width b_i, and $A_{\overline{b_i},j}$ denote a Kenyon's square of width $w(\overline{b_i}, j)$. The idea is to prove that such a partition is the one depicted in Figure 14.5, where the two small $M \times S$ rectangle areas are shown by arrows in Figure 14.5 and fully described in Figure 14.6. The intuitive idea of the proof is the following: The large squares are used to prevent the two small $M \times S$ rectangular zones from becoming neighbors. Hence, these areas must be filled separately by the remaining squares, namely, the squares A_{b_i} and the Kenyon's squares. This will be possible if and only if the b_is can be partitioned into two subsets of the same sum, which in turn will be possible if and only if the a_is can be partitioned into two subsets of same sum and same cardinal. The Kenyon's squares are introduced to fill the holes in the two rectangular zones and to obtain a true tiling of the whole area.

Altogether, the proof is technical, but the two main ideas are simple: (i) Force the A_{b_i} rectangles to lie into one of the two $M \times S$ rectangle areas, hence, separating them into two sides of the partition, and (ii) fill up these areas with a polynomial number of squares. Note that using many little squares of width 1 would have led to using a number of squares exponential in the size of the original instance. Of course, there remain several technical and painstaking constraints to check, but the main ideas are all here. Please refer to [8] for further details. □

14.3 A guaranteed heuristic

In this section, we investigate a restriction of the PERISUM(s, K) partitioning problem where one considers partitions that consist of columns of rectangles. For this restriction, one can compute the optimal solution in polynomial time, via dynamic programming. We describe here this optimal solution, and then we analyze the performance guarantee obtained when comparing to the solution of PERISUM(s, K).

14.3.1 The ColPeriSum(s) problem

Because PeriSum(s) is NP-hard, we consider a more constrained problem, ColPeriSum(s). In this new problem, we specify that the partition of the unit square must consist of columns of rectangles, as shown in the example in Figure 14.7. Formally, given p real positive numbers $s_1, \ldots, s_p$ such that $\sum_{i=1}^{p} s_i = 1$, we want to partition the unit square into $\mathbb{C}$ columns (where $\mathbb{C}$ is to be determined) of widths $c_1, \ldots, c_{\mathbb{C}}$. Each column C_i is itself partitioned in k_i rows (to be determined as well). The resulting partition of the unit square consists of $\sum_{i=1}^{\mathbb{C}} k_i = p$ rectangles. The goal is to produce such a partition that minimizes the sum of the semiperimeters of the rectangles.

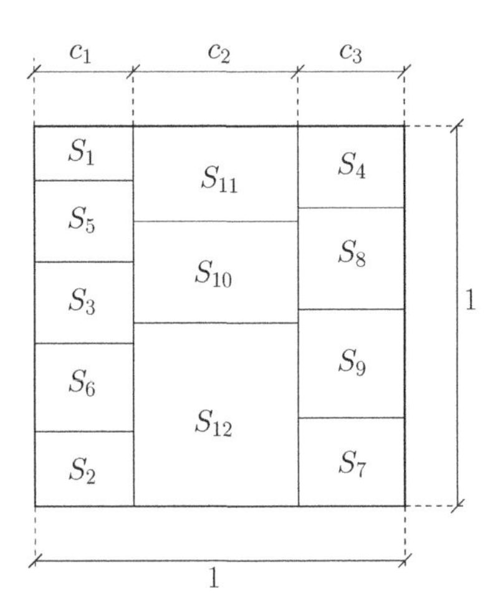

FIGURE 14.7: Column partition of the unit square with $\mathbb{C} = 3$ columns with $k_1 = 5$, $k_2 = 3$, and $k_3 = 4$ rectangles, respectively.

An optimal algorithm to solve this problem uses dynamic programming and relies on the two following ideas:

1. It renumbers variables $s_1, \ldots, s_p$ so that $s_1 \leqslant s_2 \leqslant \cdots \leqslant s_p$. Indeed, we show in the following that we can restrict to sorted sequences.

2. It iteratively constructs p functions $f_{\mathbb{C}}$ for values of $\mathbb{C}$ going from 1 to p. For $q \in \{1, \ldots, p\}$, $f_{\mathbb{C}}(q)$ is defined as the sum of the semiperimeters in an optimal partition of a rectangle with height 1 and width $\sum_{i=1}^{q} s_i$, into $\mathbb{C}$ columns and q rectangles with areas $s_1, \ldots, s_q$.

The key idea behind the algorithm is that it is straightforward to compute function $f_{\mathbb{C}}$ recursively based on function $f_{\mathbb{C}-1}$ as follows:

$$f_{\mathbb{C}}(q) = \min_{a \in [\mathbb{C}-1, q-1]} \left(1 + (q-a) \sum_{a < i \leqslant q} s_i + f_{\mathbb{C}-1}(a) \right). \qquad (14.1)$$

TABLE 14.1: The values of $f_{\mathbb{C}}(q)$ and of a_0 (separated by a "|") for our eight-rectangle example. The values in boldface indicate the optimal solution.

$\mathbb{C}$ \ q	$q=1$	$q=2$	$q=3$	$q=4$	$q=5$	$q=6$	$q=7$	$q=8$
$\mathbb{C}=1$	1.05 | 0	1.2 | 0	**1.54 | 0**	2.12 | 0	2.9 | 0	4 | 0	5.90 | 0	9 | 0
$\mathbb{C}=2$		2.10 | 1	2.28 | 2	2.56 | 2	2.94 | 3	**3.50 | 3**	4.38 | 4	5.76 | 5
$\mathbb{C}=3$			3.18 | 2	3.38 | 3	3.66 | 4	4 | 4	4.58 | 5	**5.50 | 6**
$\mathbb{C}=4$				4.28 | 3	4.48 | 4	4.78 | 5	5.20 | 6	5.88 | 7
$\mathbb{C}=5$					5.38 | 4	5.60 | 5	5.98 | 6	6.50 | 7
$\mathbb{C}=6$						6.50 | 5	6.80 | 6	7.28 | 7
$\mathbb{C}=7$							7.70 | 6	8.10 | 7
$\mathbb{C}=8$								9 | 7

In the above minimum, a corresponds to the number of rectangles in the first $\mathbb{C}-1$ columns, and thus $q-a$ corresponds to the number of rectangles in the last column. The first term in the minimum, $1+(q-a)\sum_{a<i\leqslant q} s_i$, is the contribution of the last column to the total sum of the semiperimeters. The intuitive interpretation of equation (14.1) is that it looks at all possible ways to make use of an additional column to achieve the optimal partition. We denote by a_0 the value of parameter a that achieves the minimum in equation (14.1). This recursion is initialized by setting

$$f_1(q) = 1 + q \times \sum_{i=1}^{q} s_i ,$$

which simply gives the sum of the semiperimeters of a partition of a rectangle with height 1 and width $(\sum_{i=1}^{q} s_i)$ into one column and q rectangles with areas $s_1, \ldots, s_q$.

To understand better how the algorithm works, let us apply it in an example. Consider $p = 8$ rectangles with areas (0.05; 0.05; 0.08; 0.1; 0.1; 0.12; 0.2; 0.3). The algorithm recursively computes all $f_{\mathbb{C}}(q)$ values for $1 \leqslant q, \mathbb{C} \leqslant 8$, as shown in Table 14.1. Because we wish to partition the unit square in eight rectangles, we look at column $q = 8$ and find that the optimal $f_{\mathbb{C}}(q)$ value, 5.5, is achieved for $\mathbb{C} = 3$, indicating that the optimal partition consists of three columns. Furthermore, the optimal $f_{\mathbb{C}}(q)$ value is achieved for $a_0 = 6$. Therefore, the last column of the optimal partition must contain $8 - 6 = 2$ rectangles. We now look at column $q = 6$ of the table and find out that the optimal $f_{\mathbb{C}}(q)$ value is achieved for $a_0 = 3$. Therefore, the next-to-last column of the optimal partition must contain $6 - 3 = 3$ of the remaining rectangles. Similarly, we determine that, in column $q = 3$ of the table, the optimal $f_{\mathbb{C}}(q)$ value is achieved for $a_0 = 0$. The first column of the optimal partitioning consists of all remaining $3 - 0 = 3$ rectangles, which makes sense because we know that the optimal partitioning consists of three columns. The widths of the three columns in the optimal partitioning are thus $c_1 = s_1 + s_2 + s_3$, $c_2 = s_4 + s_5 + s_6$, and $c_3 = s_7 + s_8$. This partitioning is shown in Figure 14.8.

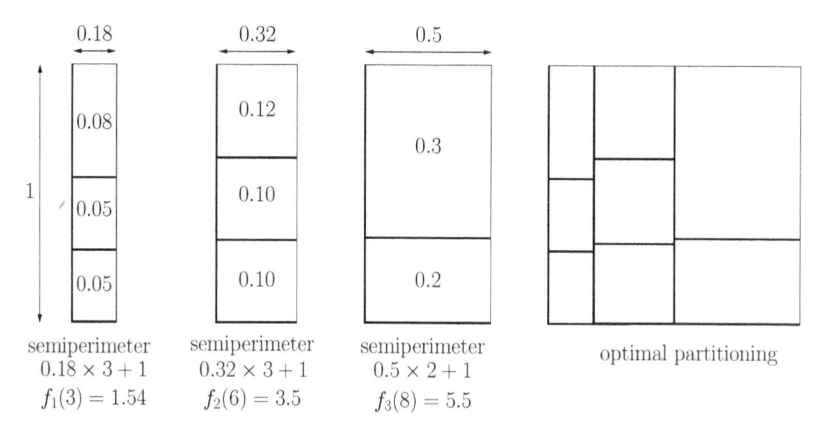

FIGURE 14.8: Optimal column partitioning for our eight-rectangle example. The thicker lines correspond to the rectangle sides that contribute to the sum of the semiperimeters.

Algorithm 14.1 makes it possible to compute the optimal partition with, in the worst case, a complexity of $O(p^3)$. The final partition corresponds to function $f_{\mathbb{C}_{opt}}(p) = \min_{1 \leqslant \mathbb{C} \leqslant p} f_{\mathbb{C}}(p)$. This partition is found by Algorithm 14.2. We saw in an example how this algorithm operates, namely, by reading Table 14.1 from right to left and by following the values in boldface. The unit square is partitioned in $\mathbb{C}_{opt}$ columns. The i-th column contains rectangles $s_{d_i}, s_{d_i+1}, \ldots, s_{d_i+k_i}$ with $d_i = k_1 + k_2 + \cdots + k_{i-1}$.

Algorithms 14.1 and 14.2 together compute the optimal solution of problem ColPeriSum(s), under the assumption that the rectangles are sorted according to their areas. There remains to be shown that one can indeed consider only ordered $s_1 \leqslant s_2 \leqslant \cdots \leqslant s_p$ sequences.

DEFINITION 14.5. A partitioning is said to be *well ordered* if for any pair of columns $\mathbb{C}_i$ and $\mathbb{C}_j$, either all elements of $\mathbb{C}_i$ are smaller than or equal to all elements of $\mathbb{C}_j$, or the other way around. Figure 14.9 illustrates this definition.

Without loss of generality, consider a partitioning into $\mathbb{C}$ columns whose number of rectangles is $k_1 \geqslant k_2 \geqslant \cdots \geqslant k_{\mathbb{C}}$ (swapping columns around does not change the quality of the partitioning). Assume that the s_i values are indexed so that $s_1 \leqslant s_2 \leqslant \cdots \leqslant s_p$; let τ be a permutation of $\{1, 2, \ldots, p\}$ such that the i-th column of the partitioning contains rectangles $s_{\tau(d_i+1)}, \ldots, s_{\tau(d_i+k_i)}$, where $d_i = k_1 + k_2 + \cdots + k_{i-1}$. Recall from equation (14.1) that the contribution of column C_i to the total sum of the

```
1  PARTITION(s_1, ..., s_p)
2  |  S = 0
3  |  for q = 1 to p do
4  |  |  S = S + s_q
5  |  |  f_1(q) = 1 + S × q
6  |  |_ f_1^cut(q) = 0
7  |  for ℂ = 2 to p do
8  |  |  for q = ℂ to p do
9  |  |  |  f_ℂ(q) = min_{a∈[ℂ−1,q−1]} ( 1 + (q − a) Σ_{a<i≤q} s_i + f_{ℂ−1}(a) )
10 |  |  |  f_ℂ^cut(q) = q − a_0    { where a_0 is the value of a that leads to
   |  |  |_     the minimum in the previous expression }
   |  |_
11 |_ return (f_*^cut)
```

ALGORITHM 14.1: Algorithm for COLPERISUM(s): Construction of the functions $f_{\mathbb{C}}$. $f_{\mathbb{C}}^{\text{cut}}(q)$ corresponds to the number of rectangles used in the $\mathbb{C}-1$ first columns, which leaves $q - f_{\mathbb{C}}^{\text{cut}}(q)$ rectangles in column $\mathcal{C}$.

semiperimeters is $1 + k_i \sum_{j=d_i+1}^{d_i+k_i} s_{\tau(j)}$. The total sum is computed as follows:

$$\begin{aligned}&\mathbb{C} + k_1 s_{\tau(1)} + k_1 s_{\tau(2)} + \cdots + k_1 s_{\tau(k_1)}\\&\quad + k_2 s_{\tau(k_1+1)} + k_2 s_{\tau(k_1+2)} + \cdots + k_2 s_{\tau(k_1+k_2)}\\&\quad + \cdots\\&\quad + k_{\mathbb{C}} s_{\tau(k_1+\cdots+k_{\mathbb{C}-1}+1)} + k_{\mathbb{C}} s_{\tau(k_1+\cdots+k_{\mathbb{C}-1}+2)} + \cdots + k_{\mathbb{C}} s_{\tau(k_1+\cdots+k_{\mathbb{C}})}\,.\end{aligned}$$

Because $k_1 \geqslant k_2 \geqslant \cdots \geqslant k_{\mathbb{C}}$, the above expression is minimized when τ is the identity, which characterizes a well-ordered partition. The proof is completed via an induction on the number of element swaps in permutation τ.

```
1  RE-BUILD(f_*^cut, ℂ_opt)
2  |  q = p
3  |  for ℂ = ℂ_opt down to 1 do
4  |  |  k_ℂ = q − f_ℂ^cut(q)
5  |  |_ q = f_ℂ^cut(q)
6  |_ return (k_1, ..., k_{ℂ_opt})
```

ALGORITHM 14.2: Reconstructing the optimal solution from functions $f_{\mathbb{C}}^{\text{cut}}$.

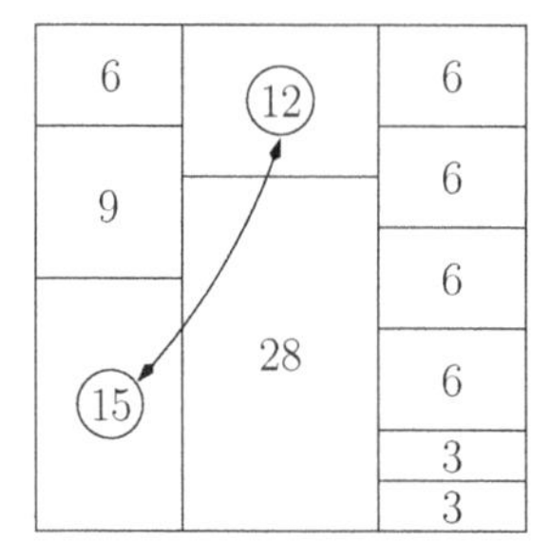

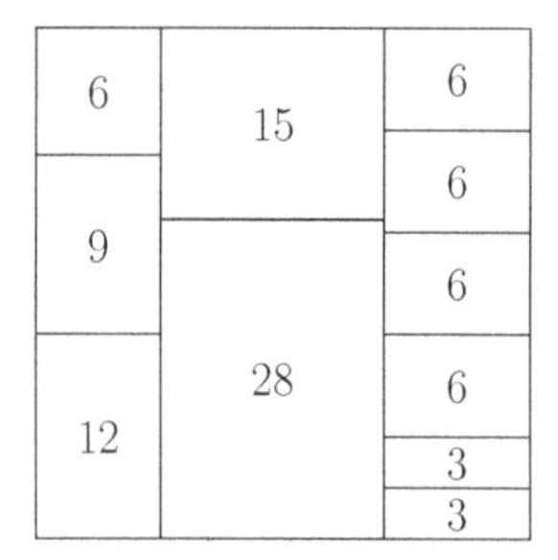

FIGURE 14.9: Two example column partitions. The one on the left is not well ordered, while the one on the right is. For clarity, rectangle areas are shown in percentage of the total area.

We conclude that, for any partition, there is a corresponding well-ordered partition that leads to a total sum of semiperimeters that is lower than or equal to that of the original partition. This completes the proof that Algorithms 14.1 and 14.2 compute the optimal solution to ColPeriSum(s).

14.3.2 Performance guarantee

In this section, we show that column partitioning leads to a good approximation of the optimal (arbitrary) partition. This is especially true when the ratio between the largest rectangle area, $\max s_i$, and the smallest one, $\min s_i$, is low.

THEOREM 14.2. *Let $r = \frac{\max s_i}{\min s_i}$. Let $\widehat{C}$ denote the sum of the semiperimeters of the rectangles in the optimal column partition, and let $LB = 2\sum_{i=1}^{p}\sqrt{s_i}$. Then, we have*

$$\frac{\widehat{C}}{LB} \leqslant \sqrt{r}\left(1+\frac{1}{\sqrt{p}}\right) = \sqrt{\frac{\max s_i}{\min s_i}}\left(1+\frac{1}{\sqrt{p}}\right).$$

Proof. We define $\mathbb{C} = \lceil \sqrt{r}\sum_i \sqrt{s_i} \rceil$. Let us consider the partition that consists of $\mathbb{C}$ columns in which the rectangles are distributed evenly across the columns. Therefore, the number of rectangles per column is either $\lfloor \frac{p}{\mathbb{C}} \rfloor$ or $\lceil \frac{p}{\mathbb{C}} \rceil$. Let $\widehat{C}^*$ denote the sum of the semiperimeters of this partitioning. We, then, have

$$\begin{aligned}\widehat{C}^* &\leqslant \left\lceil \sqrt{r}\sum_i \sqrt{s_i} \right\rceil + \left\lceil \frac{p}{\lceil \sqrt{r}\sum_i \sqrt{s_i} \rceil} \right\rceil \\ &\leqslant 2 + \sqrt{r}\sum_i \sqrt{s_i} + \frac{p}{\sqrt{r}\sum_i \sqrt{s_i}}.\end{aligned}$$

And, therefore,

$$\frac{\widehat{C}^*}{2\sum_i \sqrt{s_i}} \leqslant \frac{1}{\sum_i \sqrt{s_i}} + \frac{\sqrt{r}}{2} + \frac{p}{2\sqrt{r}(\sum_i \sqrt{s_i})^2} .$$

Furthermore,

$$\begin{aligned}\sum_i s_i = 1 &\Longrightarrow p \max s_i \geqslant 1 \\ &\Longrightarrow \min s_i \geqslant \frac{1}{pr},\end{aligned}$$

which leads to

$$\sum_i \sqrt{s_i} \geqslant p\sqrt{\min s_i} \geqslant \sqrt{\frac{p}{r}} .$$

Finally, we obtain

$$\begin{aligned}\frac{\widehat{C}^*}{2\sum_i \sqrt{s_i}} &\leqslant \sqrt{\frac{r}{p}} + \frac{\sqrt{r}}{2} + \frac{\sqrt{r}}{2} \\ &\leqslant \sqrt{r}\left(1 + \frac{1}{\sqrt{p}}\right) .\end{aligned}$$

Because $\widehat{C}$ corresponds to the best solution among all the possible column partitionings, we have $\widehat{C} \leqslant \widehat{C}^*$, which completes the proof. □

If $r = 1$, i.e., if all rectangles have the same area, column partitioning is asymptotically optimal. By contrast, if r is large, then the upper bound of Theorem 14.2 is very pessimistic.

14.3.3 Looking for a better solution

The column-based solution may not be very satisfactory in some "degenerate" artificial cases. To illustrate the point, consider the following partitioning problem into $p = 6$ rectangles of respective areas

$$(0.2488, 0.2488, 0.2488, 0.2488, 0.0024, 0.0024).$$

The absolute lower bound for this example is $LB = 2\sum_{i=1}^{6} \sqrt{s_i} = 4.19$. Consider the following solutions, which have different degrees of freedom:

1. The partitioning is constrained to be a column-based partitioning. Using the column-based algorithm, we obtain the solution depicted in Figure 14.10.

2. The partitioning is constrained to be recursively defined as follows. The unit square is divided into several columns. Each column, in turn, is divided into several rows, and so on. Of course, there are multiple choices for the number of columns, and for the number of rows within each column, and so on. In Figure 14.11, we give an example with two columns divided into two and three rows, respectively. Finally, the last row of the second column is split into two rectangles. In the example, this partitioning is optimal among recursively defined partitionings (proof by exhaustive case analysis). Note that it is shown in [9] that one can design an approximation algorithm that builds a recursive partitioning and that is a 7/4 approximation to PERISUM(s), regardless of the relative sizes of the rectangles.

3. The partitioning is constrained to be made out only of rectangles. An example is given in Figure 14.12. Note that this solution is neither column-based nor recursively defined. This partitioning is optimal among all rectangle-based partitionings.

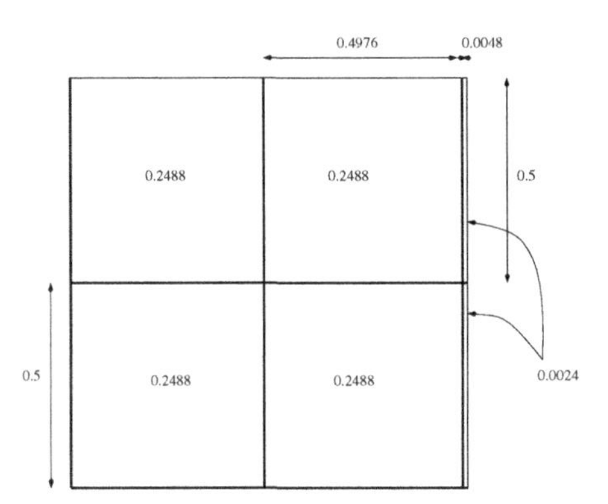

FIGURE 14.10: Optimal column-based partitioning. The sum of the semiperimeters is 5.

Clearly, the less constrained the partitioning, the better the solution. Note that the improvement over the column-based partitioning is not negligible here, roughly 16% for the rectangle-based solution. However, going back to the original motivation of the problem in terms of matrix product with different-speed processors, we point out that the rectangle areas correspond to six processors whose relative cycle-times are approximately $(1, 1, 1, 1, 100, 100)$. In a realistic experiment, we would never use a processor in conjunction with another one that is 100 times faster; this is why the previous example may be called *artificial*.

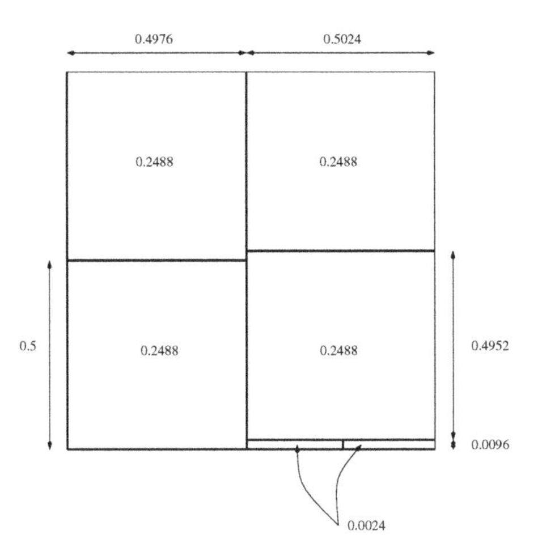

FIGURE 14.11: Optimal recursively defined partitioning. The sum of the semiperimeters is 4.51.

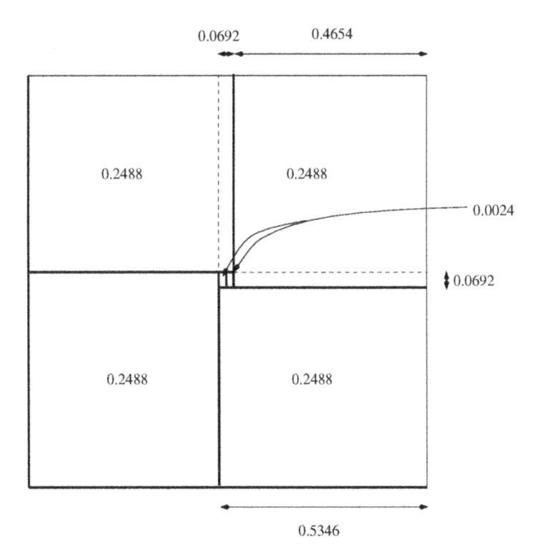

FIGURE 14.12: Optimal rectangle-based partitioning. The sum of the semiperimeters is 4.21.

14.4 Related problems

To conclude this case study, we outline several problems related to PeriSum(s) that have been considered in the literature:

- The most similar problem is the following: How do we tile the unit square into p rectangles of same area so as to minimize the maximum perimeter of these rectangles? This problem is shown to be polynomial by Kong et al. [63, 64]. The optimal solution is one of the following two arrangements: let either $m = \lfloor \sqrt{p} \rfloor$ or $m = \lceil \sqrt{p} \rceil$, and use m columns composed of $\lfloor \frac{p}{m} \rfloor$ or $\lceil \frac{p}{m} \rceil$ rectangles. This problem is motivated by a data-allocation problem, which is related to our matrix-product formulation in the following sense: Assume that we have p equal-speed processors and that we aim at minimizing the largest amount of communications made by one processor (as opposed to the total volume of communications in our original problem).

  The heterogeneous counterpart of this problem is how do we tile the unit square into p nonoverlapping rectangles of prescribed areas $s_1, \ldots, s_p$ whose sum is 1, so that the largest perimeter is minimized. Or, in terms of parallel algorithms (given p different-speed processors), how to allocate data so that the length of the largest communication is optimized? This interesting problem is NP-complete, too [9], which again shows the intrinsic difficulty of designing heterogeneous parallel algorithms.

- Another related problem is to find the minimum partition of a rectangle with interior points. Given a rectangle R and a finite set P of points located inside R, find a set of line segments that partition R into rectangles such that every point in P is on the boundary of some rectangle. The goal is to minimize the total length of the introduced line segments. This problem is shown NP-complete in [41, 42, 73], where approximation algorithms are given. The link with our problem is that the objective function is the same, but the original motivation in [41, 42] was a VLSI (very-large-scale integration) routing problem (and the constraints are quite different).

- Several other NP-complete geometric optimization problems are listed in [28]. One example is the minimum rectangle tiling problem [61]: Given an $n \times n$ array A of nonnegative numbers and a positive integer p, find a partition of A into p nonoverlapping rectangular subarrays, such that the maximum weight of any rectangle in the partition is minimized (the weight of a rectangle is the sum of its elements).

Chapter 15

Online scheduling

In all the other case studies (Chapters 11 to 14), we are given full knowledge of a problem instance, and we are interested in finding the best possible solution for that particular instance. In this chapter, however, we are focusing on online problems, i.e., on problems for which algorithmic decisions must be taken before all the characteristics of the whole instance are known. More specifically, we are considering online scheduling problems. Several *tasks* are submitted to a system that must process them. The scheduling system must decide at what time to allocate which resource to which task. The problem being *online* by nature, the system does not know beforehand how many tasks it will have to process, when they will arrive in the system, and what their characteristics will be. Hence, the scheduling system will have to make decisions with only a partial knowledge of the instance at hand. One would think that in such a context it is impossible to design optimal algorithms. However, some scheduling problems are still simple enough to enable the existence of algorithms making optimal decisions at any step of any online scenario. For more complex problems, we need a way to assess the quality of the solution produced by an online algorithm, independently of the scenario. This is the purpose of *competitive analysis*.

In Section 15.1, we show that, for some settings and objective functions, there exist optimal online algorithms. Since this is obviously not always the case, this section also motivates the need for a way to assess the performance of online algorithms. We introduce competitive analysis in Section 15.2 and show how one can establish a competitive result for an algorithm, or a bound on the performance of any algorithm. Finally, in Section 15.3, we change the framework and the target objective function. This gives us the opportunity to introduce a randomized algorithm that achieves a better performance than any deterministic algorithm could provide.

We conclude this introduction by mentioning that the online scheduling problems addressed in this chapter have little in common with the task graph scheduling problems presented in Section 6.4.4, p. 140: There are no precedence constraints, the number of tasks is unknown, and the objective function is not always makespan minimization (recall that the makespan of a schedule is its total execution time). Still, we are playing a *scheduling* game that assigns tasks to processors.

15.1 Flow time optimization

Consider a series of n tasks, $T_1, \ldots, T_n$, submitted online to a system comprising a single processor. Neither the value of n nor the release dates of these n tasks are known beforehand. Furthermore, the execution time w_j of task T_j is known only at the time when T_j arrives in the system. We denote by r_j the release date of task T_j, that is, the date at which it is submitted to the system. We denote by C_j its completion time, that is, the date at which its processing is completed. The flow time F_j (or response time) of a task T_j is the time spent in the system by the task, between its release date and its completion time: $F_j = C_j - r_j$. The first objective that we consider is the minimization of the maximum flow time: $\max_{1 \leqslant i \leqslant n} F_i$. Even in an online setting, this problem can be optimally solved:

THEOREM 15.1. *First Come First Served (FCFS) optimally solves the online minimization of the maximum-flow.*

Note that, by definition, FCFS does not use preemption (for a definition of preemption, see Exercise 9.1, p. 211). In other words, under FCFS, the processing of a task is never interrupted to be resumed after the processing of another task. This theorem is proved below for settings without preemption, but it remains valid for settings taking preemption into account (see [71]).

Proof. The proof follows a simple exchange argument illustrated in Figure 15.1. We consider any optimal schedule $\mathcal{S}$. We assume that the optimal solution differs from that of FCFS. Thus, there exist two tasks T_i and T_k such that:

- T_k is executed right after T_i (and thus $C_i < C_k$);
- T_i was released after T_k: $r_i > r_k$.

In this optimal schedule, looking at flow times, we have: $F_k = C_k - r_k > C_i - r_k > C_i - r_i$. (Note that two tasks sharing the same release date can be executed in any order. Therefore, we do not need to consider that case.)

We build from $\mathcal{S}$ a new schedule $\mathcal{S}'$ identical to $\mathcal{S}$ except that the executions of T_i and T_k are interchanged. Figure 15.1 illustrates these two schedules. Since the executions of T_i and T_k were consecutive under schedule $\mathcal{S}$, and since T_k was released earlier than T_i, the execution of T_k can start under $\mathcal{S}'$ at the time the execution of T_i started under $\mathcal{S}$. Furthermore, the execution times of the other tasks are not modified. Therefore, T_i and T_k are the only tasks whose flow times are modified by the transformation of the schedule. Their new flow times are

$$\begin{cases} F'_i = C'_i - r_i = C_k - r_i < C_k - r_k = F_k \\ F'_k = C'_k - r_k < C_k - r_k = F_k \end{cases} .$$

Therefore, $\max\{F'_i, F'_k\} < F_k \leq \max\{F_i, F_k\}$ and the maximum flow of $\mathcal{S}'$ is not greater than that of $\mathcal{S}$. One can thus transform any optimal schedule into

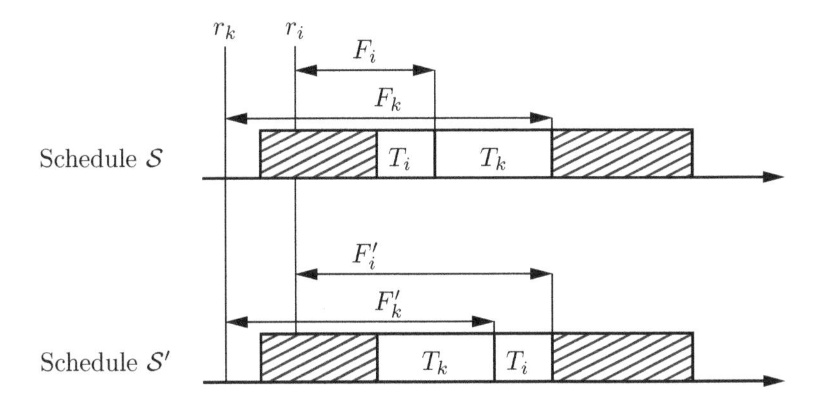

FIGURE 15.1: Exchange argument used in the proof of Theorem 15.1.

the FCFS schedule by the repeated application of this exchange argument, first to reduce the number of tasks executed before the first submitted task, then to reduce the number of tasks executed before the second submitted task, and so on. This proves that the maximum flow time of the FCFS schedule is not greater than the optimal maximum flow time. □

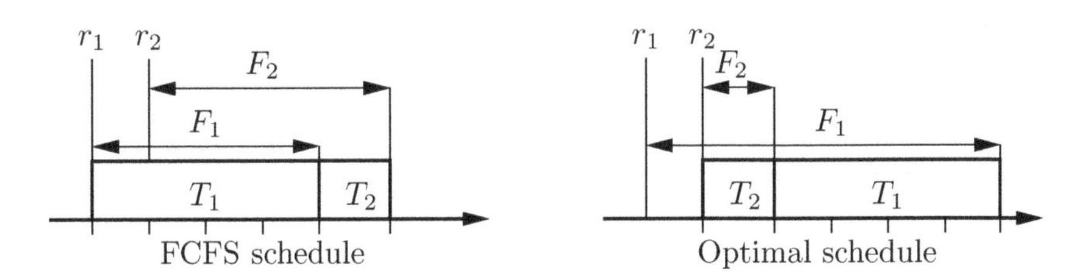

FIGURE 15.2: Example showing the suboptimality of FCFS for the minimization of the sum of flow times.

Thus, FCFS is optimal for the minimization of the maximum flow time. Another important optimization objective is the sum of flow times, namely, $\sum_{i=1}^{n} F_i$, which corresponds to the average response time for each task (dividing by the constant n does not change the optimization problem). FCFS is not optimal, however, for the minimization of the sum of flow times. Figure 15.2 shows a counterexample with two tasks: The first task is released at time 0 and has an execution time of 4; the second task is released at time 1 and its execution time is 1. FCFS achieves a sum of flow times of 8 in this example when the optimal value is 7 (when preemption is not allowed). One can easily check that FCFS is optimal for variants of the counterexample where the second task is not released before time 4, or where the execution time of the second task is at least equal to 2. Therefore, for some instances, FCFS is suboptimal, and, for some other instances, it is optimal. We would like to have a way to assess the performance of FCFS for this objective function, knowing that we are in an online setting. To achieve this purpose, the notion of competitive analysis has been introduced by Sleator and Tarjan [98].

15.2 Competitive analysis

15.2.1 Definition

In competitive analysis, the performance of an *online* algorithm $\mathcal{A}$ for a given objective function is compared to the performance of the optimal *offline* algorithm, that is, to the performance of the optimal solution when all of the instance characteristics are known beforehand.

DEFINITION 15.1 (Competitive algorithm). Let $\mathcal{A}$ be an online algorithm for a minimization problem. For any instance $\mathcal{I}$, let $C^{\mathcal{A}}(\mathcal{I})$ denote the value of the objective function for algorithm $\mathcal{A}$, and $C^{\text{OPT}}(\mathcal{I})$ that of the optimal (offline) solution. Then, algorithm $\mathcal{A}$ is said to be ϱ-competitive if there exist two constants $\varrho \geqslant 1$ and c such that, for any instance $\mathcal{I}$,

$$C^{\mathcal{A}}(\mathcal{I}) \quad \leqslant \quad \varrho \times C^{\text{OPT}}(\mathcal{I}) + c.$$

(To adapt this definition to a maximization problem, one needs only to change the inequality sign and to replace ϱ with $1/\varrho$ in the condition.) We have given here the most general definition of a competitive algorithm. This definition contains an "offset" c in addition to the competitive ratio ϱ. However, in all the competitive results presented in this chapter, this offset will be zero.

The notion of competitive algorithm can be seen as an extension of the notion of approximation algorithm (Section 8.1) for online settings. An important difference is that the complexity of an algorithm does not play any role in the definition of a competitive algorithm, contrarily to the definition of an approximation algorithm (cf. Definition 8.1, p. 180).

We now establish a competitive result for FCFS when the objective is the minimization of the sum of flow times. In the next section, we will show how one can elaborate and derive such a result.

THEOREM 15.2 ([71]). *FCFS is Δ-competitive for the online minimization of the sum of flow times, where Δ is the ratio of the largest task size to the smallest one. Furthermore, this bound is tight.*

Proof. We first show that FCFS is Δ competitive, then we show that this bound is tight. In this proof, $\mathcal{F}^{\mathcal{A}}(\mathcal{I})$ denotes the sum of flow times achieved by the schedule $\mathcal{A}$ on instance $\mathcal{I}$. $\mathcal{F}^{\text{OPT}}(\mathcal{I})$ is the optimal sum of flow times for instance $\mathcal{I}$.

Let $\mathcal{I} = \{T_1 = (r_1, w_1), \dots, T_n = (r_n, w_n)\}$ be any instance (recall that r_i is the release date of task T_i and w_i its execution time). We show by induction on n that $\mathcal{F}^{\text{FCFS}}(\mathcal{I}) \leqslant \Delta\mathcal{F}^{\text{OPT}}(\mathcal{I})$. This property obviously holds for $n = 1$. Let us assume that it has been proved for n and prove that it holds true for $n + 1$.

Consider $\mathcal{I} = \{T_1 = (r_1, w_1), \dots, T_{n+1} = (r_{n+1}, w_{n+1})\}$ an instance with $n+1$ tasks. Without loss of generality, we assume that $\min_j w_j = 1$. Therefore, Δ is the size of the largest task in instance $\mathcal{I}$. Without loss of generality, we can restrict the study to schedules such that the processor is never deliberately left idle. Furthermore, we also can restrict the study to schedules for which, at any time, the task executed is, among all of the available tasks, the task that completes at the earliest under that schedule. Indeed, transforming a schedule so that it meets this assumption can only decrease its sum of flow times. Under these assumptions, a schedule is completely defined by the order of task completions. We call this order a *priority list*. Thus, let Θ denote an optimal priority list for $\mathcal{I}$; the corresponding schedule is an optimal schedule. In the following, we denote by A_1 the set of tasks that have a higher priority than T_{n+1} and A_2 the set of tasks that have a lower priority than T_{n+1} under Θ. $t^{\text{OPT}}(k)$ denotes the remaining processing time of T_k at time r_{n+1} in the optimal schedule. This remaining processing time is equal to: i) 0 if task T_k was completed at the latest at time r_{n+1}, ii) w_k if the processing of T_k has not started by time r_{n+1}, and, otherwise, iii) w_k minus the time elapsed since the processing of T_k started. We can see that the execution of the $n+1$ tasks of $\mathcal{I}$ with the priority list Θ as the scheduling of the n first tasks of $\mathcal{I}$ in which task T_{n+1} had been inserted. Thus, we have:

$$\mathcal{F}^{\text{OPT}}(T_1, \dots, T_{n+1}) = \mathcal{F}^{\text{OPT}}(T_1, \dots, T_n) + \underbrace{w_{n+1} + \sum_{k \in A_1} t^{\text{OPT}}(k)}_{\text{Flow of } T_{n+1}} + \underbrace{\sum_{k \in A_2} w_{n+1}}_{\substack{\text{Cost induced} \\ \text{by } T_{n+1}}} .$$

We also have (using the recurrence hypothesis to establish the first inequality):

$$\begin{aligned} \mathcal{F}^{\text{FCFS}}(T_1, \dots, T_{n+1}) &= \mathcal{F}^{\text{FCFS}}(T_1, \dots, T_n) + \underbrace{w_{n+1} + \sum_{k \leqslant n} t^{\text{FCFS}}(k)}_{\text{Flow of } T_{n+1}} \\ &\leqslant \Delta\mathcal{F}^{\text{OPT}}(T_1, \dots, T_n) + w_{n+1} + \sum_{k \leqslant n} t^{\text{FCFS}}(k) \\ &= \Delta\mathcal{F}^{\text{OPT}}(T_1, \dots, T_n) + w_{n+1} + \sum_{k \leqslant n} t^{\text{OPT}}(k). \end{aligned}$$

The last equality is established by focusing at any given time t on the sum of the task remaining processing times under any priority-based schedule. Let

τ be the last date prior to t at which the processor was idle. Then, the sum of the task remaining processing times is equal to the sum of the execution times of the tasks released between τ and t minus $t - \tau$, that is, minus the amount of processing that was performed during that time interval. This is independent of the priority list. Therefore, this sum is the same under FCFS and under an optimal priority-based schedule. Now, since we have

$t^{\text{OPT}}(k) \leqslant \Delta \leqslant \Delta w_{n+1}$ and $\sum_{k \leqslant n} t^{\text{OPT}}(k) = \sum_{k \in A_1} t^{\text{OPT}}(k) + \sum_{k \in A_2} t^{\text{OPT}}(k)$, we get:

$$\begin{aligned}\mathcal{F}^{\text{FCFS}}(T_1, ..., T_{n+1}) &\leqslant \Delta \mathcal{F}^{\text{OPT}}(T_1, ..., T_n) + w_{n+1} + \sum_{k \in A_1} t^{\text{OPT}}(k) + \sum_{k \in A_2} \Delta w_{n+1} \\ &\leqslant \Delta \mathcal{F}^{\text{OPT}}(T_1, ..., T_n) + \Delta w_{n+1} + \Delta \sum_{k \in A_1} t^{\text{OPT}}(k) + \Delta \sum_{k \in A_2} w_{n+1} \\ &\leqslant \Delta \mathcal{F}^{\text{OPT}}(T_1, ..., T_{n+1}).\end{aligned}$$

We now prove that the previous bound is tight. Let us consider the following example: n tasks $T_1, \ldots, T_n$ of size Δ arrive at time 0, and, then, for $1 \leqslant j \leqslant n^2$, task T_{n+j} of size 1 arrives at time $j - 1 + \frac{1}{n}$.

A possible schedule would be to process each of the n^2 tasks $T_{n+j}, \ldots, T_{n+n^2}$ at their release dates and to wait for the completion of the last of these tasks before completing tasks $T_1, \ldots, T_n$. The sum of flow times would then be $n^2 \times 1 + (n^2 + \Delta + \frac{1}{n}) + (n^2 + 2\Delta + \frac{1}{n}) + \cdots + (n^2 + n\Delta + \frac{1}{n}) = n^3 + n^2 + \frac{n(n+1)}{2}\Delta + 1$. Therefore, the optimal sum of flow times is not greater than this value, and this value can be used as an upper bound (we do not need to prove that it is the optimal value).

Rather, FCFS schedules first the tasks of size Δ, and the sum of flow times achieved by FCFS on this instance is

$$\Delta + 2\Delta + \cdots + n\Delta + n^2\left(1 + n\Delta - \frac{1}{n}\right) = \frac{2n^3\Delta + n^2(2+\Delta) + n(\Delta - 2)}{2}.$$

Therefore, the competitive ratio $\varrho(n)$ of FCFS on this instance satisfies:

$$\varrho(n) \geqslant \nu(n) = \frac{\frac{2n^3\Delta + n^2(2+\Delta) + n(\Delta-2)}{2}}{n^3 + n^2 + \frac{n(n+1)}{2}\Delta + 1} = \frac{2n^3\Delta + n^2(2+\Delta) + n(\Delta - 2)}{2n^3 + 2n^2 + n(n+1)\Delta + 2}.$$

To conclude, we only have to remark that $\lim_{n \to +\infty} \nu(n) = \Delta$. $\square$

It is important to see that different objective functions that share the same optimal solutions may lead to different competitive ratios for nonoptimal solutions. Take, for instance, the sum of flow times $\sum_{i=1}^{n} F_i = \sum_{i=1}^{n}(C_i - r_i)$ and the sum of completion times $\sum_{i=1}^{n} C_i$. The two objectives differ by the constant (for a given instance) term $\sum_{i=1}^{n} r_i$. Therefore, any schedule optimal for the sum of flow times is also optimal for the sum of completion times, and reciprocally. Furthermore, a scheduling algorithm Θ that is ϱ-competitive for the sum of flow times is also ϱ-competitive for the sum of completion times. This is easily proved (remember that, by definition, $\varrho \geqslant 1$):

$$\begin{aligned}
& \sum_{i=1}^{n}(C_i^{\Theta} - r_i) \leqslant \varrho \sum_{i=1}^{n}(C_i^{\text{OPT}} - r_i) \\
\Rightarrow \quad & \sum_{i=1}^{n} C_i^{\Theta} \leqslant \varrho \left(\sum_{i=1}^{n} C_i^{\text{OPT}} \right) - (\varrho - 1) \sum_{i=1}^{n} r_i \\
\Rightarrow \quad & \sum_{i=1}^{n} C_i^{\Theta} \leqslant \varrho \left(\sum_{i=1}^{n} C_i^{\text{OPT}} \right).
\end{aligned}$$

The reverse, however, is not true. A competitive result on the sum of completion times does not imply the same result for the sum of flow times.

Finally, an obvious remark: Not all online algorithms have competitive ratios. For instance, take the Shortest Processing Time First algorithm. This algorithm leads to unbounded maximum flows and, therefore, is not a competitive algorithm for maximum flow minimization. To establish this result one needs only to consider an instance made of a task of size 2 released at time 0, and n tasks of size 1, the i-th task, $1 \leqslant i \leqslant n$ being released at time $i - 1$. The optimal maximum flow is equal to 3. It is obtained by scheduling first the task of size 2 and then the other tasks in the FCFS order. Shortest Processing Time First achieves a maximum flow of $n + 2$.

15.2.2 Method to establish a competitive analysis result

We now explain how to establish a bound on the competitiveness of *any* online algorithm. The idea is to start from a problematic instance. A problematic instance, for an online scheduler, is an instance in which the optimal choice for the first task, or for a handful of first tasks, depends on the potentially following tasks. Therefore, depending on whether the first task(s) is (are) followed by some other tasks, schedulers should not make the same decision.

Starting from a problematic instance, one generalizes it. One, thus, replaces the instance characteristics with parameters. The instance characteristics are the characteristics of tasks and of the processing platform. One, then, studies all possible responses by schedulers, and one adds constraints on the instance characteristics so as to force schedulers to make bad decisions. One then expresses the competitive ratio of any algorithm as a function of the instance characteristics. Finally, one looks for a set of instance characteristics that maximizes this bound on the competitive ratio.

We now illustrate this method with the scheduling of identical tasks with realease dates on a heterogeneous master-worker platform. The processing platform, illustrated in Figure 15.3, is composed of a master that sends tasks to workers that are then in charge of processing them. The master cannot send several tasks simultaneously, and each task is sent to a single worker. It takes a time c_i for the master to send a task to processor P_i. Worker P_i must wait to have completely received a task from the master before it can start processing it. This processing takes a time w_i. Each worker must process tasks one at a time and wait for the completion of the processing of one task before starting the processing of the next one. Different processors, however, can simultaneously be processing different tasks. The platform is said to be *heterogeneous* because the communication and processing times are worker dependent. The objective function is then to minimize the maximum flow time, where the flow time of a task is defined as before as the completion time minus the release date.

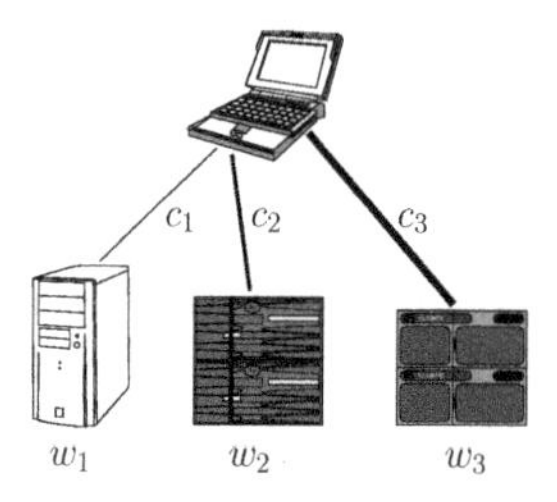

FIGURE 15.3: Example of a heterogeneous master-worker platform.

Which is the most interesting processor to process a given task? The answer depends on when tasks are submitted to the system (remember, all tasks share the same characteristics in our problem). To establish the result stated in Theorem 15.3, we build an instance in which an optimal (online) schedule executes the first submitted task on the processor that maximizes its flow time (instead of minimizing it!).

THEOREM 15.3 ([88]). *There is no online algorithm for scheduling identical tasks on a master-worker platform that achieves a competitive ratio strictly lower than $\sqrt{2}$ for the minimization of the maximum flow time.*

A weaker result is that no online algorithm is optimal for this problem. To establish such a result, we must prove that, whatever the online algorithm, there exists a scenario for which the produced schedule is suboptimal. In order to prove a result such as the one stated in Theorem 15.3, one builds a scenario that provokes the failure of any online algorithm. Hopefully, with this scenario comes a lower bound $\varrho > 1$ on the performance of any online algorithm. We will tune the value of the parameters defining the scenario in order to maximize ϱ.

In this proof, we use the *adversary* technique. Each time the scheduling algorithm makes a decision (or decides to delay making a decision), the adversary decides whether the instance will contain any more tasks and when they will be released. Each time the adversary defines the "next part" of the instance, its aim is for the last decision by the algorithm to have the worst possible consequences on the objective function.

Proof. Since we want to show how to establish a theorem such as Theorem 15.3, we assume that we want to establish a lower bound $\varrho > 1$ on the competitive ratio, but we do not know yet that $\varrho = \sqrt{2}$. We prove Theorem 15.3 by contradiction. Let us assume that there exists a deterministic online algorithm $\mathcal{A}$ whose competitive ratio is $\varrho - \varepsilon$, with $\varepsilon > 0$. We build a platform and an adversary to derive a contradiction. The platform is made up of three processors P_1, P_2, and P_3. The idea is to have one processor very slow at communicating but very quick at computing (P_1) and two other processors that are identical, quick at communicating, slow at processing, and slower than P_1 for the overall processing (communication + computation) of a single task. The idea is that:

- If a single task is submitted to the platform, the optimal solution is to schedule it on P_1.
- If several tasks are submitted to the platform, the first one should not be assigned to P_1. This is because the duration of the communication induced by the first task is potentially impacting all tasks (remember that a single communication can take place at any time) while computations can be run in parallel.

This is the structure of our problematic instance. Formally, we have the properties:

- $c_2 = c_3$ and $w_2 = w_3$ (P_2 and P_3 are identical);
- $c_1 > c_2$ (P_1 communicates slowly);
- $w_1 < w_2$ (P_1 computes quickly);
- $c_1 + w_1 < c_2 + w_2$ (P_1 is the optimal choice if only one task is submitted to the platform).

Initially, the adversary sends a single task i at time 0. $\mathcal{A}$ can execute the task i on:

- P_1. The maximum flow time is then at least equal to $c_1 + w_1$.
  Note that we wrote "at least" because nothing forbids the scheduler to wait some time before initiating the communication from the master to worker P_1. Also, nothing forbids the scheduler to impose some idle time on P_1 between the end of the reception of the first task and the beginning of its processing. We are trying to establish a bound on the performance of *any* algorithm. Therefore, almost *anything* can happen, however inefficient it may look, and our study must take that into account.
- P_2 or P_3. The maximum flow time is then at least equal to $c_2 + w_2 = c_3 + w_3$.

At time τ—the value of τ is also a parameter that will be defined later—we check whether $\mathcal{A}$ has made a decision concerning the scheduling of i, and which one. We want any online algorithm to make the wrong decision, that is, to send task i right away to worker P_1. Thanks to our hypotheses, this is the optimal behavior when a single task is submitted to the system. Let us assume that $\mathcal{A}$ does not send the first task to P_1, or at least not before time τ. Then the adversary does not send any more tasks. The behavior of $\mathcal{A}$ is then obviously suboptimal. We need only to obtain a contradiction with the assumption on $\mathcal{A}$ (namely, that $\mathcal{A}$ has a competitive ratio of at most $\varrho - \varepsilon$). So, we consider all the possible behaviors for $\mathcal{A}$ and add the corresponding constraints on ϱ, that is, we add the constraint that ϱ is smaller than the smallest competitive ratio that $\mathcal{A}$ can obtain on these instances. This will give us a contradiction because the smallest competitive ratio that $\mathcal{A}$ can obtain on these instances must be, by hypothesis, not greater than $\varrho - \varepsilon$. We have two cases to consider:

1. $\mathcal{A}$ scheduled the task i on P_2 or P_3. Then, the best possible maximum flow time is $c_2 + w_2 = c_3 + w_3$. The optimal scheduling has a maximum flow time of $c_1 + w_1$. Hence, the competitive ratio of $\mathcal{A}$, in this case, is at least equal to $\frac{c_2+w_2}{c_1+w_1}$. Since we have assumed that $\mathcal{A}$ has a competitive ratio of $\varrho - \varepsilon$, this implies:

$$\varrho - \varepsilon \geqslant \frac{c_2 + w_2}{c_1 + w_1}. \tag{15.1}$$

   Since we want to forbid the scheduling of task i on P_2 or P_3, we add the weakest possible constraint on ϱ that contradicts equation (15.1). Therefore, we decide that the lower bound ϱ on the competitive ratio of any algorithm must satisfy:

$$\varrho \leqslant \frac{c_2 + w_2}{c_1 + w_1}. \tag{15.2}$$

2. $\mathcal{A}$ did not begin to send task i. Then, the best maximum flow time that can be achieved is equal to $\tau + c_1 + w_1$. The competitive ratio of $\mathcal{A}$ is thus at least equal to $\frac{\tau+c_1+w_1}{c_1+w_1}$ and

$$\varrho - \varepsilon \geqslant \frac{\tau + c_1 + w_1}{c_1 + w_1}. \tag{15.3}$$

   To forbid this case from happening, we proceed as for the previous case and add the minimal constraint on ϱ that contradicts equation (15.3), that is:

$$\varrho \leqslant \frac{\tau + c_1 + w_1}{c_1 + w_1} = 1 + \frac{\tau}{c_1 + w_1}. \tag{15.4}$$

With the two additional constraints on the value of ϱ, equations (15.2) and (15.4), algorithm $\mathcal{A}$ has no choice but to send task i to processor P_1. Indeed, because of equations (15.2) and (15.4), any other decision would contradict the hypothesis on the competitive ratio of algorithm $\mathcal{A}$. Then, the most favorable case for algorithm $\mathcal{A}$ is to send it as soon as possible, that is, as soon as it is submitted to the system at time 0.

Now, at time τ, the adversary sends two tasks, j and k. We consider all the scheduling possibilities:

1. j and k are scheduled on P_1. Then, the best achievable maximum flow time is the maximum of the flows of tasks i, j, and k:
   - Flow of i: $c_1 + w_1$.
   - Flow of j: Task j can be sent neither before it is released (at time τ) nor before the communication link from the master is freed from the sending of task i (at time c_1). Hence, task j is sent at the earliest at time $\max\{c_1, \tau\}$. Then, the processing of j cannot start before the end of its communication or the end of the previous computation on P_1, that is, at time: $\max\{\max\{c_1, \tau\} + c_1, c_1 + w_1\} = c_1 + \max\{c_1, \tau, w_1\}$. Therefore, the flow of j is at least equal to: $c_1 + w_1 + \max\{c_1, \tau, w_1\} - \tau$ as task j was released at time τ.
   - Flow of k: Task k is sent at the earliest at the time the communication of task j ends, that is, at time $\max\{c_1, \tau\} + c_1$. (Since tasks j and k are identical, we can assume without loss of generality that task j is sent before task k.) The processing of task k can start neither before the end of its communication nor earlier than the completion time of task j. Hence, since task k was released at time τ, its flow is at least equal to: $\max\{\max\{c_1, \tau\} + 2c_1, c_1 + w_1 + \max\{c_1, \tau, w_1\}\} + w_1 - \tau = c_1 + \max\{\max\{c_1, \tau\} + c_1, w_1 + \max\{c_1, \tau, w_1\}\} + w_1 - \tau$.

   Since the flow of task k is not smaller than that of task j, which is not smaller than that of task i, the maximum flow in this case is at least equal to: $c_1 + w_1 + \max\{\max\{c_1, \tau\} + c_1, w_1 + \max\{c_1, \tau, w_1\}\} - \tau$.

   We derive the lower bounds on the maximum flow time of the other cases along the same lines.

2. The first of the two tasks (j and k) is scheduled on P_2 (or P_3) and the other one on P_1.
   - Without loss of generality, we assume task j is scheduled on P_2. Then, its flow is at least: $\max\{c_1, \tau\} + c_2 + w_2 - \tau$.
   - The flow of task k is then at least: $\max\{c_1 + w_1, \max\{c_1, \tau\} + c_2 + c_1\} + w_1 - \tau = c_1 + \max\{w_1, \max\{c_1, \tau\} + c_2\} + w_1 - \tau$.

   Then, the best achievable maximum flow time is

$$\max\{\max\{c_1, \tau\} + c_2 + w_2, c_1 + w_1 + \max\{w_1, \max\{c_1, \tau\} + c_2\}\} - \tau.$$

3. The first of the two tasks (j and k) is scheduled on P_1 (say j) and the other one on P_2 (or P_3).
   - The flow of j is at least: $\max\{c_1, \tau, w_1\} + c_1 + w_1 - \tau$ (cf. case 1).
   - The flow of k is at least: $\max\{c_1, \tau\} + c_1 + c_2 + w_2 - \tau$.

Since, by hypothesis, $c_1 + w_1 < c_2 + w_2$, then $w_1 \leqslant c_2 + w_2$ and the best achievable maximum flow time is: $c_1 + \max\{2w_1, \max\{c_1, \tau\} + c_2 + w_2\} - \tau$.

4. One of the two tasks (j and k) is scheduled on P_2 and the other one on P_3.
   - The flow of j is at least: $\max\{c_1, \tau\} + c_2 + w_2 - \tau$.
   - The flow of k is at least: $\max\{c_1, \tau\} + 2c_2 + w_2 - \tau$ (as $c_3 = c_2$ and $w_3 = w_2$).

   The best achievable maximum flow time is then: $\max\{c_1, \tau\} + 2c_2 + w_2 - \tau$.

5. The case where j and k are both executed on P_2, or both on P_3, leads to an even worse maximum flow time than the previous case. Therefore, we do not need to study it. Indeed, we are interested in the best achievable maximum flow time, and this case never leads to it.

Hence, whatever the choices made by algorithm $\mathcal{A}$, its best achievable maximum flow time is the minimum of the maximum flow time of cases 1 through 5, which we denote μ:

$$\mu = \min \begin{cases} c_1 + w_1 + \max\{\max\{c_1, \tau\} + c_1, w_1 + \max\{c_1, \tau, w_1\}\} - \tau, \\ \max \begin{Bmatrix} \max\{c_1, \tau\} + c_2 + w_2, \\ c_1 + w_1 + \max\{w_1, \max\{c_1, \tau\} + c_2\} \end{Bmatrix} - \tau, \\ c_1 + \max\{2w_1, \max\{c_1, \tau\} + c_2 + w_2\} - \tau, \\ \max\{c_1, \tau\} + 2c_2 + w_2 - \tau. \end{cases} \tag{15.5}$$

We now deal with the optimal (offline) schedule of the three tasks and want it completely different. Instead of having task i being processed by processor P_1, we want task i to be processed by P_2, then task j on P_3, and finally task k on P_1, each task being in turn scheduled as soon as possible. We now establish the maximum flow achieved by this schedule:

- Flow of task i: $c_2 + w_2$.
- Flow of task j: Its sending to P_3 can start neither before its release date nor before the end of the communication of task i. Its processing can take place as soon as P_3 has received it. Its flow is thus equal to: $\max\{\tau, c_2\} + c_2 + w_2 - \tau$ (since P_2 and P_3 share the same characteristics).
- Flow of task k: The reasoning follows the one we had for task j; its flow is: $\max\{\tau, c_2\} + c_2 + c_1 + w_1 - \tau$.

Thus, the maximum flow achieved by this schedule is equal to:

$$\max\left\{c_2 + w_2,\ \max\{c_2, \tau\} + c_2 + w_2 - \tau,\ \max\{c_2, \tau\} + c_2 + c_1 + w_1 - \tau\right\} \\ = \max\{c_2, \tau\} + c_2 + \max\{w_2, c_1 + w_1\} - \tau. \tag{15.6}$$

Therefore, the best achievable competitive ratio for algorithm $\mathcal{A}$, denoted ν, is the ratio of its best achievable maximum flow, μ, and of the maximum flow of the best schedule, as expressed by equation (15.6). Since, by hypothesis, the competitive ratio of algorithm $\mathcal{A}$ is not greater than $\varrho - \varepsilon$, $\varrho - \varepsilon \geqslant \nu$, where

$$\nu = \frac{\mu}{\max\{c_2, \tau\} + c_2 + \max\{w_2, c_1 + w_1\} - \tau}. \tag{15.7}$$

The problem, now, is to find values of c_1, c_2, w_1, w_2, and τ that maximize the value of ν while satisfying the constraints set on ϱ, i.e., equations (15.2) and (15.4). In other words, we want to solve the following problem:

MAXIMIZE ν UNDER THE CONSTRAINTS

$$\begin{cases} 0 \leqslant w_1 < w_2 \\ 0 \leqslant c_2 < c_1 \\ c_1 + w_1 < c_2 + w_2 \\ \nu \leqslant \dfrac{c_2 + w_2}{c_1 + w_1} \\ \nu \leqslant 1 + \dfrac{\tau}{c_1 + w_1} \\ \nu \leqslant \dfrac{c_1 + w_1 + \max\{\max\{c_1, \tau\} + c_1, w_1 + \max\{c_1, \tau, w_1\}\} - \tau}{\max\{c_2, \tau\} + c_2 + \max\{w_2, c_1 + w_1\} - \tau} \\ \nu \leqslant \dfrac{\max\{\max\{c_1, \tau\} + c_2 + w_2, c_1 + w_1 + \max\{w_1, \max\{c_1, \tau\} + c_2\}\} - \tau}{\max\{c_2, \tau\} + c_2 + \max\{w_2, c_1 + w_1\} - \tau} \\ \nu \leqslant \dfrac{c_1 + \max\{2w_1, \max\{c_1, \tau\} + c_2 + w_2\} - \tau}{\max\{c_2, \tau\} + c_2 + \max\{w_2, c_1 + w_1\} - \tau} \\ \nu \leqslant \dfrac{\max\{c_1, \tau\} + 2c_2 + w_2 - \tau}{\max\{c_2, \tau\} + c_2 + \max\{w_2, c_1 + w_1\} - \tau}. \end{cases} \tag{15.8}$$

System (15.8) seems too formidable to be analytically solved. To circumvent this difficulty, we start by solving it numerically. We can indeed solve it with any available software package. The obtained numerical solution gives us the shape of the solution. In our case, this leads to: $w_1 = 0$, $\tau = c_2 = 1$, $1 < c_1$, and $c_1 = w_2$. We can then simplify the system of equations (15.8). We then obtain:

MAXIMIZE ν UNDER THE CONSTRAINTS

$$\begin{cases} 1 < c_1 \\ \nu \leqslant 1 + \dfrac{1}{c_1} \\ \nu \leqslant \dfrac{3c_1 - 1}{1 + c_1} \\ \nu \leqslant \dfrac{2c_1}{1 + c_1} \\ \nu \leqslant \dfrac{3c_1}{1 + c_1} \\ \nu \leqslant \dfrac{2c_1 + 1}{1 + c_1}. \end{cases}$$

By deleting the redundant constraints, we obtain the following equivalent system:

$$\text{MAXIMIZE } \nu \text{ UNDER THE CONSTRAINTS}$$
$$\begin{cases} 1 < c_1 \\ \nu \leqslant 1 + \dfrac{1}{c_1} \\ \nu \leqslant \dfrac{2c_1}{1+c_1}. \end{cases}$$

Therefore, we look at maximizing the minimum between $1 + \frac{1}{c_1}$ and $\frac{2c_1}{1+c_1} = 2 - \frac{2}{1+c_1}$. The former expression being decreasing with c_1 while the second is increasing, the maximum is obtained when both expressions are equal. This leads to $c_1 = 1 + \sqrt{2}$ and $\nu = \sqrt{2}$.

Therefore, we have proved that there is no optimal online algorithm for the minimization of the maximum flow time for the scheduling of independent tasks on heterogeneous master-worker platforms. Furthermore, we also have proved that any competitive algorithm for this problem, if there exists any, has at least a competitive ratio of $\sqrt{2}$. □

In the proof of Theorem 15.3, we have used the technique of the adversary. At first sight, this technique seems to imply that the entity that submits tasks must have knowledge of the algorithm behavior to design the difficult instances. In fact, this is not the case. The adversary does not need to monitor the algorithm for its decisions. It needs only to envision what decisions it can make and to build a difficult instance for each possible decision. In the context of Theorem 15.3, only two instances are needed. The first instance contains a single task sent at time 0. The second instance contains three tasks: one sent at time 0 and two sent at time 1. The adversary does not need to monitor the algorithm because the algorithms that we considered in this proof are *deterministic*, i.e., they always make the same decisions for the same instances. In the next section, we also consider randomized, and, thus, nondeterministic, algorithms.

15.3 Makespan optimization

In this section, we consider the variant of the online scheduling problem that is sometimes called *online-list* [89]. The tasks are presented one by one to the scheduler, which must schedule each task on a processor before seeing the next submitted task.

15.3.1 List scheduling algorithms

A *list schedule* is a schedule such that no processor is deliberately left idle. As soon as a processor is available, the first free task—first according to some priority list—is scheduled on it. However, simple as these greedy algorithms may seem to be, these algorithms are quite efficient as shown by Theorem 15.4 and Lemma 15.1.

THEOREM 15.4 ([45]). *Any list scheduling algorithm is $\left(2 - \frac{1}{m}\right)$-competitive for the online minimization of the makespan on m processors, and this bound is tight.*

In the definition of list schedules, we introduced the term "free" tasks. This is because list algorithms are often used when tasks can depend on each other, and because Theorem 15.4 holds in this broader context. If task T_j depends on task T_i, then the execution of task T_i must have been completed before the execution of task T_j can start. The set of precedence constraints naturally defines a directed (precedence) graph. This graph is obviously acyclic. A task is said to be free if it does not depend on any task, or if all the tasks that it depends upon have already been completed.

Proof. In this proof, we consider a set of potentially dependent tasks and their precedence graph G. We first need to establish a preliminary result. There exists a precedence path ϕ in G whose weight $w(\phi)$ satisfies

$$Idle \leqslant (m-1) \times w(\phi),$$

where *Idle* is the cumulated idle time of the m processors during the whole execution of the list schedule, and where the weight of a path in the precedence graph is equal to the sum of the execution times of its constitutive tasks.

Let T_{i1} be a task whose execution terminates at the end of the schedule. Let t_1 be the latest time before the start of the execution of $T_{i,1}$ such that there exists an idle processor during the time interval $[t_1, t_1 + 1[$ (let $t_1 = 0$ if such a time does not exist).[1] Why is this processor idle? Since the schedule is a list schedule, no task is free at t_1; otherwise, the idle processor would start executing a free task at that time. Therefore, there must be a task $T_{i,2}$ that is a (direct or indirect) predecessor of $T_{i,1}$ and that is being executed at time t_1; otherwise, $T_{i,1}$ would have been started at time t_1 by the idle processor. Because of the definition of t_1, we know that all processors are active between the end of the execution of $T_{i,2}$ and the beginning of the execution of $T_{i,1}$. We start the construction again from $T_{i,2}$ so that we obtain a task $T_{i,3}$ such that all processors are active between the end of $T_{i,3}$ and the beginning of $T_{i,2}$. Iterating the process, we end up with r tasks $T_{i,r}$, $T_{i,r-1}$, ..., $T_{i,1}$ that belong to a precedence path ϕ of G and such that all processors are always active

[1]Here we assume, without loss of generality, that all values are integers. If needed, we can always scale them up for this to be true.

except perhaps during the execution of the tasks constituting ϕ. In other words, the idleness of some processors can occur only during the execution of these r tasks, during which at least one processor is active (the one that executes the task). Hence,

$$\mathit{Idle} \leqslant (m-1) \times \sum_{j=1}^{r} w(T_{i,j}) \leqslant (m-1) \times w(\phi).$$

We now use this property to establish the competitive ratio. Let $\mathcal{M}$ be the makespan of the considered list schedule and n the number of tasks. Then, since a processor is either busy or idle at any time,

$$m \times \mathcal{M} = \mathit{Idle} + \sum_{i=1}^{n} w_i.$$

Now, take the precedence path ϕ that we previously constructed. We have that $w(\phi)$ is a lower bound on the optimal makespan $\mathcal{M}^{\text{OPT}}$, because the makespan of any schedule is greater than the weight of all precedence paths in G (simply because precedence constraints must be met). Furthermore, $\sum_{i=1}^{n} w_i \leqslant m \times \mathcal{M}^{\text{OPT}}$ (with equality only if all m processors are active all the time). Putting this together, we get

$$m \times \mathcal{M} = \mathit{Idle} + \sum_{i=1}^{n} w_i \leqslant (m-1) \times w(\phi) + m \times \mathcal{M}^{\text{OPT}} \leqslant (2m-1) \times \mathcal{M}^{\text{OPT}},$$

which proves the competitive ratio.

We now prove that this bound is tight. Let K be an arbitrarily large integer. We build a set of dependent tasks G for which any list schedule has a makespan $\mathcal{M} \approx \frac{2m-1}{m}\mathcal{M}^{\text{OPT}}$. There are $2m+1$ tasks, whose execution times are as follows: $w_i = K(m-1)$ for $1 \leqslant i \leqslant m-1$; $w_m = 1$; $w_i = K$ for $m+1 \leqslant i \leqslant 2m$; and $w_{2m+1} = K(m-1)$. The only existing precedence constraints are as follows:

- All the tasks T_i, for $m+1 \leqslant i \leqslant 2m$, depend on task T_m;
- Task T_{2m+1} depends on each of the tasks T_i, for $m+1 \leqslant i \leqslant 2m$.

There are exactly m entry tasks, i.e., tasks that do not depend on any other tasks, tasks T_i for $1 \leqslant i \leqslant m$. Hence, any list schedule schedules them at time 0. At time 1, the execution of T_m completes and the free processor (the one that executed T_m) will be successively assigned $m-1$ of the m free tasks $T_{m+1}, T_{m+2}, \ldots, T_{2m}$. Note that this processor starts the execution of the last of its $m-1$ tasks at time $1+K(m-2)$ and terminates it at time $1+K(m-1)$. Therefore, the remaining m-th task will be executed at time $K(m-1)$ by another processor. Only at time $K(m-1)+K = Km$ will task T_{2m+1} be free, which leads to $\mathcal{M} = Km + K(m-1) = K(2m-1)$.

However, the tasks can be scheduled in only $Km+1$ time units. The key is to keep $m-1$ processors idle deliberately while executing task T_m at time

0 (which is forbidden in a list schedule). Then, at time 1, each processor executes one of the m tasks T_{m+1}, T_{m+2}, ..., T_{2m}. At time $1 + K$, one processor starts executing T_{2m+1} while the other $m - 1$ processors execute tasks T_1, T_2, ..., T_{m-1}. This defines a schedule with a makespan equal to $1 + K + K(m - 1) = Km + 1$, which is optimal because it is equal to the weight of the precedence path from T_m to T_{m+1} to T_{2m+1}. Hence, we obtain the ratio $\frac{\mathcal{M}}{\mathcal{M}^{\text{OPT}}} \geqslant \frac{K(2m-1)}{Km+1} = \frac{2m-1}{m} - \frac{2m-1}{m(Km+1)} \underset{m\to+\infty}{\longrightarrow} \frac{2m-1}{m}$. This concludes the proof. □

Theorem 15.4 establishes a competitive ratio for any list scheduling algorithm. One can then wonder whether some scheduling algorithms could achieve a better ratio. To establish such a result, one can either exhibit an algorithm with a lower competitive ratio or establish a lower bound on the competitive ratio of any algorithm. We will start with the latter option.

To establish lower bounds, we use the notion of *prefix* instances. A problem instance $\mathcal{I}_1$ with n tasks is a prefix of an instance $\mathcal{I}_2$ if the first n tasks of $\mathcal{I}_2$ are exactly the tasks of $\mathcal{I}_1$ in the same order. Then, under the *online-list* model, any (deterministic) scheduler will take the exact same decisions for each of the first n tasks of $\mathcal{I}_2$ than for its counterpart task in $\mathcal{I}_1$. Indeed, as a scheduler must have scheduled a task before it can discover the characteristics of the next task in the instance, the set of the first n tasks of instance $\mathcal{I}_2$ is undistinguishable from instance $\mathcal{I}_1$. We use this property to prove the following lemma.

LEMMA 15.1 ([34]). *If the platform contains two or three processors (i.e., $m = 2$ or $m = 3$), then any list scheduling algorithm achieves the best possible competitive ratio for the online minimization of the makespan.*

Proof. To establish this result, we just have to show that if $m \in \{2, 3\}$, any online algorithm has a competitive ratio of at least $2 - \frac{1}{m}$. Since we are under the *online-list* model, in all the instances used in this proof, we assume that all tasks are available from the start, and each instance is fully characterized by a sequence of task sizes. Indeed, under the *online-list* model, the scheduler must decide where (and when) to execute a task before it can discover the characteristics of the next task. Here, once assigned on a processor, tasks are executed as soon as possible.

$m = 2$. We consider two instances: $\mathcal{I}_1 = (1, 1)$ and $\mathcal{I}_2 = (1, 1, 2)$. In other words, $\mathcal{I}_1$ is made of two unitary tasks and $\mathcal{I}_2$ of two unitary tasks and then one task of size 2. Because of the *online-list* model, up to the second task included, scheduling algorithms take the exact same decisions when processing instances $\mathcal{I}_1$ and $\mathcal{I}_2$.

An algorithm that schedules both tasks of $\mathcal{I}_1$ on the same processor achieves a ratio of at least 2 on $\mathcal{I}_1$ and thus a ratio greater than our target ratio of $2 - \frac{1}{m} = \frac{3}{2}$. Therefore, we need only to focus on the other algorithms, that is, those that schedule both tasks of $\mathcal{I}_1$ on two different processors. Such an algorithm achieves a ratio of at least $\frac{3}{2}$ on $\mathcal{I}_2$, since the optimal makespan for $\mathcal{I}_2$ is achieved by scheduling the task of size 2 alone on a processor, leading to a makespan of 2.

We have shown that any algorithm achieves either a ratio of 2 on $\mathcal{I}_1$ or a ratio of $\frac{3}{2}$ on $\mathcal{I}_2$, hence, the desired lower bound.

$m = 3$. We consider three instances: $\mathcal{I}_1 = (1, 1, 1)$, $\mathcal{I}_2 = (1, 1, 1, 3, 3, 3)$, and $\mathcal{I}_3 = (1, 1, 1, 3, 3, 3, 6)$. Because of the *online-list* model, up to the third (respectively sixth) task included, scheduling algorithms take the exact same decisions when processing instances $\mathcal{I}_1$ and $\mathcal{I}_2$ (resp. $\mathcal{I}_1$, $\mathcal{I}_2$, and $\mathcal{I}_3$).

Any algorithm that schedules two tasks of $\mathcal{I}_1$ on the same processor achieves a ratio of at least 2 and thus greater than $2-\frac{1}{m} = \frac{5}{3}$. Therefore, we need only to focus on the other algorithms, the ones that schedule the tasks of $\mathcal{I}_1$ (and, thus, the first three tasks of $\mathcal{I}_2$ and of $\mathcal{I}_3$) on three different processors. Such an algorithm that schedules two tasks of size 3 of $\mathcal{I}_2$ on the same processor achieves a ratio of at least $\frac{7}{4} > \frac{5}{3}$. Therefore, we need only to focus on algorithms that schedule on each processor exactly one unitary task and one task of size 3 when dealing with $\mathcal{I}_2$ and, thus, with $\mathcal{I}_3$. Such an algorithm achieves a makespan of at least 10 on $\mathcal{I}_3$. The optimal schedule for $\mathcal{I}_3$ is to schedule the three unitary tasks and one task of size 3 on one processor, the two remaining tasks of size 3 on a second processor, and the task of size 6 on the last processor, hence, achieving a makespan of 6. Therefore, any schedule that achieves a competitive ratio strictly smaller than $\min\{2, \frac{7}{4}\}$ on $\mathcal{I}_1$ and $\mathcal{I}_2$ achieves a ratio of at least $\frac{10}{6} = \frac{5}{3}$ on $\mathcal{I}_3$. This concludes the proof.

□

Therefore, no deterministic online algorithm can achieve a better competitive ratio than list scheduling algorithms when the computing platform comprises two or three processors. When the number of processors is strictly greater than four, it is possible to design online algorithms with lower competitive ratios than those of list scheduling algorithms [7]. Rather than focus on the design of such algorithms, we look for *randomized* algorithms that achieve lower competitive ratios than list scheduling algorithms.

15.3.2 Randomized optimization of makespan

In this section, we consider randomized algorithms, i.e., algorithms that can make random decisions. The principles and advantages of randomized algorithms have been presented in Section 8.4. We recall them here.

Let us consider an online scheduling problem and an incomplete instance $\mathcal{I}$ for which any online scheduling algorithm has to make a choice between two potential schedules. The first schedule will deliver a strictly better or a strictly worse performance than the second schedule, depending on the tasks yet to be submitted (if any). By definition, a deterministic algorithm will always choose the same schedule when facing instance $\mathcal{I}$. A randomized algorithm will toss a coin to decide which schedule to pick. The worst-case performance of a randomized algorithm is obviously the worst performance between those of the deterministic algorithms that uses the same schedules. What matters for randomized algorithms is their *average* performance. The hope is that not always using the same schedule will lead to an average performance better than the worst case of a deterministic algorithm.

We start our study of randomized algorithms for the online minimization of the makespan by giving a lower bound on the competitive ratio of any randomized algorithm.

THEOREM 15.5 ([7]). *There is no randomized scheduling algorithm for the online minimization of the makespan whose competitive ratio is strictly lower than $\frac{4}{3}$.*

Proof. We consider an instance with m processors, a randomized scheduling algorithm $\mathcal{A}$ whose competitive ratio is ϱ, and two instances denoted $\mathcal{I}_1$ and $\mathcal{I}_2$. $\mathcal{I}_1$ is composed of m unitary tasks. $\mathcal{I}_2$ is identical to $\mathcal{I}_1$ except that it contains a $(m+1)$-th task of size 2. Therefore, instance $\mathcal{I}_1$ is a prefix of instance $\mathcal{I}_2$. With m unitary tasks to schedule, algorithm $\mathcal{A}$ can:

1. Schedule the m unitary tasks on the m different processors. Algorithm $\mathcal{A}$ makes such a choice with probability p. If the submitted instance is

   $\mathcal{I}_1$: The makespan of $\mathcal{A}$ is 1 and so is its ratio.

   $\mathcal{I}_2$: The makespan of $\mathcal{A}$ is at least 3, while the optimal makespan is 2 (by assigning two unitary tasks to the same processor and having one processor free to process the task of size 2), hence, a ratio of at least $\frac{3}{2}$.

2. Schedule at least two of the m unitary tasks on the same processor, freeing at least one processor. Algorithm $\mathcal{A}$ makes such a choice with probability $1-p$. If the submitted instance is

   $\mathcal{I}_1$: The makespan of $\mathcal{A}$ and its competitive ratio are at least equal to 2.

   $\mathcal{I}_2$: The makespan of $\mathcal{A}$ can be 2, which is optimal (hence, a ratio of 1).

To compute (a lower bound on) the competitive ratio of $\mathcal{A}$, we take into account the different choices that algorithm $\mathcal{A}$ can make and the respective probabilities of these choices. If the submitted instance is $\mathcal{I}_1$, the competitive ratio ϱ of algorithm $\mathcal{A}$ satisfies:

$$\varrho \geqslant p \times 1 + (1-p) \times 2 = 2 - p.$$

Note that here the lower bound on the competitive ratio is a function of the probability of the random choice. If the submitted instance is $\mathcal{I}_2$, the competitive ratio ϱ of algorithm $\mathcal{A}$ satisfies:

$$\varrho \geqslant p \times \frac{3}{2} + (1-p) \times 1 = 1 + \frac{p}{2}.$$

Overall, we have

$$\varrho \quad \geqslant \quad \max\left\{2-p,\ 1+\frac{p}{2}\right\} \quad \geqslant \quad \frac{4}{3}$$

since the maximum is minimized for $p = \frac{2}{3}$. This concludes the proof. □

Note that in the proof of Theorem 15.5, letting $p = 0$ or $p = 1$ enables to also model deterministic algorithms. Hence, the proof and the theorem are valid for both randomized *and* deterministic algorithms.

One should remark that Theorem 15.5 states a lower bound for the competitive ratio of randomized algorithms that is *strictly lower* than the best achievable competitive ratio $\frac{3}{2}$ for any *deterministic* online algorithm, as stated by Lemma 15.1. We now introduce an algorithm that achieves this lower bound for two processors.

A very significant problem for online scheduling algorithms is the arrival of a large task when the load is perfectly balanced. In such a case, the makespan is increased by the size of the task; the larger the size, the worse the imbalance, and the worse the lower bound for the algorithm competitive ratio. The solution to avoid this problem is for scheduling algorithms to purposefully maintain some imbalance between processors. The problem, then, is to keep the imbalance large enough to be able to cope with the arrival of large tasks, while keeping the imbalance small enough for the algorithm to achieve a low competitive ratio. This is what Algorithm 15.1 [7] tries to achieve for platforms with two processors.

With two processors

In the two-processor framework, the imbalance is the difference between the current makespan of both processors. Algorithm 15.1 randomly schedules the next task so that the imbalance, in expectation, is exactly equal to one third of the size of all already scheduled tasks (Step 10). To achieve this goal, Algorithm 15.1 adjusts the probability to assign the next task to the most loaded processor or to the least loaded one. If the algorithm is unable to achieve this goal, the next task is deterministically assigned to the least loaded processor (Step 17). In order to be able to compute the expectation of the imbalance, Algorithm 15.1 records all the schedules that it may have built so far (each one as a triplet of the probability of occurrence of the schedule, the loads of the least loaded processor, and that of the most loaded one). Therefore, Algorithm 15.1 has an exponential complexity. We will conclude this section by showing how it can be converted into a polynomial-time complexity algorithm.

THEOREM 15.6 ([7]). *Algorithm 15.1 is exactly $\frac{4}{3}$-competitive.*

Proof. We know from Theorem 15.5 that, if Algorithm 15.1 has a competitive ratio, then it is at least equal to $\frac{4}{3}$. Therefore, we only have to prove that it has a competitive ratio and that this ratio is not greater than $\frac{4}{3}$.

```
1  Schedule task T_1 on the first processor
2  S ← {(1, 0, w_1)}   { The only schedule has a probability 1, the minimum
       load is 0 (on processor 2), and the maximum load is w_1 (on
       processor 1) }
3  for t = 2 to n do
4      U_min ← 0    { U_min is the expectation of the imbalance if task T_t is
           scheduled on the least loaded processor }
5      U_max ← 0    { U_max is the expectation of the imbalance if task T_t is
           scheduled on the most loaded processor }
6      foreach (π, W_min, W_max) in S do
7          U_min ← U_min + π · |(W_min + w_t) − W_max|
8          U_max ← U_max + π · |(W_max + w_t) − W_min|
9      S' ← ∅
10     if ∃p ∈ [0, 1] such that p · U_min + (1 − p) · U_max = (1/3) Σ_{i=1}^{t} w_i then
11         Schedule T_t on the processor with lower makespan with
           probability p
12         foreach (π, W_min, W_max) in S do
13             (W'_min, W'_max) = sort{W_min + w_t, W_max}
14             S' ← S' ∪ {(pπ, W'_min, W'_max)}
               (W'_min, W'_max) = (W_min, W_max + w_t)
15             S' ← S' ∪ {((1 − p)π, W'_min, W'_max)}
16     else
17         Schedule T_t on the processor with lower makespan
18         foreach (π, W_min, W_max) in S do
19             (W'_min, W'_max) = sort{W_min + w_t, W_max}
20             S' ← S' ∪ {(π, W'_min, W'_max)}
21     S ← S'
```

ALGORITHM 15.1: Randomized algorithm for the minimization of the makespan on two processors.

Algorithm 15.1 maintains an invariant on the imbalance in average; the imbalance is always equal to at least one third of the sum of all scheduled tasks. We establish this invariant by induction. The invariant obviously holds after the scheduling of the first task (the imbalance is equal to the size of the first task). We suppose that the invariant holds after the scheduling of task T_i. The invariant obviously holds if the test at Step 10 is positive for task T_{i+1}; for such a task, the imbalance is exactly equal to one third of this sum. We then say that the invariant is *tight*. Let us suppose now that the test is negative. Because the invariant held after the scheduling of task T_i, placing task T_{i+1} on the most loaded processor would increase the imbalance by w_{i+1}, and the new imbalance would be strictly greater than one third of the sum of the sizes of the tasks from T_1 to T_{i+1}. Therefore, since the test at Step 10 is negative, placing T_{i+1} on the least loaded processor also leads to an imbalance strictly larger than one third of the sum of the sizes and, hence, the invariant holds.

The proof on the upper bound of the competitive ratio is a proof by induction on the scheduled tasks. We prove that the upper bound holds after the scheduling of the first task, and that if the upper bound holds after the scheduling of task T_i, then it also holds after the scheduling of T_{i+1}. The proof relies on the notion of *large* tasks. A *large* task is a task whose size is at least equal to the sum of the sizes of the previous tasks. Formally, task T_i is large if and only if $w_i > \sum_{j=1}^{i-1} w_j$. Note that T_1 is a large task. To prove the induction, we consider several cases and prove the following properties:

1. The upper bound holds whenever the invariant is tight, i.e., whenever the test at Step 10 is positive.
2. If the invariant is tight after the scheduling of task T_i but not after the scheduling of task T_{i+1}, then T_{i+1} is a large task.
3. The upper bound holds after the scheduling of a large task (and, hence, after the scheduling of task T_1).
4. The upper bound holds when the invariant is not tight and the task is not a large task (this property covers all remaining cases).

We now prove these properties one by one:

1. We assume that the test at Step 10 is positive for some task T_i. In other words, there exists a value p such that $p \cdot \mathcal{U}_{\min} + (1-p) \cdot \mathcal{U}_{\max} = \frac{1}{3} \sum_{j=1}^{i} w_j$. Let M_i be the expectation of the makespan of Algorithm 15.1 after task T_i is scheduled. This is the average of the makespans of all the possible schedules, the schedules being weighted by their respective probabilities. Let U_i be the expected (average) imbalance at that time. U_i is the expectation of the difference of the makespans of the two processors, i.e., $U_i = \frac{1}{3} \sum_{j=1}^{i} w_j$. Finally, let $\mathcal{O}_i$ be the optimal makespan when scheduling the set of the first i tasks: $\mathcal{O}_i \geqslant \frac{1}{2} \sum_{j=1}^{i} w_j$. One of the processors is busy from the start up to the completion of the whole schedule, while the other remains idle at the end for a time equal, by

definition, to the imbalance. Therefore, we have

$$M_i = \frac{1}{2}\left(U_i + \sum_{j=1}^{i} w_j\right) = \frac{1}{2}\left(\frac{1}{3}\sum_{j=1}^{i} w_j + \sum_{j=1}^{i} w_j\right) = \frac{2}{3}\sum_{j=1}^{i} w_j \leqslant \frac{4}{3}\mathcal{O}_i,$$

and the upper bound on the competitive ratio is satisfied.

2. We now assume that the invariant is tight after the scheduling of task T_i but not after the scheduling of task T_{i+1}. We do not prove anything on the competitive ratio for this property, but proving that T_{i+1} is a large task will then allow us to prove easily the upper bound.

   Let m be the number of possible schedules after the scheduling of task T_i. Let p_j and u_j, respectively, be the probability and imbalance of the j-th of these schedules. Let $S = \sum_{j=1}^{i} w_j$. Then, since the invariant is tight after the schedule of T_i, $\sum_{j=1}^{m} p_j u_j = \frac{1}{3}S$.

   As seen when proving the invariant, the invariant is not tight if and only if the expected imbalance is strictly greater than one third of the sum of the task sizes when placing the new task, here T_{i+1}, on the least loaded processor. Therefore, we focus on the expected imbalance, $U(w_{i+1})$, when T_{i+1} is placed on the least loaded processor. We want to assess under which condition(s) this imbalance is strictly greater than one third of the sum of task sizes, $\frac{1}{3}(S + w_{i+1})$, when the invariant is tight after the scheduling of T_i. We consider the function:

$$f(w_{i+1}) = U(w_{i+1}) - \frac{1}{3}\sum_{j=1}^{i+1} w_j = U(w_{i+1}) - \frac{1}{3}(S + w_{i+1})$$

   and study its sign; the invariant is not tight if and only if f is strictly positive. If the invariant for a schedule is u_j, then after placing T_{i+1} on the least loaded processor, the imbalance becomes $|u_j - w_{i+1}|$. Therefore,

$$f(w_{i+1}) = \left(\sum_{j=1}^{n} p_j |u_j - w_{i+1}|\right) - \frac{1}{3}(S + w_{i+1}).$$

   Because the invariant is tight after the scheduling of task T_i, $f(0) = 0$. Note that each u_j is smaller than the sum of the sizes of the tasks from 1 to i, i.e., S. In order to ease the study, we add two new indices to the set of u_js: $u_0 = 0$ and $u_{m+1} = +\infty$. We also assume, without loss of generality, that the schedules are ordered by nondecreasing imbalance: $u_1 \leqslant u_2 \leqslant \cdots \leqslant u_m$. Let k be any integer such that $0 \leqslant k \leqslant i$ and let us assume that w_{i+1} belongs to the interval $[u_k, u_{k+1}[$. We can then rewrite f as follows:

$$\begin{aligned}
f(w_{i+1}) &= \left(\sum_{j=1}^{k} p_j(w_{i+1} - u_j)\right) + \left(\sum_{j=k+1}^{m} p_j(u_j - w_{i+1})\right) \\
&\quad - \frac{1}{3}(S + w_{i+1}) \\
&= \left(\sum_{j=1}^{k} p_j - \sum_{j=k+1}^{m} p_j - \frac{1}{3}\right) w_{i+1} \\
&\quad - \left(\sum_{j=1}^{k} p_j u_j\right) + \left(\sum_{j=k+1}^{m} p_j u_j\right) - \frac{1}{3}S.
\end{aligned}$$

Function f is a continuous, piecewise affine function. The coefficient of w_{i+1} is nondecreasing from an interval $]u_k, u_{k+1}[$ to the next, going from $-\frac{4}{3}$ (case $k = 0$) to $\frac{2}{3}$ (case $k = m$). Therefore, on the interval $[0, +\infty[$, the function f is strictly decreasing, may be constant, and then is strictly increasing, at the latest starting from S. We have seen that $f(0) = 0$, and we have also $f(S) = 0$. Therefore, f is nonpositive on $[0, S]$ and strictly positive on $]S, +\infty[$. In other words, if the invariant is tight after the scheduling of task T_i, it is also tight after the scheduling of task T_{i+1} if and only if $0 \leqslant w_{i+1} \leqslant S$, i.e., if and only if T_{i+1} is not a large task, hence, the result.

3. Assume that T_i is a large task. Let $S = \sum_{j=1}^{i-1} w_j$. From the study of the previous case, the invariant cannot be tight after the scheduling of T_i, and T_i is placed on the least loaded processor. If U was the imbalance before T_i was scheduled, the imbalance after its scheduling is $w_i - U$ as $w_i \geqslant S \geqslant U$. Let E be the expectation of the makespan after the scheduling of T_i. By definition of the imbalance, the expectation of the makespan is equal to one half of the sum of the task sizes plus the imbalance. Therefore,

$$E = \frac{1}{2}((S + w_i) + (w_i - U)) = \frac{1}{2}S + w_i - \frac{1}{2}U.$$

We then use the lower bound of the imbalance given by the invariant:

$$E = \frac{1}{2}S + w_i - \frac{1}{2}U \leqslant \frac{1}{2}S + w_i - \frac{1}{2}\frac{1}{3}S = \frac{1}{3}S + w_i \leqslant \frac{4}{3}w_i$$

because T_i is a large task ($S \leqslant w_i$). Obviously, the optimal makespan cannot be smaller than the size of the largest task, that is, T_i. Hence, the property holds for large tasks.

4. We now consider the remaining cases. The task T_i considered is not a large task because large tasks are covered by point 3 (i.e., the upper

bound holds when T_i is a large task). The invariant was not tight after the scheduling of T_i, as this case was covered by point 1. Then, since T_i is not a large task, thanks to point 2, we know that the invariant was not tight after the scheduling of task T_{i-1} (otherwise, we would have a contradiction). Let T_r be the last large task scheduled before T_i. We know, thanks to point 3, that the upper bound held after the scheduling of T_r. By definition of T_r and T_i, none of the tasks $T_{r+1}, \ldots, T_i$ is large, and for none of these tasks is the invariant tight (because of point 2). We extend this sequence of tasks as much as possible. Let T_l be the last task such that none of the tasks $T_{r+1}, \ldots, T_l$ is large, and for none of these tasks is the invariant tight. We show that the makespan does not increase when scheduling tasks T_{r+1} through T_l. Since the upper bound held after the scheduling of T_r (because it is a large task by definition), this will conclude the proof.

Let m be the number of possible schedules after the scheduling of task T_{r-1}. Let p_j and u_j, respectively, be the probability and imbalance of the j-th of these schedules. Let $S = \sum_{j=1}^{r-1} w_j$. Since T_r is a large task, we have $w_r \geqslant S \geqslant u_j$, for $1 \leqslant j \leqslant m$. Indeed, the imbalance can never be greater than the sum of the sizes of all the scheduled tasks. Therefore, the average imbalance after T_r is scheduled, U, is equal to:

$$U = \sum_{j=1}^{m} p_j |w_r - u_j| = \sum_{j=1}^{m} p_j (w_r - u_j) = w_r - \sum_{j=1}^{m} p_j u_j .$$

Because of the invariant, we know that $\sum_{j=1}^{m} p_j u_j \geqslant \frac{S}{3}$, and, hence, we have $w_r - S \leqslant U \leqslant w_r - \frac{S}{3}$.

Note that $S + w_r$ is the sum of the sizes of all tasks up to task T_r included. Therefore, the invariant exactly states that b is nonnegative, where

$$b = \frac{3}{4}\left(U - \frac{S + w_r}{3}\right) . \tag{15.9}$$

Moreover,

$$0 \leqslant b \leqslant \frac{3}{4}\left(\left(w_r - \frac{S}{3}\right) - \frac{S + w_r}{3}\right) = \frac{1}{2}(w_r - S) \leqslant w_r - S .$$

By definition, for any $j \in [r+1, l]$, the invariant is not tight after the scheduling of task T_j. Therefore, task T_j is scheduled on the processor that is the least loaded after the scheduling of task T_{j-1}.

If $\sum_{j=r+1}^{l} w_j \leqslant b$, then, since $b \leqslant w_r - S \leqslant U$, all tasks are scheduled on the processor the least loaded after the scheduling of T_r (and thus they are all scheduled on the same processor) and the makespan does not increase when scheduling tasks T_{r+1} through T_l, hence, ensuring the upper bound after the scheduling of task T_i.

To complete the proof, we show that the case $\sum_{j=r+1}^{l} w_j > b$ can never happen. Let the index t be defined by: $\sum_{j=r+1}^{t-1} w_j \leqslant b$ and $\sum_{j=r+1}^{t} w_j > b$. As previously, tasks T_{r+1} through T_t are all scheduled on the same processor, the least loaded one after the scheduling of T_r. Let $\sum_{j=r+1}^{t} w_j = b + \Delta$. Instead of scheduling tasks T_{r+1} through T_t, let us suppose that a task of size b and then a task of size Δ are scheduled. Both tasks are deterministically scheduled on the processor the least loaded after the scheduling of T_r. The expected imbalance after the scheduling of these two tasks is obviously equal to the expected imbalance after the scheduling of tasks T_{r+1} through T_t.

By definition of t, $\Delta > 0$. Furthermore, by definition, task T_t is not large. Therefore,

$$w_t < \sum_{j=1}^{t-1} w_j = S + w_r + \sum_{j=r+1}^{t-1} w_j \leqslant S + w_r + b.$$

After the (fictitious) task of size b is scheduled on the least loaded processor, the imbalance is, according to equation (15.9),

$$U - b = \left(\frac{4}{3}b + \frac{S + w_r}{3}\right) - b = \frac{S + w_r + b}{3}.$$

Therefore, the invariant is tight after the scheduling of the (fictitious) task of size b. Since $\Delta < S + w_r + b$, the (fictitious) task of size Δ is not a large task. According to point 2, the invariant is tight after the scheduling of the task of size Δ. From what precedes, the invariant, therefore, is tight after the scheduling of task T_t. This contradicts the hypothesis on task T_t (recall that $r + 1 \leqslant t \leqslant l$, and by definition the invariant is not tight after the scheduling of one of the tasks T_{t+1} to T_l). Therefore, the case $\sum_{j=r+1}^{l} w_j > b$ cannot occur, which concludes the proof.

□

Algorithm 15.1 can be adapted to have a polynomial-time complexity while retaining is competitive ratio of $\frac{4}{3}$. To do so, the number of possible schedules, and thus of memorized schedules, is limited to at most i after the scheduling of i tasks. This is done as follows. We suppose there are m possible schedules after the scheduling of tasks T_1 to T_i. If the test at Step 10 is negative, that is, if the invariant is not tight after the scheduling of task T_{i+1}, then task T_{i+1} is deterministically scheduled on the least loaded processor, and there are m possible schedules after the scheduling of task T_{i+1}.

We now suppose that the test at Step 10 is positive (the invariant is tight). We number the m possible schedules in an arbitrary order. We assign T_{i+1} to the least loaded processor for each of these schedules. Because the invariant is tight, we know that the imbalance is then not greater than one third of the sum of task sizes. We then consider one by one the m schedules, and, for each schedule, we move task T_{i+1} from the least loaded to the most loaded processor. We do that as long as, after the move, the expected imbalance remains not greater than one third of the sum of all task sizes. Let k be the last schedule for which we moved T_{i+1}. We know that $k < m$ because if T_{i+1} is scheduled on the most loaded processor for each schedule, then the imbalance is strictly greater than one third of the sum of task sizes. Then, for schedules 1 through k, T_{i+1} is deterministically allocated to the most loaded processor; for schedules $k+2$ through m, T_{i+1} is deterministically allocated to the least loaded processor; and for schedule $k+1$, T_{i+1} is allocated to the least loaded processor with the probability p that enables the invariant to be tight. Hence, the number of schedules has been increased by only one.

15.4 Conclusion

The aim of this chapter was to introduce the notion and analysis of online algorithms. Competitive analysis was first introduced by Sleator and Tarjan [98]. The interested readers will find several additional randomized algorithms for makespan minimization in Seiden's article [96], along with references of deterministic algorithms achieving competitive ratios that are lower than those of list scheduling algorithms. More generally, we refer readers to the surveys on online scheduling by Albers [2] and by Pruhs, Sgall, and Torng [89].

References

[1] R. Agarwal, F. Gustavson, and M. Zubair. A high performance matrix multiplication algorithm on a distributed-memory parallel computer, using overlapped communication. *IBM Journal of Research and Development*, 38(6):673–681, 1994.

[2] S. Albers. Online algorithms: a survey. *Mathematical Programming*, 97:3–26, 2003.

[3] V. L. Arlazarov, E. A. Dinic, M. A. Kronrod, and I. A. Faradzev. On economical construction of the transitive closure of a directed graph. *Soviet Mathematics - Doklady*, 11:1209–1210, May 1970.

[4] S. Arora and B. Barak. *Computational Complexity: A Modern Approach.* Cambridge University Press, 2009.

[5] G. Ausiello, P. Crescenzi, G. Gambosi, V. Kann, A. Marchetti-Spaccamela, and M. Protasi. *Complexity and Approximation.* Springer, 1999.

[6] H. Aydin and Q. Yang. Energy-aware partitioning for multiprocessor real-time systems. In *Proceedings of the 17th IEEE International Parallel & Distributed Processing Symposium (IPDPS)*, pages 113–121. IEEE Computer Society Press, 2003.

[7] Y. Bartal, A. Fiat, H. Karloff, and R. Vohra. New algorithms for an ancient scheduling problem. *Journal of Computer and System Sciences*, 51(3):359–366, 1995.

[8] O. Beaumont, V. Boudet, F. Rastello, and Y. Robert. Matrix multiplication on heterogeneous platforms. *IEEE Transactions on Parallel and Distributed Systems*, 12(10):1033–1051, 2001.

[9] O. Beaumont, V. Boudet, F. Rastello, and Y. Robert. Partitioning a square into rectangles: NP-completeness and approximation algorithms. *Algorithmica*, 34:217–239, 2002.

[10] O. Beaumont, A. Legrand, L. Marchal, and Y. Robert. Steady-state scheduling on heterogeneous clusters. *International Journal of Foundations of Computer Science*, 16(2):163–194, 2005.

[11] A. Benoit, V. Rehn-Sonigo, and Y. Robert. Replica placement and access policies in tree networks. *IEEE Transactions on Parallel and Distributed Systems*, 19(12):1614–1627, 2008.

[12] A. Benoit, P. Renaud-Goud, and Y. Robert. Power-aware replica placement and update strategies in tree networks. In *Proceedings of the 25th IEEE International Parallel and Distributed Processing Symposium (IPDPS)*. IEEE Computer Society Press, 2011.

[13] A. Benoit and Y. Robert. Mapping pipeline skeletons onto heterogeneous platforms. *Journal of Parallel and Distributed Computing*, 68(6):790–808, 2008.

[14] J. L. Bentley. *Programming Pearls*. Addison-Wesley, 1986.

[15] P. Berlioux, M.-P. Cani, A. Lux, R. Mohr, D. Naddef, and J.-L. Roch. Algorithmique et Recherche Opérationnelle. Cours 2e année ENSIMAG, 1998/99.

[16] D. Bertsimas and D. Gamarnik. Asymptotically optimal algorithm for job shop scheduling and packet routing. *Journal of Algorithms*, 33(2):296–318, 1999.

[17] M. Best, P. van Emde Boas, and H. Lenstra. A sharpened version of the Aanderaa-Rosenberg conjecture. Research Report ZW 30/74, Mathematisch Centrum, Amsterdam, NL, 1974.

[18] L. S. Blackford, J. Choi, A. Cleary, E. D'Azevedo, J. Demmel, I. Dhillon, J. Dongarra, S. Hammarling, G. Henry, A. Petitet, K. Stanley, D. Walker, and R. C. Whaley. *ScaLAPACK Users' Guide*. SIAM, 1997.

[19] S. H. Bokhari. Partitioning problems in parallel, pipeline, and distributed computing. *IEEE Transactions on Computers*, 37(1):48–57, 1988.

[20] D. Brélaz. New methods to color the vertices of a graph. *Communications of the ACM*, 22:251–256, April 1979.

[21] P. Brucker. *Scheduling Algorithms*. Springer, 2007.

[22] A. P. Chandrakasan and A. Sinha. JouleTrack – A Web Based Tool for Software Energy Profiling. In *Proceedings of the Design Automation Conference (DAC)*, pages 220–225. IEEE Computer Society Press, 2001.

[23] J.-J. Chen and T.-W. Kuo. Multiprocessor energy-efficient scheduling for real-time tasks. In *Proceedings of the International Conference on Parallel Processing (ICPP)*, pages 13–20. IEEE Computer Society Press, 2005.

[24] I. Cidon, S. Kutten, and R. Soffer. Optimal allocation of electronic content. *Computer Networks*, 40:205–218, 2002.

[25] S. A. Cook. The complexity of theorem-proving procedures. In *Proceedings of the 3rd ACM Symposium on Theory of Computing (STOC)*, pages 151–158. ACM Press, 1971.

[26] D. Coppersmith and S. Winograd. Matrix multiplication via arithmetic progressions. In *Proceedings of the Nineteenth Annual ACM Symposium on Theory of Computing (STOC)*, pages 1–6, 1987.

[27] T. H. Cormen, C. E. Leiserson, R. L. Rivest, and C. Stein. *Introduction to Algorithms, Third Edition.* The MIT Press, 2009.

[28] P. Crescenzi and V. Kann. A compendium of NP optimization problems. World Wide Web document, URL: http://www.nada.kth.se/~viggo/wwwcompendium/.

[29] P. D'Alberto and A. Nicolau. Adaptive Strassen's matrix multiplication. In *Proceedings of the 21st International Conference on Supercomputing*, ICS'07, pages 284–292. ACM, 2007.

[30] S. Dasgupta, C. Papadimitriou, and U. Vazirani. *Algorithms.* McGraw-Hill, 2008.

[31] E. Dijkstra. *A Discipline of Programming.* Prentice-Hall, 1976.

[32] P. J. Downey, B. L. Leong, and R. Sethi. Computing sequences with addition chains. *SIAM Journal on Computing*, 10(3):638–646, 1981.

[33] T. Erlebach. Approximation algorithms for edge-disjoint paths and unsplittable flow. In *Efficient Approximation and Online Algorithms*, volume 3484 of *Lecture Notes in Computer Science.* Springer, 2006.

[34] U. Faigle, W. Kern, and G. Turán. On the performance of on-line algorithms for partition problems. *Acta Cybernetica*, 9(2):107–119, 1989.

[35] S. Fortune, J. Hopcroft, and J. Wyllie. The directed subgraph homeomorphism problem. *Theoretical Computer Science*, 10(2):111–121, 1980.

[36] G. Fox, S. Otto, and A. Hey. Matrix algorithms on a hypercube I: matrix multiplication. *Parallel Computing*, 3:17–31, 1987.

[37] D. Froidevaux, M. Gaudel, and D. Soria. *Types de données et algorithmes.* McGraw-Hill, 1990.

[38] M. R. Garey and D. S. Johnson. *Computers and Intractability, a Guide to the Theory of NP-Completeness.* W. H. Freeman and Company, 1979.

[39] GLPK: GNU Linear Programming Kit. http://www.gnu.org/software/glpk.

[40] G. H. Golub and C. F. V. Loan. *Matrix Computations.* Johns Hopkins, 1989.

[41] T. F. Gonzalez and S. Zheng. Improved bounds for rectangular and guillotine partitions. *Journal of Symbolic Computation*, 7:591–610, 1989.

[42] T. F. Gonzalez and S. Zheng. Approximation algorithm for partitioning a rectangle with interior points. *Algorithmica*, 5:11–42, 1990.

[43] M. T. Goodrich and R. Tamassia. *Algorithm Design.* John Wiley & Sons, Inc., 2002.

[44] R. Graham, E. Lawler, J. Lenstra, and A. R. Kan. Optimization and approximation in deterministic sequencing and scheduling: a survey. *Annals of Discrete Mathematics*, 5:287–326, 1979.

[45] R. L. Graham. Bounds for certain multiprocessing anomalies. *Bell System Technical Journal*, XLV(9):1563–1581, 1966.

[46] V. Guruswami, S. Khanna, R. Rajaraman, B. Shepherd, and M. Yannakakis. Near-optimal hardness results and approximation algorithms for edge-disjoint paths and related problems. *Journal of Computer and System Sciences*, 67:473–496, 2003.

[47] P. Hansen and K.-W. Lih. Improved algorithms for partitioning problems in parallel, pipeline, and distributed computing. *IEEE Transactions on Computers*, 41(6):769–771, 1992.

[48] N. J. Higham. Exploiting Fast Matrix Multiplication Within the Level 3 BLAS. *ACM Transactions on Mathematical Software*, 16(4):352–368, 1990.

[49] D. S. Hochbaum and D. B. Shmoys. Using dual approximation algorithms for scheduling problems: theoretical and practical results. *Journal of the ACM*, 34:144–162, 1987.

[50] D. S. Hochbaum and D. B. Shmoys. A polynomial approximation scheme for scheduling on uniform processors: using the dual approximation approach. *SIAM Journal on Computing*, 17(3):539–551, 1988.

[51] J. Hromkovic. *Algorithmics for Hard Problems: Introduction to Combinatorial Optimization, Randomization, Approximation, and Heuristics.* Springer, 2004.

[52] M. A. Iqbal. Approximate algorithms for partitioning problems. *International Journal of Parallel Programming*, 20(5):341–361, 1991.

[53] M. A. Iqbal and S. H. Bokhari. Efficient algorithms for a class of partitioning problems. *IEEE Transactions on Parallel and Distrbuted Systems*, 6(2):170–175, 1995.

[54] T. Ishihara and H. Yasuura. Voltage scheduling problem for dynamically variable voltage processors. In *Proceedings of the International Symposium on Low Power Electronics and Design (ISLPED)*, pages 197–202. ACM Press, 1998.

[55] K. Kalpakis, K. Dasgupta, and O. Wolfson. Optimal placement of replicas in trees with read, write, and storage costs. *IEEE Transactions on Parallel and Distributed Systems*, 12(6):628–637, 2001.

[56] K. Kalpakis, K. Dasgupta, and O. Wolfson. Steiner-Optimal Data Replication in Tree Networks with Storage Costs. In *Proceedings of the 2001 International Symposium on Database Engineering & Applications (IDEAS)*, pages 285–293. IEEE Computer Society Press, 2001.

[57] M. Karlsson and C. Karamanolis. Choosing Replica Placement Heuristics for Wide-Area Systems. In *Proceedings of the 24th International Conference on Distributed Computing Systems (ICDCS)*, pages 350–359. IEEE Computer Society, 2004.

[58] R. M. Karp. Reducibility among combinatorial problems. In R. E. Miller and J. W. Thatcher, editors, *Complexity of Computer Computations*, pages 85–103. Plenum, 1972.

[59] R. W. Kenyon. Tiling a rectangle with the fewest squares. *Journal of Combinatorial Theory A*, 76:272–291, 1996.

[60] S. U. Khan and I. Ahmad. RAMM: a game theoretical replica allocation and management mechanism. In *Proceedings of the International Symposium on Parallel Architectures, Algorithms and Networks ISPAN'05*. IEEE Computer Society Press, 2005.

[61] S. Khanna, S. Muthukrishnan, and M. Paterson. On approximating rectangle tiling and packing. In *Proceedings of the 9th Annual ACM-SIAM Symposium on Discrete Algorithms (SODA)*, pages 384–393. ACM Press, 1998.

[62] D. E. Knuth. *The Art of Computer Programming; volumes 1-3.* Addison-Wesley, 1997.

[63] T. Y. Kong, D. M. Mount, and W. Roscoe. The decomposition of a rectangle into rectangles of minimal perimeter. *SIAM Journal on Computing*, 17(6):1215–1231, 1988.

[64] T. Y. Kong, D. M. Mount, and M. Wermann. The decomposition of a square into rectangles of minimal perimeter. *Discrete Applied Mathematics*, 16:239–243, 1987.

[65] D. C. Kozen. *The Design and Analysis of Algorithms.* Springer, 1992.

[66] L. Kronsjö. *Computational complexity of sequential and parallel algorithms.* John Wiley & Sons, Inc., 1986.

[67] J. B. Kruskal. On the Shortest Spanning Subtree of a Graph and the Traveling Salesman Problem. *Proceedings of the American Mathematical Society*, 7(1):48–50, 1956.

[68] V. Kumar, A. Grama, A. Gupta, and G. Karypis. *Introduction to Parallel Computing.* The Benjamin/Cummings Publishing Company, Inc., 1994.

[69] J. D. Laderman. A noncommutative algorithm for multiplying 3x3 matrices using 23 multiplications. *Bulletin of the American Mathematical Society*, 82(1), 1976.

[70] E. Lawler. *Combinatorial Optimization — Networks and Matroids.* Dover Publications, 1976.

[71] A. Legrand, A. Su, and F. Vivien. Minimizing the stretch when scheduling flows of divisible requests. *Journal of Scheduling*, 11(5):381–404, 2008.

[72] Y.-F. Lin, P. Liu, and J.-J. Wu. Optimal placement of replicas in data grid environments with locality assurance. In *Proceedings of the International Conference on Parallel and Distributed Systems (ICPADS).* IEEE Computer Society Press, 2006.

[73] A. Lingas, R. Y. Pinter, R. L. Rivest, and A. Shamir. Minimum edge length partitioning of rectilinear polygons. In *Proceedings 20th Annual Allerton Conference on Communication, Control, and Computing*, pages 53–63, 1982.

[74] T. Loukopoulos, I. Ahmad, and D. Papadias. An overview of data replication on the Internet. In *Proceedings of the International Symposium on Parallel Architectures, Algorithms and Networks ISPAN'02.* IEEE Computer Society Press, 2002.

[75] U. Manber. *Introduction to Algorithms: A Creative Approach.* Addison-Wesley, 1989.

[76] Maple. http://www.maplesoft.com/products/maple/.

[77] A. R. Meyer and L. J. Stockmeyer. The equivalence problem for regular expressions with squaring requires exponential space. In *Proceedings of the 13th Annual Symposium on Switching and Automata Theory*, SWAT'72, pages 125–129. IEEE Computer Society, 1972.

[78] M. Mitzenmacher and E. Upfal. *Probability and Computing: Randomized Algorithms and Probabilistic Analysis.* Cambridge University Press, 2005.

[79] B. Monien and E. Speckenmeyer. Ramsey numbers and an approximation algorithm for the vertex cover problem. *Acta Informatica*, 22(1):115–123, 1985.

[80] R. Motwani and P. Raghavan. *Randomized Algorithms.* Cambridge University Press, 1995.

[81] B. Olstad and F. Manne. Efficient partitioning of sequences. *IEEE Transactions on Computers*, 44(11):1322–1326, 1995.

[82] C. H. Papadimitriou. *Computational Complexity.* Addison-Wesley, 1994.

[83] J. Pattillo, A. Veremyev, S. Butenko, and V. Boginski. On the maximum quasi-clique problem. *Discrete Applied Mathematics*, 161(1–2):244–257, 2013.

[84] D. Pearson. A polynomial-time algorithm for the change-making problem. *Operations Research Letters*, 33(3):231–234, May 2005.

[85] B. Peis, M. Skutella, and A. Wiese. Packet routing: complexity and algorithms. In *7th International Workshop on Approximation and Online Algorithms (WAOA 2009)*, LNCS 5893. Springer, 2009. Extended version: Technical Report 003-2009, Technical University Berlin.

[86] A. Pinar and C. Aykanat. Fast optimal load balancing algorithms for 1D partitioning. *Journal of Parallel and Distributed Computing*, 64(8):974–996, 2004.

[87] A. Pinar, E. K. Tabak, and C. Aykanat. One-dimensional partitioning for heterogeneous systems: Theory and practice. *Journal of Parallel and Distributed Computing*, 68:1473–1486, 2008.

[88] J.-F. Pineau, Y. Robert, and F. Vivien. The impact of heterogeneity on master-slave scheduling. *Parallel Computing*, 34(3):158–176, 2008.

[89] K. Pruhs, J. Sgall, and E. Torng. Online scheduling. In J. Y.-T. Leung, editor, *Handbook of Scheduling: Algorithms, Models, and Performance Analysis*. CRC Press, 2004.

[90] K. Pruhs, R. van Stee, and P. Uthaisombut. Speed scaling of tasks with precedence constraints. *Theory of Computing Systems*, 43:67–80, 2008.

[91] L. Qiu, V. N. Padmanabhan, and G. M. Voelker. On the Placement of Web Server Replicas. In *Proceedings of INFOCOM*, pages 1587–1596. IEEE Computer Society Press, 2001.

[92] G. J. E. Rawlins. *Compared to what?: an introduction to the analysis of algorithms*. Computer Science Press, Inc., 1992.

[93] A. Schrijver. *Theory of Linear and Integer Programming*. John Wiley & Sons, Inc., 1986.

[94] A. Schrijver. *Combinatorial Optimization: Polyhedra and Efficiency*, volume 24 of *Algorithms and Combinatorics*. Springer, 2003.

[95] P. Schuurman and G. J. Woeginger. Approximation schemes — a tutorial, 2001. www.win.tue.nl/~gwoegi/papers/ptas.pdf.

[96] S. Seiden. Randomized algorithms for that ancient scheduling problem. In F. Dehne, A. Rau-Chaplin, J.-R. Sack, and R. Tamassia, editors, *Algorithms and Data Structures*, volume 1272 of *Lecture Notes in Computer Science*, pages 210–223. Springer, 1997.

[97] J. Shallit. What this country needs is an 18-cent piece. *Mathematical Intelligencer*, 25(2):20–23, 2003.

[98] D. D. Sleator and R. E. Tarjan. Amortized efficiency of list update and paging rules. *Communications of the ACM*, 28(2):202–208, 1985.

[99] D. D. Sleator and R. E. Tarjan. Self-adjusting binary search trees. *Journal of the ACM*, 32(3):652–686, 1985.

[100] A. Srinivasan and C.-P. Teo. A constant-factor approximation algorithm for packet routing and balancing local vs. global criteria. *SIAM Journal on Computing*, 30(6):2051–2068, 2000.

[101] V. Strassen. Gaussian Elimination is not Optimal. *Numerische Mathematik*, 13(4):354–356, 1969.

[102] X. Tang and J. Xu. QoS-Aware Replica Placement for Content Distribution. *IEEE Transactions on Parallel and Distributed Systems*, 16(10):921–932, 2005.

[103] V. V. Vazirani. *Approximation Algorithms*. Springer, 2001.

[104] C.-M. Wang, C.-C. Hsu, P. Liu, H.-M. Chen, and J.-J. Wu. Optimizing Server Placement in Hierarchical Grid Environments. In *Proceedings of the 1st International Conference on Grid and Pervasive Computing GPC'07*, volume 1900 of *Lecture Notes in Computer Science*, pages 1–11, 2007.

[105] H. Wang, P. Liu, and J.-J. Wu. A QoS-aware Heuristic Algorithm for Replica Placement. In *Proceedings of the 7th International Conference on Grid Computing (GRID2006)*, pages 96–103. IEEE Computer Society Press, 2006.

[106] D. B. West. *Introduction to Graph Theory*. Prentice-Hall, 1996.

[107] H. Whitney. On the abstract properties of linear dependence. *American Journal of Mathematics*, 57(3):509–533, 1935.

[108] H. S. Wilf. *Algorithms and Complexity*. A. K. Peter, 1985. Available at http://www.cis.upenn.edu/~wilf/.

[109] O. Wolfson and A. Milo. The multicast policy and its relationship to replicated data placement. *ACM Transactions on Database Systems*, 16(1):181–205, 1991.

[110] J.-J. Wu, Y.-F. Lin, and P. Liu. Optimal replica placement in hierarchical data grids with locality assurance. *Journal of Parallel and Distributed Computing*, 68(12):1517–1538, 2008.

[111] W. Yu, H. Hoogeveen, and J. K. Lenstra. Minimizing makespan in a two-machine flow shop with delays and unit-time operations is NP-hard. *Journal of Scheduling*, 7(5):333–348, 2004.

[112] M. Yue. A simple proof of the inequality $FFD(L) \leqslant \frac{11}{9}OPT(L) + 1$, $\forall L$, for the FFD bin-packing algorithm. *Acta Mathematicae Applicatae Sinica (English Series)*, 7:321–331, 1991.

Index